To Glen and Eli...
with good wishes
Jean Andrews

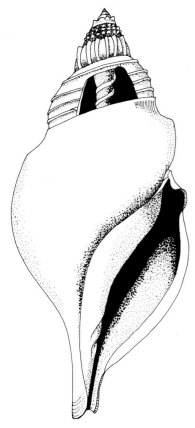

SEA SHELLS
OF THE TEXAS COAST

by Jean Andrews

SHELL PHOTOGRAPHS BY JEAN BOWERS GATES

UNIVERSITY OF TEXAS PRESS, AUSTIN AND LONDON

International Standard Book Number 0–292–70137–3
Library of Congress Catalog Card Number 70–162690
© 1971 by Jean Andrews
All Rights Reserved
Type set by Southwestern Typographics, Inc., Dallas
Printed by Steck-Warlick Company, Austin
Bound by Universal Bookbindery, Inc., San Antonio

SEA SHELLS OF THE TEXAS COAST

NUMBER 5

THE ELMA DILL RUSSELL SPENCER FOUNDATION SERIES

To the memory of my daughter, Jinxy,

whose spirit soared as the gulls

over the islands she so loved

CONTENTS

Foreword by Robert H. Parker xi
Preface xiii
Introduction xv

PART ONE

1. The Texas Coast 3
   Historical Background, *p. 3*
   Physical Description, *p. 8*

2. General Features of the Mollusk 31

3. Systematic Descriptions 51
   Class AMPHINEURA, *p. 53*
   Class GASTROPODA, *p. 54*
   Class SCAPHOPODA, *p. 145*
   Class BIVALVIA, *p. 147*
   Class CEPHALOPODA, *p. 228*

PART TWO

4. A Shell Collecting Trip 233

5. Beachcombing 262

Glossary 271

Bibliography 277

Index 287

MAPS

1. Galveston to Freeport 234–235

2. Freeport to the Colorado River 236–237

3. Colorado River to Matagorda Island 238–239

4. Cedar Bayou to Padre Island 242–243

5. Padre Island 244–245

6. Port Mansfield to the Rio Grande 248–249

FOREWORD

Seldom in my experience as a marine biologist have I come upon such a compendium of information on a single subject as is this study, so complete yet still on a popular level. Books on the sea, sea shells, beaches, and beachcombing have been written *ad nauseum*; many of them are excellent and informative, while others written in haste contain much misinformation. Laymen who write about the sea usually express themselves well, but do not know the subject thoroughly enough to satisfy the scientist. On the other hand, the scientist may be very familiar with his subject but unable to get it across to the layman. The author of this book is neither an uninformed layman nor an academically trained biologist, yet her treatment of this difficult, many-faceted subject is both authoritative and interesting.

Two summers ago I received a letter from Miss Andrews asking if I might check identifications of some small shells for her. She had gone to several other malacologists in the region for help and felt that one more opinion might assist in solving her problems. I had some misgivings as to this project, since I am not really a professional malacologist, although I am frequently besieged by questions from other collectors. She did indeed have problems and they involved the identification of tiny mollusks that most experts dare not tackle. Between the two of us, through many long hours into the night in a stuffy laboratory, some of these questions on identifications were partially solved. The fact that Miss Andrews had done a tremendous job on her own before coming to me (and other scientists) impressed me greatly. Once I had realized that the questions regarding the "right" names were scientific rather than idly curious, my interest was drawn to the entire contents of her book. It was soon evident that here was a really intelligent approach toward presenting a guide to mollusks as well as to the barrier islands of Texas, neither of which had been discussed adequately before. I was perhaps more interested in the subject than I wished to admit, as I had once intended writing (and had actually started) a similar volume on Texas mollusks, dealing with the shells of both the coastal and the offshore waters.

No stone (perhaps shell would be better) was left unturned in order to establish the accuracy or validity of the species identifications and the proper names that should be applied to the species covered in this guide. There were many times when both of us were groggy from checking old volumes of original descriptions for the proper name of one of the Texas shells that has had half a dozen names applied to it since the early 1800's. This was truly detective work, as any devoted taxonomist knows, but there is considerable satisfaction in coming to the conclusion that a certain scientific name is (one hopes) the last word in correctness.

One of the most valuable portions of this work is the beautiful set of mollusk photographs and the line drawings that go with the text. Miss Andrews is not only a competent writer but also a professional artist and an accomplished photographer, as the illustrations attest. As she points out, however, many of the photographs are the result of the efforts of others. One of the major problems in identifying Texas coastal sea shells is the lack of adequate illustrations. No single text now extant can be used to identify all the shallow-water shells of Texas. Some species may be found in R. Tucker Abbott's fine book *American Seashells*, while others only in its West Indian counterpart, *Caribbean Seashells*. This dilemma arises because the Texas coast lies within both the temperate and the tropic zones. Some of the species have never before been illustrated. The systematic portion of this book does bring together all the information on coastal species, so that most amateur shell collectors will be able to identify their finds. This brings to mind another reason why I am eager to see this volume completed and available to all. Classes in marine biology from inland Texas schools will now have some means of identifying the shells and other objects they so often collect on Texas beaches. I might add that the illustrations alone should be valuable to the oil company macropaleontologist who works with these same shells in the older, oil-bearing sands of the Texas coast.

Perhaps the greatest surprise to the reader of this book will be the tremendous range of subject matter relating to the Texas coast. Where else can one find a recipe for "Coquina Chowder" and a scientific analysis of tides, or how to clean and carry home dead shells and the detailed history of barrier-island settlement in the same book? Although many subjects are covered here, they are held together by a common thread, the author's great love for the barrier islands of the western Gulf of Mexico.

One immediately senses her fondness for those long, deserted, sandy isles when first visiting Miss Andrews's beautiful home. One portion of her grounds (hidden to the casual visitor) is devoted to accumulations of every conceivable type of float, bottle, and other flotsam and jetsam thrown up by the waves. Boxes of driftwood are stacked alongside nets, large shells, and other rare objects. I doubt if any other serious beachcomber has ever accumulated a larger or more fascinating treasure trove. Inside the house are many marine paintings (including her own), small collections of shells, translucent screens inlaid with shells and various beach-drift objects, and other mementoes, all relating to the Gulf

barrier beaches. Possibly no other woman has traveled up and down those islands as much as she has. For instance, she was a member of the first party that drove the entire length of the Mexican coastal barriers, an almost impossible feat in any type of vehicle. Actually, her four-wheel-drive car had to be transported across various passes or inlets by two small Indian canoes, one for each set of wheels.

Although I suspect that Miss Andrews's interest in shells and the barrier islands began as a hobby, there is no doubt that it soon became serious. As her list of acknowledgments indicates, she has consulted with most of the major authorities in the marine biology, malacology, and natural history of the islands. She has spent many long hours in the library of the Institute of Marine Science, The University of Texas at Port Aransas, and also in the main library of The University of Texas at Austin, and, through me, she has obtained at least one hundred books and journals, which she used in her supportive research. Although not educated primarily as a biologist, she nevertheless devoted much of her graduate study to science. In fact, her master's thesis forms part of the material of this book. What counted in the compilation of material and ideas in this book was not Miss Andrews's courses in biology, but her strong determination to find out about things not normally taught in formal courses and to get the necessary "words of wisdom" directly from authoritative sources. She has consulted the experts and utilized this information in a systematic way. What follows is a tribute to determination, intelligence, and love for the subject, her children, and nature itself.

ROBERT H. PARKER
Fort Worth, Texas

PREFACE

Have you ever picked up a jewel-like sea shell on the sparkling sands of a lonely beach? Few can resist such an impulse. Sea shells have captured the imagination of man from his beginning and have played important roles in his daily life as well as being adornments and objects to admire.

A century ago shell collecting was such an active sport that auctions were held regularly in the larger cities, and collectors traveled great distances to try to claim their choices. The sport lost its popularity, but after American servicemen had served lonely hours on isolated beaches of Pacific atolls during World War II and casually dropped the shells they picked up into their barracks bags, to rediscover them when they returned home, much new interest in the hobby was aroused. It is now one of the most popular pastimes in this mobile world.

Many of us unable to be on a Pacific isle must content ourselves with what is at hand. To those who visit Texas beaches in search of a shell or some other magical object washed ashore by the waves, this book is addressed.

The sea, like space, is a new frontier. More and more we are aware of its importance in our daily lives and in our future. Men are discovering ways to live in its hostile environment and to harvest its produce. Yet we still know little regarding our mollusks other than that they exist. Data are hard to come by. Casual collecting can add some details, but a continuous, systematic observation can provide the answers to many questions: What do they eat? How do they move? How long do they live? Why are some found in the bay and not on the Gulf beach?

This study is a compilation from the literature. I make no claim to having done the research cited from published articles. What I have attempted to do is to find the answer to questions that occurred to me as I roamed the beaches of Texas. Many of the answers have been difficult to locate; some may be imprecise. When this project was begun at least seven books totaling close to two hundred dollars in cost were needed simply to identify the most common shells found in Texas.

If this work seems to be slanted toward the southern half of the coast it is because this is the area where I live and collect. The other part is just not my "stomping ground"; however, I have driven every strip of the Gulf beach accessible by a four-wheel-drive vehicle from Sabine Pass to within sight of Tampico, Mexico.

I take responsibility for all errors found in this study. Many persons have worked to keep them to a minimum, and I am greatly indebted to them. My thanks go to Dr. William A. Burns, former director, Witte Museum; Dr. William J. Clench, malacologist, Harvard University; Dr. James X. Corgan, geologist, Austin Peay State University, Tennessee; Jean and Bob Gates, photographers; Dr. Harold Harry, marine biologist, Texas A&M University; Janet and Ed Harte, for reading and advice; Dr. Henry Hildebrand, marine biologist, University of Corpus Christi; Fred Jones, botanist, Welder Wildlife Foundation, Sinton, Texas; Dr. Myra A. Keen, Geology Department, Stanford University; Carol and Dan Kilgore, botanist and historian who read and advised; Royal Mills, biology teacher who helped with the field work; Dr. Donald Moore, marine biologist, University of Miami; Dr. Robert H. Parker, marine ecologist, Texas Christian University; Dr. W. Armstrong Price, oceanographer; Gerald Rote and Bill Gasser, photographers, Southwest Foundation for Research and Education; Dr. Harold Vokes, Geology Department, Tulane University; Manette Wilson, collector; Dr. Donald Wohlschlag, former director, The University of Texas Marine Science Institute; Carl Young, collector; and Dick and Gloria Swantner, for a quiet place in the hills to read. A generous grant from the Alice G. K. Kleberg Trust Fund aided greatly in preparing the manuscript for publication.

Without the help of my seafaring son, Robin, many of the ways of the sea would still be a mystery to me.

Over the years he has progressed from bucket toter to tide-pool explorer, jeep driver, boatman, tutor in navigation and oceanography, and companion on many underwater explorations. I shall never forget that day in the Caribbean when he discovered the sea floor literally covered with queen conchs and rushed to the surface to hasten my descent into that bewitching world beneath the surface of the sea.

PUBLISHER'S NOTE

There are two conventional methods of posing a gastropod shell for illustration. Most authors place it with the axis vertical and apex uppermost, but in France and a few other countries the shell is placed with its apex lowermost. In this volume we have sometimes taken the liberty of departing from convention in order to give the reader the advantage of the largest image possible within the limits of the format. It is felt that this change in orientation makes possible additional detail which outweighs any disadvantages.

INTRODUCTION

Before civilization dawned, early man used sea shells for implements, food, adornment, and fetishes, and as a medium of exchange. Kitchen middens and grave sites throughout the world bear witness to the dependency of man on mollusk.

Much has been written about the way man has employed the shell. Peter Dance, in his book *Shell Collecting*, and Roderick Cameron, in *Shells*, devote much space to this interesting relationship. They cite the use of mollusks by primitive man as axes, spears, utensils, ornaments, knives, and trumpets as well as a staple of diet. The ancient Phoenician empire was dependent for its existence as a world power on a dye, Tyrian purple, produced by a mollusk of the Murex family.

Although the use of shells as ornaments and tools was most important, their employment as a medium of exchange cannot be overlooked. This practice has not become completely obsolete, for even in recent times England imported tons of cowries for export to West Africa, there to be exchanged for native products.

In North America, Indians manufactured shell money, or wampum, by a laborious process. Wampum consisted of strings of cylindrical shell beads, each about a quarter of an inch in length and half that in breadth. The beads were of two colors, white and purple, the latter being more valuable. *Mercenaria mercenaria*, a common clam, furnished the largest part of the material for wampum. The trade routes of North America can be traced by marine shells transported by trade into areas far from their native habitats.

Museums, as well as curio shops, are replete with items decorated with shell. In the priceless pre-Columbian trumpet displayed in a Mexico City museum, one is thrilled to recognize the large spired horse conch from the Gulf. Beads, bracelets, rattles, drums, and trumpets from carved shells all found popularity with

early man. However, primitive man is not alone, for what woman can resist a beautiful cameo carved from the colorful shell of the cassis snail. Cameo carving is still an important industry in Italy. Mounted shells and shells transformed into silver-rimmed boxes, or worked into precious jewelry as they were in the sixteenth and seventeenth centuries, are staging a resurgence in popularity and today are to be found in expensive specialty shops throughout America.

Perhaps the veneration of the chank shell by the Hindus of India is one of the strangest uses of a shell. These shells are considered sacred, being symbolic of the many-armed Hindu god, Vishnu, and all images of Vishnu bear a chank shell in one hand.

Not only did the Renaissance artist depict shells in his paintings but he also found a practical use for them as containers for his paint after he had mixed his pigments. Georges Cuvier used sepia from cephalopods he had dissected to make the drawings used to illustrate his anatomy of the Mollusca in 1817. While sepia ink from the Mediterranean cuttlefish is the most highly prized cephalopod ink, artists and writers have used ink from various cephalopods for at least two thousand years to record their work. It is not permanent, however, and modern substitutes have largely replaced true sepia ink.

The mollusk appears frequently in art and literature, as throughout history shells were often used as symbols. The Crusaders adapted the Jacob's scallop from Europe as their emblem when they invaded the Middle East. Afterward they made pilgrimages to the tomb of the Apostle James, supposedly at Santiago de Compostela in Spain, where they found scallop shells on the nearby beaches. They carried them home as a sign that they had reached their goal. Soon people began selling scallop shells at other places and finally, in order to stop the racket in shells, the Pope decreed that anyone who sold scallop shells at any place but Santiago de Compostela would be excommunicated.

This same scallop shell, *Pecten jacobeus*, has become the familiar trade mark of a Dutch petroleum company that began long ago importing shells to sell to collectors and to be traded to natives in outlying colonies for use as money. As a sideline, the company began selling petroleum products. The origin of the company is but a memory, even though today the tankers owned by the company are each named for and decorated with a sea shell.

Pearls produced by the family Pteriidae, the pearl oyster, have been prized by man since time began. Except for use as food, perhaps the most important industry employing shell is the pearl industry. The iridescent, nacreous layer or "mother-of-pearl" from certain mollusks is used in the manufacture of buttons, knife handles, inlays, studs, and brooches. Natural pearls result from a foreign body entering the shell and causing an irritation to the animal. In its attempt to protect itself by coating the irritant with a smooth covering of nacre, the mollusk produces a pearl. Perfect pearls formed this way are highly treasured. The production of cultivated pearls is a large industry in Japan, where foreign particles are inserted into the living mollusk, forcing it to produce a pearl. Other shells produce pearls and "mother-of-pearl" but those of the Pteriidae are of the greatest commercial value.

In Texas it is possible to trace similar uses of shell by man. Around Copano Bay, to name but one coastal site, numerous mounds of oystershell abound, affirming the importance of this delicacy in the diet of the Karankawas who lived there. Around these kitchen middens and mounds may be found awls and drills made from the columella of the *Busycon contrarium*, beads from *Oliva sayana*, and scrapers from the *Busycon* and the *Macrocallista nimbosa*. Due to the lack of rock, arrowheads in this region were occasionally made of shell. Cabeza de Vaca, acting as a trader for the coastal tribes, found shells a valuable item of exchange.

In modern Texas, as in early times, the most important use of the mollusk is as food. The commercial oyster in Texas is *Crassostrea virginica* (Gmelin). Much research has been done on this oyster, and a bulletin on the subject by the Texas Game and Fish Commission, now the Parks and Wildlife Service, is worthy of study. Several important game fishes depend on oyster beds for their subsistence, and whatever endangers the oys-

ters endangers their dependents, including sportsmen and sporting goods dealers.

Not only is the oyster important as food, but also its discarded shell is of great value as building material. Shell, primarily oyster, is used in an unaltered state in the building of railway and highway roadbeds, although controversies have arisen in recent years over dredging for this purpose. The high calcium content of oystershell (98–99% as compared to 96% in limestone) makes it a good source of lime in the manufacture of Portland cement. Chemical companies along the coast use it in the production of caustic soda and other chemicals. In the Rockport area many roofs are covered with shell, as are driveways and sidewalks. Veritable mountains of oystershell can be seen being transported by barges on the Intracoastal Canal.

Before the Civil War, a material called "shellcrete" was used in building many of the early homes in Corpus Christi. The Centennial House, 411 North Broadway, built in 1848–49, is the oldest building standing and the first two-story home erected in Corpus Christi. Its foundation walls are made of this material. To produce shellcrete, oystershell was gathered along the bay front and then burned in shallow pits to be converted into lime, which, while still in a solid form, was placed in barrels to be air-slaked. This lime was used with an aggregate of crushed shell, sand, and water and was formed in wooden molds to make building blocks. Old buildings of this construction have walls twelve to eighteen inches thick that can be demolished only with dynamite.

Mollusks have always been considered a delicacy, and their cultivation for food dates back to Roman times, when runners sped fresh oysters from the Mediterranean to the Caesars and from the Pacific to the Moctezumas. In America, however, the mollusk is used less as a form of food than in other countries of the world with the exception of England. Many mollusks found in the Gulf waters are edible, but none except the oyster is found in commercial quantities. The *Donax*, or coquina shell, of the island surf can become a delicate chowder, while the adductor muscle of the *Atrina*, or pen shell, rivals the Eastern scallop (see Chapter 4 for methods of preparation). The *Mercenaria* of wampum fame also produces a succulent chowder when enough can be gathered. Clam digging is not a productive pastime on the Texas coast, because of the limited supply in waters within easy access.

Shells are sometimes harmful to man. Some, like the *Teredo*, or shipworm, destroy wooden structures in the sea by burrowing into the center of exposed wood. The shipworms were so devastating in 1917 in San Francisco that ferry ships collapsed, and warehouses and freight cars crashed into the bay, with destruction reaching catastrophic proportions. Losses in this one disaster were estimated at $25 million. Damage by this family of mollusks probably exceeds the total income from the sale of mollusks for food and all other purposes. The onslaught of boring mollusks can be retarded by the application of creosote and other chemicals, but there is no sure defense against the attack. As yet, the *Teredo* is not a major problem on the Texas coast, but their control is a continuous struggle.

In some areas of the world diseases carried by shells are a serious menace. The carrier is usually a freshwater mollusk. Several species that carry schistosomiasis are under study at the Southwest Foundation for Research and Education in San Antonio, Texas. Contamination is another danger, but state health regulations regarding the handling and packaging of seafoods make poisoning by contaminated mollusks a negligible problem.

A minor industry on the Texas coast is the manufacture of curios incorporating shells, most of which are shipped in from the Pacific and Florida. Many lovely natural forms are converted into grotesque animals, lamps, or ashtrays to remind a tourist of his visit to the shores of Texas. Often a shell that never saw the waters of Texas when living tenders "Greetings from Padre Island."

The severe extremes of the physical factors found in Texas tend to make it an inhospitable environment for sea shells. Still a profusion of molluscan life can be found. Admittedly, the Texas coast does not rival the western coast of Florida but it has enough specimens to keep the enthusiast busy for many a day. More than 150,000 shells are known in the world, of which 6,000 species are found in North American waters. Known to be living in the bays of Texas and within a mile or two offshore are approximately 400 to 600 species. An exciting aspect of shell collecting in Texas is the possibility of having a complete collection of all known local shells or even of being the proud discoverer of a species not previously known to be an inhabitant of these shores.

PART ONE

1. The Texas Coast

Texas, the home of great ranches with their cowboys and oil wells, is seldom thought of as a land of windswept beaches and soaring sea gulls. The second state of the United States in land area, Texas ranks third in length of coastline, containing one twelfth of the total in the country. The land along this coastline—extending between the Sabine River at latitude 30° N and longitude 93° 31′ W, and the Rio Grande, at latitude 25° 50′ N and longitude 97° W—is a vast coastal plain only slightly above sea level. The principal ports on the coast—Corpus Christi, Galveston, and Brownsville—are respectively 35, 20, and 57 feet above sea level. (Houston is an inland port.) The Gulf of Mexico, which forms this coastline, deepens gradually, and along most of the coast lie long barrier islands of sand, enclosing lagoons and bays. Between these islands and the mainland runs the Gulf Intracoastal Waterway. Four hundred and twenty-three miles of the total eleven hundred of this waterway lie in Texas waters, bringing an inexpensive form of transportation to the booming industries of Texas.

Historical Background

Fossil remains of man on the Texas coast date back eight thousand years. However, the first European known to view the Texas shore was probably Alonso de Piñeda while on a mapping expedition for Francisco de Garay, the governor of Jamaica. In 1519 Piñeda sailed along the coast of the Gulf of Mexico from Florida to Veracruz searching for the Strait of Anian, which would lead the treasure-hungry Spaniards to the riches of Cathay and India. While on this journey he careened his four vessels for a needed overhaul at what is now the Rio Grande, the southern boundary of Texas, but which showed on his map as the Río de las Palmas. There, in the fall of 1519, he spent forty days and noted numerous Indian villages along the lower Rio Grande. These same Indians drove Diego de Camargo away in 1520 when he attempted to found a colony near Piñeda's landing site.

Pánfilo de Narváez was commissioned to sail from the Río de las Palmas to the Cape of Florida and to establish two colonies along his route. On his staff, as treasurer, was Alvar Núñez Cabeza de Vaca, whose chronicle of the events following the shipwreck of the expedition in 1528 gives us our earliest account of life in the area now known as Texas. It is thought the survivors landed somewhere on Galveston Island. Starvation soon reduced the Spaniards to cannibalism. De Vaca survived the hardships and passed eight years as captive, trader, and medicine man among the Indians, leaving to us an exciting record of his experiences.

The Karankawa Indians of the Texas coast were a primitive tribe, without clothing, arts, or houses, and

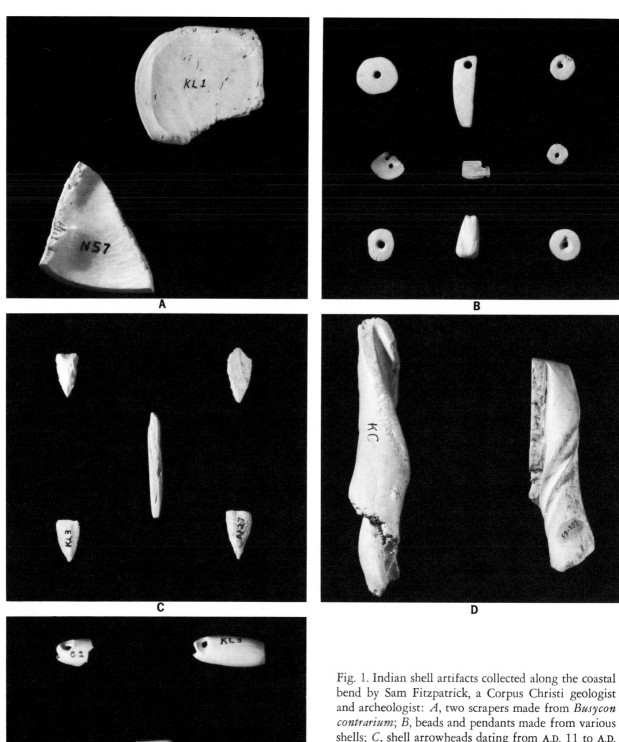

Fig. 1. Indian shell artifacts collected along the coastal bend by Sam Fitzpatrick, a Corpus Christi geologist and archeologist: *A*, two scrapers made from *Busycon contrarium*; *B*, beads and pendants made from various shells; *C*, shell arrowheads dating from A.D. 11 to A.D. 1500; *D*, archaic *Busycon* columella gouges; *E*, beads of olive shells (the fish type is very rare).

practicing ritualistic cannibalism. Having come to this region in about A.D. 1400, they were well established and had lived primarily on fish and oysters for many generations when the first white man sighted these shores. Eventually driven out by the gun and sword, they left the European colonist an untouched source of wealth. These Indians were reported by early settlers to be as treacherous and inhospitable as the islands they inhabited. Their primary locale between 1536 and 1821 was from Galveston Island to the northern tip of Padre Island. Shell mounds, shell implements, and primitive arrowheads are still found to attest their former presence (fig. 1).

The Karankawas, "water walkers," were a vile-smelling group as a result of their practice of smearing their bodies with alligator fat or any other odorous material they could find to ward off the tormenting mosquitos. Although usually hungry, they did not resort to cannibalism when threatened with starvation. Contrary to popular opinion resulting from early Spanish reports these coastal Indians were not cannibals in the true sense of the word but practiced the custom of eating human flesh for reasons of revenge or for magical purposes. They were lazy and preferred to suffer hunger and nakedness in order to be at liberty in the inland woods or on the beach.

Even though these savages were said to be the largest American Indians, they were no match for their invaders. They were exterminated by the usurpers of their territory. The remnants eventually drifted into Mexico where they were absorbed by local tribes. Kuykendall reports (1903, p. 253): "In the year 1855 the once formidable tribe of Karankawas had dwindled to six or eight individuals who were residing near San Fernando, State of Tamaulipas, Mexico."

Their neighbors to the east and into Louisiana across the Sabine River were the Atakapans. The French officer Simars de Bellisle left us a reliable account of this people after he was captured in 1719 and lived among them for three years. Closely related to the Karankawas, they were, as the translation of their name indicates, "maneaters." Their territory was more hospitable than that of the larger Karankawas, but they, too, moved in seasonal rhythms, gathering roots and nuts, hunting and fishing. Both tribes led a beachcombing, scavenger type of existence, and neither survived civilization.

The coast south of the last Karankawa camp was largely uninhabited because of its inaccessibility. Occasional bands of Coahuiltecans would prowl its south-

ern limits from their territory to the west. Inhabiting the country inland from Galveston Bay to the Rio Grande and west to San Antonio, they had few usable natural resources in their southern zone and were constantly on the move looking for food. It was this never-ending search that brought them to the coast at times in attempts to supplement their usual diet of cacti, mesquite beans, nuts, sotol, and agave. Ritualistic cannibals, they probably did not resort to cannibalism for food but only ate captives and their own people who died of natural causes.

The Spaniards, prospering from their conquest of the fabled Aztecs in Mexico, were sending vast treasure fleets back home with their loot. In 1553 vessels laden with such a cargo, as well as a thousand proud conquistadors and their families, sailed from Veracruz for Spain via Havana. Before reaching the Bahama passage they were hit by a tropical hurricane that sank four ships on the spot and drove thirteen across the Gulf to crash in the pounding surf off Padre Island. Only three finally straggled to safety.

At Devil's Elbow, about forty miles from the southern tip of Texas, three hundred men, women, and children staggered ashore. They saved much of the food supply from the wrecked vessels and had been on the beach for six days when a hundred Karankawa Indians arrived bearing fresh fish and making signs of peace. While the survivors were eating the meal prepared for them, the savages attacked. Crossbows salvaged from the wreckage drove the Indians away but not until many of the Spaniards were dead.

The next day they headed south down the island, abandoning most of their supplies because they thought they were near the Pánuco River and a Spanish outpost. Five days later when they arrived at Brazos Santiago Pass they built rafts and crossed but in doing so lost the weapons overboard. The Indians attacked again at the Pass and only two hundred disaster victims managed to reach the Rio Grande, half of them wounded. Thinking that they could appease the Indians so that the savages would go away, the foolhardy Spaniards gave them all their clothing, leaving themselves exposed to the vicious sun and wind.

In desperation they stumbled on, dogged at every step by the Indians, who killed all stragglers. Only two survived to tell the tale. One of these men reached the Pánuco in spite of festering wounds from arrows that were to remain in his body until he was carried to Mexico City. The other was found a year later at the site of the catastrophe when Spanish ships returned to salvage the wrecks. Twelve of the ships were located and most of the treasure recovered but the precious cargo of the thirteenth has defied man's search until this day. ¿Quien sabe?

Only hapless victims of the sea came to the Texas coast to fall prey to the merciless Karankawas until René Robert Cavelier, Sieur de La Salle, discoverer of the Mississippi and a hero of France, arrived in 1685. He claimed that the purpose of his journey was the establishment of a colony near the mouth of the Mississippi, but it is believed by some research students that he was on a mission to conquer New Spain and that the stated plan was only a cover. In order to facilitate a French invasion he needed to be based nearer Spanish territory. He sailed on until he reached Matagorda Bay. One of his four vessels went aground there and the others sailed back to France after bitter arguments between La Salle and one of his captains.

Between 180 and 220 unskilled artisans and soldiers were in the remaining group. They built a fort on a high bluff on the west bank of Garcitas Creek about five or six miles upstream. La Salle named the fort St. Louis and the nearby bay La Vaca. He made two long overland expeditions from the fort in search of rumored Spanish settlements. His people were dying and hopeless by the time he attempted a third march, and he was murdered near the Brazos River by one of his men.

In the meantime the Spanish had heard of the French colony in Texas and had begun to send expeditions by land and sea attempting to locate it. When Alonso de León, coming north from Mexico, arrived at the ill-fated French fort in 1689 he found only the bodies of the few who had remained to fall victims of the Karankawas.

De León's penetration of the country as far as the Guadalupe River marked the beginning of the European settlement of Texas. The Spaniards realized they must establish colonies to hold the land and so began their period of mission building.

As years went by the French, Spanish, and Americans sailed up and down the Gulf coast, landing or being shipwrecked. To the southern end of what is now called Padre Island, in the year 1804 came a Spanish seafaring priest named Nicolás Balli (pronounced Baa-yee). He and his alleged nephew, Juan José Balli, lived there until the Padre went to Matamoros, Mexico, to establish a church. The two men established the Santa Cruz ranch about twenty-six miles north of Point Isabel and in 1829 received a grant to the land. In 1852 a special act of the Texas legislature made the grant permanent. It is for this adventurous priest, Padre Nicolás Balli, that the longest of the Texas islands is named.

The offshore islands, worlds of incessant change, linked with piracy and tales of sunken and buried treasure and lost Spanish galleons, have inspired reams of romantic lore. Some fact, some fiction. It is a fact that Louis Aury, a famous pirate, called Galveston his base until 1816. He was succeeded by Jean Lafitte, who raided Spanish shipping and sold slaves from Galveston until 1821.

Colonists from the United States began to come to Texas early in the nineteenth century and by 1836 they were strong enough to fight a bloody war and win their independence from Mexico, which had only a dozen years earlier overthrown the Spanish yoke. The Treaty of Velasco, ending this conflict, was signed May 14, 1836, at the temporary capital located on the beach at the mouth of the Brazos River, now Surfside.

For ten years following the Texas Revolution both the new Republic of Texas and Mexico claimed the territory between the Nueces River and the Rio Grande. Into this void sailed the army of the United States in the form of several thousand troops led by General Zachary Taylor. On July 26, 1845, the first U.S. flag in Texas was flown on St. Joseph Island, where the troops had landed. In his journal (1847, p. 14), Captain W. S. Henry recorded the event with these words: " . . . on top of the loftiest sand hill was erected a pole, from the top of which was unfurled the star spangled banner. It floats over a rich acquisition, the most precious Uncle Sam has yet added to his crown."

The stars and stripes were waving over the Karankawa's old hunting grounds, and hunting grounds they were. The men marveled at the fish and game so easily taken and at how rapidly melons and potatoes would grow. Deer and wild mustang, ducks and geese, oysters and fish of every description abounded. Foul-tasting but potable water could be had by digging a few feet into the sand and shoring the hole with a barrel. These young soldiers and their camp women frolicked in the surf before moving across the bay and settling down to the business of training for a war.

The camp was set up in what is now the northern part of downtown Corpus Christi. From this base the soldiers fought the mosquitos and the elements for seven months and eleven days. During this period Texas statehood was made official, and in December 1845 the Republic became the Lone Star State.

After sending a reconnoitering party down Padre Island, Taylor decided it would be better to march his men to a site opposite Matamoros. In their trek across the "Wild Horse Desert" on the trail used by the Mexicans as they retreated from San Jacinto, the men suffered from heat, dust, and lack of water. Soon after their arrival at Fort Brown, they engaged and overwhelmingly defeated the larger Mexican army on the flats nearby in two battles—the first, Palo Alto, May 8, 1846, and the second, on the following day at the Resaca de la Palma. From there Taylor and his young West Point lieutenants, among them Ulysses S. Grant, went on to win the territory for the United States.

With the Mexicans whipped and Texas in the Union, ranchers and planters from the States began to move in. Gone were the treasure fleets and pirates, but perhaps the real treasure of these coastal islands is the grass and abundant fresh water that have made ranching profitable since the days of Balli.

This grass and water made life possible for the family of John V. Singer, which was shipwrecked on the southern tip of Padre in 1847 (*Writer's Round Table*, p. 158). He landed near the Padre's old ranch headquarters, now covered by the shifting sand hills. The family built a home from their wrecked schooner and soon prospered by ranching and salvaging until they were ordered from the island by the Confederate army in 1861.

The War between the States brought considerable activity to the coast of Texas. The Union navy attempted to blockade the almost limitless coast in order to cut off Confederate shipments of cotton and hides. Numerous skirmishes took place between the two forces, but lack of communications and remoteness made achieving this blockade an almost impossible task. The shanty town of Bagdad sprang up on the south side of the mouth of the Rio Grande, and the beach was covered with bales of cotton awaiting shipment to Europe.

Attesting the isolation of this area at that dramatic time in history is the fact that the last battle of the Civil War was fought at Palmito Hill near the southern tip of the coast, on May 12–13, 1865, thirty-four days after Lee had surrendered at Appomattox. Ironically the Confederates were the victors. We wonder if, when word of the battle reached the Federal commander in chief, Ulysses S. Grant, he remembered the place where he had fought as a young lieutenant, only a stone's throw from Palmito Hill. Twenty years later it was still in isolation.

Reports made soon after the war by the hardy engineer R. E. Halter brought about the establishment of a series of lighthouses along the coast, and make interesting reading for those who marvel at the fiber of the men who defied the elements on these islands long ago.

It was cheaper and easier to have shallow-draft lighters built in the East and shipped by schooner to Texas than to attempt to have them made in Corpus Christi or Galveston in the mid–nineteenth century. Traffic behind the islands went by flat-bottomed scows through the bays to Indianola or through Aransas Pass into Corpus Christi along a channel that had been dredged to a depth of six feet.

Between 1840 and 1880 an active business sprang up in the coastal area in slaughtering the wild mustangs and cattle for their hides, tallow, and bones. Packing houses dotted the coast from Galveston to Flour Bluff, and salt was in demand to pack the hides in. The trade boomed, then died. Also, during this period the area supported a large green turtle fishery; in 1890 more than 83,000 pounds were produced. This creature no longer inhabits the Texas bays.

Near where La Salle had landed, on Lavaca Bay, the prime entry point for colonists coming to Texas from over the world developed. It was called Indianola. In 1875 Indianola, grown to six thousand population, was heavily damaged by a severe tropical storm; finally, in 1886, the town was wiped from the map by a hurricane. Here, from this flat, featureless shore, countless numbers of pioneering souls streamed to the frontiers of Texas. There were the Germans who followed the rivers, establishing along the banks settlements that bear their imprint to this day. Here too, came much of the building material for early Texas homes, along with the camels for the ill-fated dream of using these beasts as transportation for the army in the "deserts" of West Texas.

Gone too are many other settlements. Throughout history northers and hurricanes, like those at Indianola, have made the barrier islands of Texas inhospitable for permanent settlements, although several have survived. The city of Galveston has clung to the sands of Galveston Island since the early nineteenth century in spite of having been the victim of the worst natural disaster ever to strike the continental United States. The hurricane of 1900 left about six thousand dead in the havoc that had visited the island.

Port Aransas, to the south, is the only other incorporated town on the islands. A large resort settlement flourishes on the southern tip of Padre Island, a colony is growing near the northern end of Padre, private ranches thrive on mid–St. Joseph and southern Matagorda islands, and the air force practices bombing at a range on northern Matagorda. All these establishments receive repeated attacks from Mother Nature, but a better understanding of the environment, more efficient hurricane warning systems, and increased knowledge of construction principles are easing the blows.

After the building of the lighthouses was completed, between 1852 and 1856, and the activity of the Civil War had subsided, the beaches of Texas lapsed into their early ways, unmolested except for a few cattlemen who came after the Ballis and the Singers, fattening their cattle on the lush grasses of their unfenced domains.

The Singers were followed by the Kings, Kenedys, and Dunns on Padre. Each island along the coast was a series of large privately owned ranches—kingdoms populated by cattle, deer, coyotes, turkeys, cranes, herons, and the skittish ghost crab. The barbed-wire entanglements that covered the beaches in World War II only made these kingdoms more secure for a while.

Each of the barrier islands has been privately owned at least once. Today, however, 69.5 miles of Padre have become a national seashore. The plan to conserve the region, begun in 1954 by the National Park Service, is now a reality. Development is progressing slowly, but it is reassuring to know that a part of this coastline will be kept as it was when the Karankawas, the Spaniards, the French, the pirates, Taylor's troops, the Confederates, and the cattle barons roamed its shores. It is hoped that this seashore, along with the large national wildlife refuge on Aransas Bay near Matagorda Island, will be a haven for the abundant wildlife of the area. Where the shore birds soared over the heads of the tall Karankawas, the giant turtles, and the wild mustangs, modern man is striving to reestablish the almost extinct whooping crane before he too is driven from his native shores.

PHYSICAL DESCRIPTION

The 370 miles of coastline plus the bays and tidewater flats make an impressive 624 miles of beach for the shell collector in Texas to investigate. The barrier-island chain has been described as the longest in the world. Located in a semitropical climate, these islands have a minimum width of a few hundred feet to a maximum of nearly four and one-half miles. The southern ends of the islands are usually narrower than the northern ends. The white sand beach slopes gently beneath the Gulf surf. The slope of the continental shelf is only about six feet per mile. Beyond the beach is a ridge of

sand dunes that reaches a height of fifty feet in some places. Behind the dunes the terrain breaks into a scattering of lower dunes and finally into a tidal flat that eventually disappears beneath the waters of the bays and lagoons.

Naturally, there are variations. Let us begin at the Sabine River and make our way southwestward to the Rio Grande. (Reference to the maps in Chapter 4 will aid in following this description of the coast.) We are concerned here only with the shallow waters and not the offshore shelves and their fauna. However, these extremely broad, muddy shelves extending out into the Gulf have caused the beach collecting to be less productive than in some other localities of the United States. This fact, coupled with the lack of natural rock formations and coral reefs close to shore, contributes to the brevity of our species list. Nevertheless, the list is rapidly being added to due to the growing interest in the Texas coast for serious students of marine biology. Actually, the greatest part of the coast is relatively inaccessible to all except the well-equipped and endowed researcher.

Perhaps of less interest to the collector is that part of the coast between the jetties at Sabine Pass and Galveston. Because longshore currents here carry large amounts of silt from the Mississippi, the water is seldom the clear blue that it is off the Florida and southern Texas coasts. Salt marsh extends from the Sabine to MacFadden Beach near Galveston, and oyster grass (*Spartina alterniflora*) grows in the intertidal zone, which is chiefly marsh clay.

State Highway 87, from Sabine Pass to Bolivar Pass, closely follows the beach ridge, which has been stabilized by grasses, and the beach slope is almost imperceptible. The near-shore Gulf bottom is largely clay covered with a few inches of sand. Toward Bolivar Pass the beach on this narrow peninsula separating East Bay from the Gulf of Mexico begins to slope gently, and a wide sand flat occurs at the pass. This flat is a favorite spot for collectors in this area.

At Bolivar Pass we find the second jettied inlet. The area adjacent to the pass encompasses the city of Galveston, which is protected by a ten-mile-long sea wall. Continuing westward, this offshore island is separated from a barrier spit by San Luis Pass, which was long an obstacle to driving along the beach from Galveston to Freeport. A new toll bridge now spans the gap. Along this route the beach slopes very gently and the foreshore area (area between the tide lines) is wide. Low dunes, seldom more than six feet in height, have formed along the beach ridge on the backshore. San Luis Pass is the only channel opening from West Bay to the Gulf. There is a wide sand flat here that becomes a special haunt of the shell collector at low tides.

In the vicinity of Freeport the coast again becomes a part of the mainland and the Velasco jetties act to divert the Brazos River so that its channel can be utilized by the many industrial plants located here. In good weather the beach is crowded in spite of the fact that the Gulf is generally muddy from waters entering it from the Brazos and San Bernard rivers. From here to Sargent Beach, the next point readily accessible to the collector, the beach is eroding badly. A road has been completed to the mouth of the San Bernard and a proposed bridge will soon add another access to the Gulf shore. Between the San Bernard and Sargent Beach are several places where fossil marsh clay is exposed in the intertidal zone. At Sargent Beach, sandstone and beach rock almost completely replace the sand. Instead of a beach ridge, there is a cut bank two to three feet in height. The beach slopes gently away from the bank. Some cottages on stilts line the beach and are separated from the Intracoastal Waterway by a marshy area.

From Sargent Beach, near Brown Cedar Cut, south to the Colorado River, the surrounding land is low and is cut by bayous on its bay shores. The beach nearest Brown Cedar Cut is covered with heavy shell, and the beach ridge has a greater variety of plants than is found to the south. Brown Cedar Cut prevents a four-wheel-drive vehicle from traveling the beach between the Colorado and Farm Road 457 at Sargent Beach. Those who drive northeast on the beach from the Colorado River along Matagorda peninsula will find a wide, flat beach and low dunes.

The bayside is cut by numerous bayous. An amazing amount of driftwood litters the beach ridge; most of this wood floats down the Colorado River. A log jam that accumulated over a period of at least fifty years once choked the river for a distance of forty-six miles upstream. This jam was dynamited and cleared between 1926 and 1930. In 1936 a man-made channel was dredged across the barrier peninsula to the Gulf. This diversion of the major source of fresh water into Matagorda Bay changed the bay from an estuarine environment to one with a salinity more typical of the Gulf of Mexico.

Matagorda Peninsula is isolated and difficult to reach except at the point near the Colorado River where State Highway 60 follows the east bank of the river. Fishing

craft are unable to use the river as an exit to the Gulf except on very calm days, due to the shallow water over the longshore sand bars; otherwise the area would be more extensively developed. The new Matagorda ship channel, maintained by the Corps of Engineers, cuts the peninsula about three miles north of Pass Cavallo. From this pass to the Rio Grande, the Gulf coast is a chain of barrier islands that front a series of shallow coastal bays.

In general the islands' outer shorelines are almost straight. The beach is corrugated with temporary serrations, or cusps, sometimes several hundred feet apart, which are caused by storm action but may be smoothed out by later storms. According to Hedgpeth (1953), the inner shoreline is irregular, scalloped by washover fans or prolonged into spits, which, on the Texas coast, usually point to the southwest. Some of the spits may persist as "flying bars" of small chain islands, for example, the Bird Islands of the Laguna Madre and Mud Island in Aransas Bay. In less humid regions to the south, there is a development of clay dunes fifteen to twenty feet high.

The northernmost two islands in the main barrier chain are privately owned and can only be reached by boat or light plane. Matagorda Island is approximately thirty-four miles long and is bounded on the north by Pass Cavallo and on the south by Cedar Bayou. Matagorda Island receives more rain than the islands to the south and, except at the southern end, dunes are not a conspicuous part of the landscape. The sea oats (*Uniola paniculata*) so familiar to the visitor on Padre Island are replaced here by wiregrass (*Spartina patens*) as the dominant plant on the beach ridge. The beach is fine sand and slopes very gently. Fences cut the beach at frequent intervals on the southern two thirds and small bomb craters from the Air Force practice range on the island pock the wide, flat beach on the northern portion. The rusting hulk of a large ship rests in the surf about mid-island. The absence of beachcombers on this island has permitted driftwood, bottles, and other wreck to accumulate in quantities on the beach. The area behind the dunes on this most isolated of the barrier beaches is a wonderland. Here man has developed the natural resources without despoiling nature. The acres of lush grasslands are crossed by shell-paved roads from which can be seen incredible quantities of wildlife mingling with the vast herds of cattle. Small, fenced-in wooden tepees, serving as quail covers, dot the flat terrain, which is devoid of the brush these birds naturally choose for protection and nesting.

St. Joseph Island is bounded on the north by Cedar Bayou and on the south by Aransas Pass, an inlet jettied in 1887, through which the ocean-going freighters bound for the port of Corpus Christi must pass. The island is approximately twenty miles long and inhabited only by a few ranch employees. Previously, in winter and early spring, commercial beach seiners from Port Aransas, using nets over twenty feet in length, worked the outer beach. This practice is against the law in or on any of the waters of the Gulf of Mexico within one mile of the Horace Caldwell Pier located on Mustang Island and the Bob Hall Pier on Padre Island and within one thousand feet of the shoreline of Padre Island in Nueces County. The owners of St. Joseph could and often did bar this practice because there is no public access to the island. Early in 1969 they chose to put a stop to the commercial seining.

The southernmost five miles of St. Joseph Island comprise an open flat where the dunes have not developed. The natural shifting of Aransas Pass during the ages before it was stabilized by the jetties occurred over this stretch of beach. The lighthouse that stands on Harbor Island was built in 1852 to guard the pass. During the Civil War a village of two thousand people flourished on St. Joseph opposite the lighthouse. Through the years following the destruction of the village during that war, the pass gradually shifted to the south, creating a vast flat and leaving the lighthouse nothing to guard. It was deactivated in 1952 and the old pass site is called North Pass.

In Lydia Ann Channel, the western boundary of St. Joseph, the wreckage of a World War II Victory ship, *The Worthington*, lies next to the island's bay shore opposite the old light. This ship, torpedoed by German submarines in the Gulf, was being towed to Corpus Christi for salvage. Considered a hazard to navigation by port authorities it was denied entry, and the owners anchored it across from the lighthouse where it was sunk to protect navigation. It is now a favored spot for anglers. On the shore nearby, fragments of the shell-crete foundations of the former village can be found.

The northern end of the island is covered with sand dunes in the foreshore area. It differs from Mustang Island in that the small active drifts common at the base of the dunes on Mustang are rare or absent on most of St. Joseph. Vegetation is similar to that on Padre Island but the backshore growth is denser with large clumps of mesquite trees.

Mustang Island is separated from the southern tip of St. Joseph by the Aransas Pass jetties. The fishing-resort town of Port Aransas is located on the northern end and is connected with the town of Aransas Pass by a ferry and causeway and with Corpus Christi by a road (completed in 1954) down the center of the island to the John F. Kennedy Causeway. The University of Texas maintains its Marine Science Institute at Port Aransas.

The foreshore of Mustang Island is a gently sloping beach of fine sand and is passable to passenger cars at times during low tides. This fine sand is one of the first things a visitor notices, for it is more highly sorted than most beach sand in this country. The backshore, that area between high tide and the base of the dunes, is approximately 150 feet wide. The fore dunes are nearly all stabilized by sea oats. However, the small dunes at their seaward base are in their first stages of development and are more or less active. Mustang is not always joined to Padre Island at its southern boundary as it has been since the dredging of the ship channel at Port Aransas closed Corpus Christi Pass. These islands become separated intermittently when storm tides wash the old pass open. A bridge on the island road spans this pass. They will be continuously separated when the Mustang Island Water Exchange Pass, which was dedicated February 6, 1971, is completed and bridged at the island road.

Padre Island, a low, barren, stormswept strip of sand for 113 miles, is the longest barrier island in the world. It separates the Laguna Madre from the Gulf of Mexico. There are few landmarks for the beach traveler. Causeways built at the northern end in 1950 and the southern end in 1954 link it to the mainland. Nueces County in the north and Cameron County to the south have developed parks at either extremity. Twelve miles south of Nueces County Park, the Padre Island National Seashore Area has been established and is being developed with federal funds. A causeway is proposed in Willacy County to link the island to Port Mansfield, 78 miles south of Bob Hall Pier. We shall designate this pier in Nueces County Park "mile zero" when indicating the location of the prominent features south of it on Padre Island (see Map 5).

The topography of Padre Island, like that of the other barrier islands, is shaped by the wind and the sea. The most conspicuous features are the sand dunes and the washovers. Washovers are open, flat areas or shallow channels formed by the spilling of water across the barrier island during the times of storm tides. Sand soon closes the seaward opening and the remaining areas appear to be dry washes. There are no washovers on the northern end of Padre Island, but beginning about 50 miles from the pier and extending to the southern end they become numerous. Vegetation invades the washovers only very slowly and, since there is little or no vegetation, dunes do not form.

The first plant to invade the shore is sea purslane. Drs. Hildebrand and Gunter (1955) have described a dune's development as being based on this plant. As the purslane grows, sand collects around its base and builds a mammilliform mound or coppice dune. After this mound reaches a certain height, and if it is not dissolved by a storm tide, it is invaded by other plants, usually sea oats. However, in very wide washovers, such as the old North Pass on St. Joseph Island, where there are no other plants adjacent, coppice dunes up to six feet in height are built solely on sea purslane. Normally other plants would crowd out the purslane before the coppice dune was three feet high. The dunes on Padre do not reach a great height, probably no more than thirty to forty feet. Behind the beach dunes are grass flats and smaller dunes. The shore along the Laguna Madre is a poorly defined area of mud flats that merge with the waters of the lagoon.

About 8 miles south of the pier the Chevron offshore oil field can be seen. The entrance to the seashore area is at the 12-mile point. A paved road has been constructed down the center of the island to these facilities at Malaquite Beach and is projected for the entire island. In the meantime the beach traveler in a passenger car can safely go only about 20 miles or to the beginning of Little Shell, a section that is difficult to delimit but has been so called by fishermen because of the large deposit of small clam shells on the foreshore. At the long-closed Yarborough Pass (28.3 miles) the beach becomes streaked with blackish lines caused by magnetite, or iron oxide, in the sands. About three miles south of this pass, Big Shell commences and extends 10 to 12 miles farther south. In Big and Little Shell the steep, high-tide crest and the soft shell on the backshore make passage difficult. Big Shell differs from Little Shell in being composed only of large heavy shell. Mostly, however, the shells are fragmented and the fragments form ridges and cusps on an otherwise steep beach. Old, badly worn, heavy clam shells almost form a pavement as one nears the Mansfield jetties where an outcropping of clay is encountered in the foreshore.

The barrier islands are paralleled by two or three rows of sand bars with troughs between them that are called longshore bars and longshore troughs. One dif-

Fig. 2. Big Shell, the "graveyard of ships." A shrimp boat grounded in the pounding surf at the point on Padre Island where a convergence has produced a historic hazard to navigation.

ference in the foreshore of mid–Padre Island is that the longshore bars do not parallel the shore, but form what the fishermen call "blind guts" along the shore. The absence of the bars also accounts for the area called Big Shell. Without the bars the full force of the waves strikes the beach, fragmenting all but the large heavy shell so characteristic of this stretch of shore. The probable reason for the absence of sand bars is the abrupt curve that the coastline makes in this mid-region. It not only changes direction in relation to the prevailing winds but also becomes the meeting place of the longshore currents.

The same approximate area is referred to by Leipper (1954) as a "graveyard of ships" (fig. 2). This 100-square-mile area of current convergence is 20 miles east and 40 miles south of Aransas Pass. If one examines the wrecked boats along the shore, he is impressed by the fact that they are most numerous in the vicinity of Yarborough Pass, which lies in the convergence region. A beached shrimp boat being pounded by the waves is a familiar sight in this area. At times it is possible to see an abrupt change in the color of the Gulf, as if there were a line of demarcation in the sea with cloudy, blue water to the north of it and clear, aqua water to the south.

Three and two-tenths miles below Yarborough Pass, which is recognized by the radio tower at an oil com-

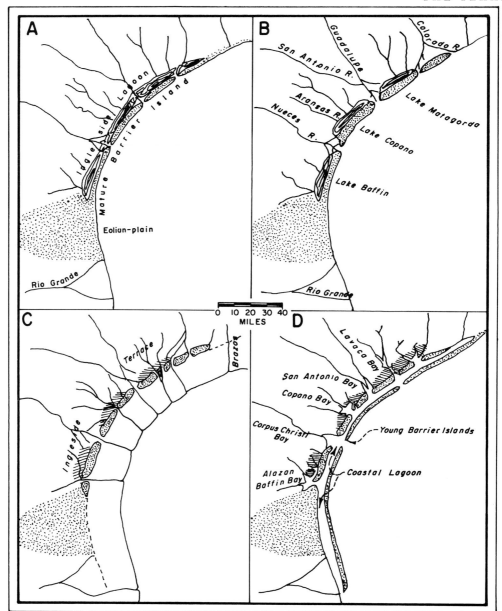

Fig. 3. Theoretical stages in the development of the barrier islands of Texas (after Hedgepeth 1953): *A*, master streams flow directly into Gulf; *B*, period of delta building; *C*, fall in sea level entrenches streams; *D*, rise in sea level widens the estuaries of master streams into bays.

pany camp near the bay side, one can find a ranch road behind the dunes. This simple trail extends 13 miles down the center of the island and reenters the beach near a large, privately owned beach cottage at the 44.4-mile point. This road is inundated following rains and storms.

The only other significant landmark seen before arriving at the Mansfield jetties is the boiler of the *Nicaragua*, a six-hundred-ton Mexican coastal steamer that ran aground during a storm on October 18, 1918. The crew of this hapless vessel was rescued five days later. According to local tradition the ship was running guns to be used in the Mexican revolution, but it was probably a banana boat on its regular run. Her remains protrude from the surf 65.5 miles south of the pier and a small Park Ranger station can be seen here back in the dunes.

A point on the beach about 2.8 miles north of the Mansfield jetties is a haunt of the metal-detector-equipped treasure hunter because a Spanish galleon lies ¾ miles offshore. An out-of-state company began salvage operations in 1969 on the 415-year-old treasure ship, an action that may put an end to the thrill of catching the gleam of an ancient doubloon in the sand. Laws are now being proposed in the state legislature to protect such treasures, and the Texas Antiquities Committee is making plans for the preservation and display of the artifacts.

The jetties, 77.6 miles south of the pier and 36 miles north of Brazos Santiago Pass, were completed in May 1962. Prior to that it was possible to drive the entire length of the island. There is talk of a causeway to connect this section of the island to the mainland, but at present these jetties are accessible only by boat or four-wheel-drive vehicle. Here the water is generally clear because no rivers empty into the Laguna Madre, which is opened by this cut.

South of the Mansfield jetties the driving is difficult until one reaches the pavement about five miles north of Isla Blanca Park; then it is an easy drive to the Port Isabel jetties and the well-developed resort facilities at this end of Padre. Fine hotels and cottages as well as more modest motels and trailer parks dot the sands of this almost tropical part of the island.

Brazos Island, the last and smallest in the chain, can be reached by going inland and taking State Highway 4, which goes directly to Boca Chica Beach. As you go through the town to Port Isabel the old Point Isabel Lighthouse (1852–1905) can be seen. The pass this historic light guarded is said to have been named El Paso de Brazos de Santiago (the Pass of the Arms of St. James) by Piñeda in 1519.

The road passes near the site of the last battle of the Civil War fought on the barren flats of Palmito Hill. At present there is no development at the beach although hundreds of people visit it every summer to fish and swim. The Boca Chica inlet, which is often closed, forms the southern boundary of Brazos Island. From this inlet to the mouth of the fabled Rio Grande (10 miles from the jetty) the beach becomes progressively more silty and muddy. In the immediate vicinity of the river, which empties directly into the Gulf, marsh grasses grow into the edge of the water. The flow of the Rio Grande has been greatly reduced by dams along its course; therefore, the supply of fresh water entering the Gulf has a negligible effect on the ecology of the region.

The most conspicuous and biologically significant feature of the shoreline are the passes we have described as we proceeded down the coast: Sabine, Bolivar, San Luis, Matagorda, Cavallo, Cedar Bayou, Aransas, Corpus Christi, Port Mansfield, and Brazos Santiago. These passes become stabilized at the extreme southern ends of the bays into which they open. Related to these inlets are the tidal deltas built up by deposition of sediment carried from the direction of the barrier island northeast of the inlet. Thus the delta is built up at the northern end of the island, that is, south of the inlet.

Harbor Island at Port Aransas and the islands at Pass Cavallo are examples of the tidal delta. Pass Cavallo is probably the most stable natural pass on the Texas coast. We will recall that La Salle entered Matagorda Bay through Pass Cavallo, which has been shown on charts since 1690, and that the German pioneers of Texas braved its treacherous shoals to reach Indianola.

The barrier islands, formed between two and five thousand years ago, earn their name because they make a barrier between the Gulf of Mexico and a series of shallow coastal bays and lagoons (fig. 3). The bottoms of these bays are for the most part mud; the Laguna Madre, however, has a predominately sandy bottom. Behind the coastal bays is a system of inner bays, of lower salinity than the coastal or outer bays, which are in turn lower in salinity than the Gulf. These coastal bays are connected to the Gulf of Mexico by the series of passes previously listed. Seven of these passes—Sabine, Bolivar, Matagorda, Aransas, Port Mansfield, and Brazos Santiago—have been stabilized as navigable channels with jetties maintained by the United States Army Corps of Engineers.

Across the bays from the islands the shores are wind-tide flats and the surface is only a few feet above sea level. The greatest fluctuation of water level in the bays is caused by the wind and these changes in water level are called "windtides." The windtide, especially after the passage of a cold front, leaves large areas of shoreline exposed and these exposed flats are referred to as "windtide flats."

Indications are that this area, now a part of the mainland, was once a chain of barrier islands. The bays that separate the ancient from the new barrier islands vary from five to six miles in width. These lagoons and large embayed (indented) river mouths are dotted with small islands and spoil banks, thrown up from the dredging of the Intracoastal Waterway, and have become nesting grounds for countless thousands of birds.

The seasonal temperature of the area rivals that of Florida and the sea breezes make even the hot Texas summers pleasant. The average reading in the spring is 73.2 degrees, while the summer average is 81.8, fall 77.4, and winter 65.5. According to the *Texas Almanac*, temperatures of 32 degrees or lower occur only four years out of ten in Brownsville, the southernmost tip of the coastal range. Killing freezes, those of long duration, occur about once every ten years. August is the hottest month and late spring and early fall are the wettest times.

Perhaps the fear of hurricanes with their great property damage has been the largest single factor in retarding the development of this beautiful, semitropical coastal area. Tropical cyclones, the largest and most destructive storms that affect the Texas coast, have been occurring in the Gulf of Mexico since long before the time of man. Ancient languages spoken in the area had a word for them: Mayan, *hunrakin*; Guatemalan, *hurakan*; Carib Indian, *aracan*, *urican*, and *huiranvucan*.

Of the approximate ninety-nine tropical cyclones affecting Texas during the period 1766–1961, only sixteen have been designated "major," that is with maximum winds of 101 to 135 miles per hour; however, two-thirds of them have had winds over 74 miles per hour. Mass mortality of near-shore marine fauna due to hurricanes is significant on the Texas coast.[1]

Flora and Fauna

Vegetation on the coast of Texas is sparse because of the wind and the high salinity of both air and soil. On the barrier islands, the plant growth is confined to the dune area and the sand barrens behind bare beaches. On and near the dunes you will see the silvery beach croton, *Croton punctatus* Jacq.; waving sea oats, *Uniola paniculata* Linné; the large, purple flowers of the trailing goatfoot morning-glory or railroad vine, *Ipomoea pes-caprae* Linné; the smaller white-flowered *Ipomoea stolonifera* Gmelin; the spreading fleshy sea purslane, *Sesuvium portulacastrum* Linné; the yellow beach evening primrose, *Oenothera drummondii* Hook, which opens in late afternoon; the dune sunflower, *Helianthus argophyllus* Torr. & Gray, with silky, silver leaves; wiry dune sedge *Fimbristylis spa-*

dicea Vahl; and several other varieties of sand-strand vegetation.

In late August and early September when the sea oats are at their prime, they wave their heavy heads from atop the dunes, casting long shadows across the rippled, white sands to make an unforgettable sight. They do not extend through the eastern half of the Texas coast.

On the barrens of the islands and across the bays on the shores of the mainland the saltwort, *Salicornia bigelovii* Torr., whose succulent stems take on a reddish tinge in the fall, the round-leaved pennywort, *Hydrocotyle bonariensis* Lamarck, and the fleshy, green *Batis maritima* Linné sparsely cover the salt-glinting ground. In more marshy areas the flora is dominated by clumps of a grass the Mexicans call *sacahuiste*, *Spartina spartinae* Merr., which forms a dull green background for the bright wild flowers that dot the area seasonally. Indian blanket, *Gallardia pulchella* Foug, the yellow aster, *Machaeranthera phyllocephala* Shinners, and a daisy, *Borrichia frutescens* Linné, are but a few of the latter. Cat tails, *Typha domingensis* Pers., stand guard in ditches and low spots.

Isolated motts (an old term derived from the Spanish for clump, *mota*) of live oak, *Quercus virginiana* Mill, offer a rare patch of shade and their wind-sculptured forms add variety to an otherwise flat terrain. The sweet bay, *Persea borbonia* Speng., whose leaves add zest to a good seafood gumbo, the false willow, *Baccharis angustifolia* Michx., and the groundsel, *Baccharis halimifolia*, to the east, are some of the first woody invaders of a salt marsh. Scattered about with the usual chaparral-type (thorny brush) vegetation of the shell ridges one sees the stately Spanish dagger,

---

[1] During the ten-year period in which I have been working on the material for this book three major hurricanes have struck the Texas coast: Carla on September 11, 1961, Beulah on September 20, 1967, and Celia on August 3, 1970. The first two had a marked effect on the marine life of the coast, but the last primarily affected man and his structures. Finding myself alone in my home on that afternoon in August with the edited copy of this manuscript and the only carbon copy, I became frantic looking for a way to protect them. As my windows broke and water poured through the ceiling, the light fixtures, and each slight orifice, I tried to recall scenes of other natural disasters. The vision I conjured up as my home shook and leaked was of fallen walls with a lone refrigerator standing in their midst. In haste and desperation I pulled the contents from my refrigerator and replaced them with the manuscript and its copy—there they remained until I remembered them several days later. When the door was opened the light did not come on as usual, but the papers were dry and safe; that is quite a bit more than I could say about the rest of my home. Celia will be long remembered by those caught in its unpredictable, devastating path.

Fig. 4. Salt marsh flats: *A*, marsh grass at low tide near Pass Cavallo; *B*, close-up of marsh with *Littorina* snails, *Abra* clams, and pelican feather; *C*, close-up of marsh grass with *Littorina* on leaf and *Mercenaria* in sand.

Yucca treculeana Carr., whose massive shafts of white flowers begin to bloom in February; the prickly pear, or nopal, *Opuntia lindheimeri* Engelm., the fruit of which formed an important staple in the diet of the Karankawas during summer and fall; and the Turk's-cap, *Malvaviscus drummondii* T.&G., which adds a dash of red to the fall landscape.

There are other plants that are submerged in the shallow waters of the bays and marshes but whose importance to the ecology of the area cannot be underestimated. Not only do they provide cover and protection in the nursery grounds of game fish and mollusks but they also are responsible for the development of the coastal prairie. As the plants die and decay, the substratum is built up, the water becomes shallower, and reed-swamp plants take over. These are succeeded by marsh grasses, which in turn are followed by the prairie grasses. Here the most prominent submerged grasses are widgeon grass, *Ruppia maritima* Linné; a type of eelgrass, *Diplanthera wrightii* Aschers; turtle grass, *Thalassia testudinum* Koenig & Sims; and manatee grass, *Syringodium filiforme* Kutzing.

The grass prairies and salt marshes that once extended to the water's edge are being invaded by thorny brush. This brush may be deplored by stockmen of the region but this extension of subtropical jungle plants has saved thousands of acres from erosion and complete denudation. These vast coastal prairies and salt marshes are not as extensive as they once were due to overgrazing by early pioneers, drainage, and cultivation (fig. 4). It is an ecological tragedy because a salt marsh has a higher rate of primary production, that is, the manufacture of plant material by photosynthesis, than any other crop in the world except sugar cane. *Spartina*, without cultivation by man, yields three times as much as the best wheat lands. The shoreline marshes, wastelands to many, actually produce much of the nourishment for many forms of life in the coastal seas. Destroying the marshes by draining or filling to provide industrial sites can bring disaster to the fauna of shore and ocean.

The coast of Texas is a birdlover's paradise. It would be foolish in a book of this nature to try to elaborate on a subject that has been so well covered by Roger Tory Peterson in his *Field Guide to the Birds of Texas*. It is common knowledge that the area falls in one of the major flyways of this hemisphere, and Mrs. Connie Hagar in Rockport has done more than any other one person to record the birds' activities.

While shelling on our coasts one can see avian hordes that include the giant blue heron, roseate spoonbills, snowy egrets, reddish egrets, cattle egrets, wood ibis, white pelicans, laughing gulls, herring gulls, curlews, avocets, terns, skimmers, and cormorants. Racing at the water's edge are the smaller species, such as sanderlings, sandpipers, willets, turnstones, and dowitchers. Behind the dunes one might find a huge flock of the giant sandhill cranes or on a coastal roadside in the spring watch a pair of prairie chickens perform their ritual mating dance.

A still greater thrill can be had between October and March when your boat glides near the almost legendary whooping cranes as they feed along the edge of the Intracoastal Waterway behind Matagorda Island. The quail and turkeys are almost unbelievably abundant. The turkeys on St. Joseph were introduced by the late owner of the ranch, who built weird roosts with metal flanges to protect them from preying coyotes, since no natural roosts exist on the island barrens.

On rare occasions a giant frigate bird strays our way from faraway Yucatán to soar effortlessly over the Texas shore. It is not an uncommon sight while searching for a shell to come upon a speckled egg in an indentation in the sand. If it is that of a black skimmer, the mother bird will make shrieking divebomb attacks on the unwary beachcomber. At times the small spoil islands in the bays appear to be paved with eggs, some of which are laid so carelessly that they roll off the low banks and break.

Such landbirds as the mockingbird, falcon, cardinal, barn swallow, hummingbird, meadow lark, horned lark, and warblers, to name but a few, inhabit the land. Ducks raft on the bays by the thousands and geese fly in honking formations overhead. It is a birdwatcher's paradise.

The mammals and reptiles are not so numerous. The wily coyote comes to the water's edge in the evening to scavenge for food. Spotted ground squirrels, kangaroo rats, grasshopper mice, jack rabbits, pocket gophers, and deer hide under clumps of grass during the day. Cattle stand amid the sea oats on top of the dunes or bunch together in the breeze at the water's edge through the night to avoid the mosquitos behind the dunes. Lizards, turtles, and rattlesnakes sun themselves on a winter's day on the lee side of a dune. The tracks of these animals tell tales of hunting and being hunted on the smooth, white sand.

In 1860 W. S. Gilbert, who made a coastal survey for the United States government, in his description of the San Antonio Bay area wrote, "The scattered 'motts' within the same limits afford a refuge for the wild animals that infest the country. Panthers, tiger cats and mustangs abound" (p. 355). Perhaps panthers and mustangs no longer roam the region, but nature, unspoiled, is still visible at every turn on the barrier islands of Texas.

The Coastal Waters

The previous sections describe the shores one walks in search of a sea shell; in this section the many factors influencing the variety and number of marine animals will be examined. The tides, waves, and currents, the salinity, temperature, and turbidity, and the substratum all have an influence on the marine fauna. The combined effects of variations in these factors make the long Texas coast a relatively inhospitable environment for marine mollusks.

TEMPERATURE. Perhaps the greatest controlling factor in the distribution of marine animals is temperature. The temperature of the ocean has a very limited range; consequently, comparatively small changes in temperature have a very decided effect on marine organisms. In the ocean, temperature creates as much of a barrier to the dispersal of animals as a mountain range does to land animals. The maximum and minimum temperatures each animal can tolerate limit marine animals to certain regions of the seas.

Texas is renowned for its "blue northers." In reality these are rapidly moving cold fronts that sweep across the land, dropping temperatures as much as thirty to forty degrees in a few hours. An officer in Taylor's army described a typical norther (Henry 1847, p. 45) —"Hast thou, dear reader, ever felt a norther? Heard tell of one? No. Well, your northern cold is nothing to it. It comes 'Like a thief in the night,' and all but steals your life. I'll venture to say there is no part of the U.S. cursed with such a variable one in the winter. Oh Texas, if we have not 'fought, bled and died for you,' we have done as Dick Riker (peace to his ashes) did, 'suffered some.'" In the century and a quarter since that was written the northers have not changed and the beach is a miserable place to be during one.

The extremely rapid onset of the northers, accompanied by sharp drops in water temperature, occasionally causes massive mortality of marine life. Texas bays are invaded by such subtropical animals as the commercial shrimp, *Penaeus setiferus* and *P. aztecus*, in the spring and summer; in the fall and winter they return to the Gulf. The lingering strays seldom survive a Texas winter.

Biologists have determined that if the damaging cold waves are preceded by other, lesser freezes their destructive effects are lessened. Some animals escape to deeper water if the onset of the cold wave is slow. However, the slow-moving or attached mollusk has little opportunity to escape. The marine life in the shallow bays must be able to withstand temperature changes far greater than animals in corresponding positions in deeper water on the outer coast.

According to Pulley (1952) and Moore (1961) the coast from South Texas to Cape Hatteras, North Carolina (excepting southern Florida), is considered a single biogeographic area, the Carolinian Province (fig. 5). The climate in this zone is warm temperate; the tropical element is scanty. The line of demarcation between the Carolinian and the Caribbean provinces is at a point near 27° N latitude. Benthic (bottom-dwelling) forms south of this line are usually more West Indian in nature and are seldom found alongshore north of it (Hedgpeth 1953).

Only those forms able to withstand wide extremes of temperature during the year can survive along the Texas coast. Cold winter winds greatly lower the temperature of inshore and bay waters but the waters of the continental shelf may never get below 65° or 70°F. This stable offshore temperature results in a shelf fauna almost entirely West Indian in character (Pulley 1952), while the shoreline fauna in the northern Gulf is similar to that of the Carolinas. The broad range of environmental factors that characterize the estuarine, bay, and coastal waters of Texas produces a mixed invertebrate fauna containing both temperate Atlantic and tropical Caribbean species, with a very low endemic component (Hedgpeth 1953).

CLIMATE AND SALINITY. The surface climate of the Texas coast has a direct effect on the water temperature and the salinity of the bays and lagoons. From Sabine to Galveston we find a humid climate with a slight surplus of moisture (fig. 6). The next climatic zone is the moist subhumid, extending from Galveston to Port Lavaca. Here moisture supply and loss are in balance. This condition is not always true in the case of the next zone, from Port Lavaca to Corpus Christi, a dry, subhumid belt. The transition between the dry, subhumid region and that of the semiarid portion is

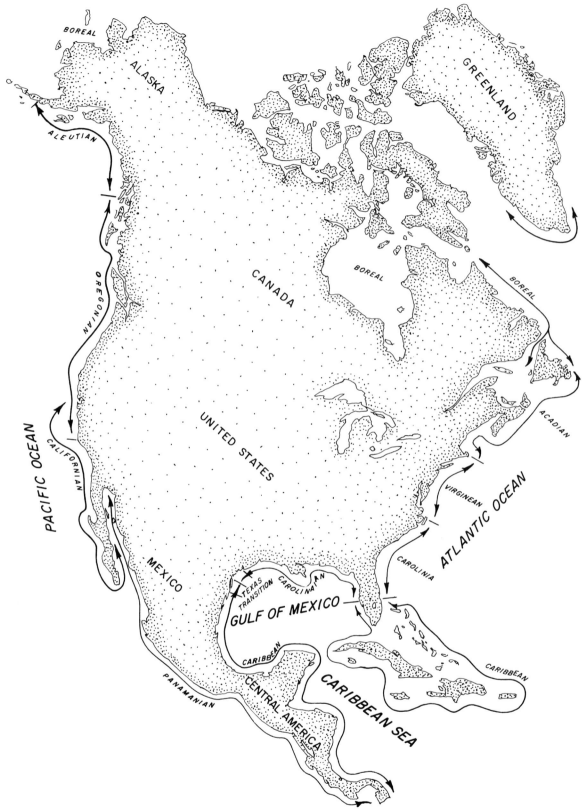

Fig. 5. Faunal provinces.

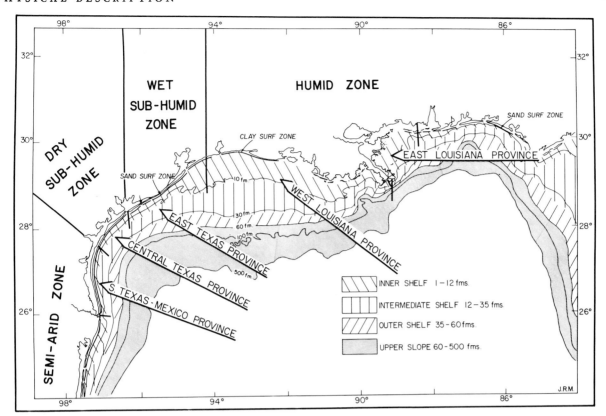

Fig. 6. Areal distribution of macrofaunal assemblages, northern Gulf of Mexico, and climatic zones (map modified from Hedgpeth 1953). Faunal provinces, extending generally to depths of about eleven fathoms, are based on breaks in mollusk distribution and related to climatic and physical boundaries (Parker 1960, based partially on data from Pulley 1953).

very gradual. The remainder of the coast is characterized by a permanent moisture-deficient climate, almost desert (Hedgpeth 1953).

Along the Texas coast these climatic zones are roughly correlated with rainfall but in the bays this correlation breaks down due to the drainage of rivers into the bay systems. If there has been extensive rain and flooding along the watersheds of the rivers that drain into a bay system, the salinity will be lowered in spite of the fact that there has been little or no rain in the bay area itself. At present the annual rainfall on this coastline ranges between 15 and 40 inches per year, with, of course, a trend toward a drier, more arid climate in South Texas.

Normal Gulf salinity is usually 36 parts per thousand (36 $^o/oo$).[2] Salinity conditions are nearly stable at Port Isabel, where the monthly averages range from 30.4 to 37.5 $^o/oo$. This fact permits shells that are only found offshore in other less stable areas of the coast to move in closer to shore here. Variation is the rule, how-

ever, in most of the other tidal waters of the Texas coast. Salinities can rapidly decrease after heavy rains and flooding along the rivers. Hurricane Beulah, in 1967, is an extreme example of the devastation to sessile (sedentary or attached) marine life produced by sustained flooding and reduced salinities over a long period. The animals that inhabit estuarian waters are more adaptable than those of other more stable regions but they too will succumb if the water stays "fresh" for too long a period. It is usually about eighteen months after the water returns to its normal salinity before the bottom fauna becomes reestablished.

A rapid change in the proportions of salt in the

[2] In referring to the degree of salinity of a body of water the following terms are used:

Oligohaline	0/00–1/00
Mesohaline A	2/00–10/00
Mesohaline B	10/00–18/00
Polyhaline	18/00–30/00
Marine	30/00–40/00
Metahaline (hypersaline)	above 40/00

water will interfere with the mollusk's process of osmosis, during which fresh water moves through a membrane into a region of greater concentration and in doing so sets up a considerable pressure known as osmotic pressure. When a marine mollusk is placed in fresh water, the fresh water rushes into its cells, which have a higher concentration, causing expansion and rupture of the cell tissues. If the change is gradual the mollusk might be able to adjust its enzyme output and thus prevent immediate death due to breakdown of its membranes. Such bivalves as *Barnea* (the angel wing), which are unable to completely close their shells, are particularly susceptible.

If certain marine invertebrates are not forced into a sudden salinity change they are able to make considerable adjustment. For example, the dilution of sea water by fresh water from rains or from a limited amount of flooding of the rivers may have little effect on marine animals. A reverse process of gradually raised salinities due to evaporation and drought may also have little effect. But, if great amounts of fresh water are introduced into a bay, or if portions of it become cut off and evaporation goes on until the salinity becomes exceedingly high, all the strictly marine animals will die.

A look at the map of Texas shows at least nine rivers emptying into the waters of this area. Extended droughts, alternation of the drainage pattern by reclamation projects, or agricultural diversions along these rivers cause low river runoff with accompanying increases of the salinity in landlocked bays. Normally these bays are less saline than the open Gulf; consequently, when the salinity increases they are invaded by many marine or open-Gulf species of invertebrates, and the low-salinity or brackish-water mollusks disappear. As the salinity of the bay water approaches Gulf salinity, either from a fresh or from a hypersaline condition, the number of species present increases and the number of individuals of each species decreases (Keith & Hulings 1965).

Flooding by the rivers emptying into the bays, bringing an influx of fresh water on the one hand and summer evaporation on the other, causes variable and often extremely low and high salinities. Normal salinity for Aransas Bay is about 19 to 30 $^0/_{00}$ as compared with 31 to 36 $^0/_{00}$ for the open Gulf. It consistently rises during the summer months and was recorded as high as 42 $^0/_{00}$ in August 1951. Port Isabel, a more stable area, has no river drainage into its lagoons, and the flow of the Rio Grande (10 miles south) has been reduced to but a trickle due to upstream damming. The Baffin Bay of the Laguna Madre is one of the saltiest bodies of water on earth. During the late thirties it was rivaled only by the Dead Sea and the Great Salt Lake.

Mass mortality due to killing cold spells in the winters, high salinity in the summer droughts, and flooding of the bays or estuaries have been characteristic features of this area since Pleistocene times (about 1 million years ago); yet man and animal on the Texas coast still find the adjustment difficult.

CURRENTS. Currents, or movements of water, are related to the productivity of the sea. Currents help to distribute not only nutrients for both plants and animals, but also the animals themselves. Of primary concern to the area under consideration are the inshore currents. As yet they have not been studied in sufficient detail but it would seem that the Mississippi Current follows the coast southwestward, getting narrower toward Tamaulipas, Mexico, and is the principal force in governing the productivity of the region. With respect to the current system of the Caribbean and western Atlantic as a whole, the currents of the Gulf of Mexico, exclusive of the main system of the Florida Current, comprise an eddy (fig. 7).

It would be well to consider briefly the general nature of the currents that may be expected in the Gulf. Swerdrup (1942) lists three different groups of currents, each of which is represented in the Gulf of Mexico. These are as follows:

1. Currents that are related to the distribution of density in the sea
2. Currents that are caused directly by the stress that the wind exerts on the sea surface
3. Tidal currents and currents associated with internal waves

No body of water of comparable size has generated as much difference of opinion regarding its current patterns as has the Gulf of Mexico. It is a field under study, and only a few general remarks can be made about the Gulf currents.

In the northern hemisphere the Coriolis force (force due to the rotation of the earth) acts toward the right of the flow of water which results from the movement of water of higher density into water of lower density. Temperature, salinity, and atmospheric pressure determine the relative density of water. Hence, in this hemisphere the more dense water is on the left of a person standing with his back to the current.

Surface or wind currents involving frictional and Coriolis force will become better understood when

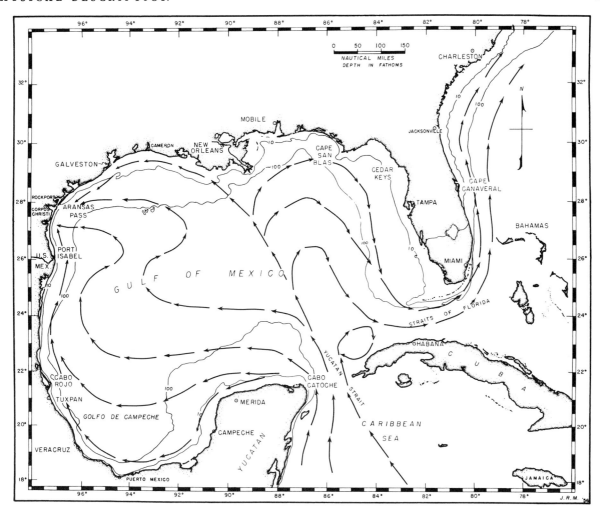

Fig. 7. Surface currents in the Gulf of Mexico, based on studies by Dale F. Leipper (Parker 1960).

studies being made from data gathered on the oil-drilling platforms off the Gulf coast are analyzed.

Tidal currents are horizontal movements of water that have the effect of raising the sea level at a given location. On the Gulf the bays and lagoons have restricted outlets or passes to the sea. All the water necessary to bring about a change in level of the bays must flow through these passes. At certain stages of the tide these tidal currents become quite strong through the passes.

The sands that pile up on the south side of the jetties are part of the evidence that the prevailing set of inner coastal currents is westward and southward, or counter-clockwise, in the northwestern part of the Gulf, at least to Big Shell on Padre Island, where there seems to be a convergence with a northward current along the

shore. Mounds of driftwood and storm debris at Big Shell attest this convergence.

One does not think of the currents of the Gulf of Mexico without a reference to the Gulf Stream. On either side of the equator is a zone where the wind blows constantly in one direction. These winds, trade winds, drive surface waters before them. South of the equator these wind-blown waters form the South Equatorial Current and north of the equator, the North Equatorial Current. The latter is driven across the Atlantic by northeast trade winds to be divided into two currents. Half of the North Equatorial Current joins the South Equatorial waters near the Windward Islands and continues northwestward across the Caribbean through the Yucatán Channel into the Gulf of Mexico. Here a pressure wall turns the current to the northeast out through the Straits of Florida. This current is now the Gulf Stream. As the warm and salty waters of this stream flow along the southern Atlantic coast of North America they are reinforced by the remainder of the North Equatorial Current. Opposite Cape Hatteras the Gulf Stream begins its long journey eastward across the Atlantic.

The action of the Gulf Stream on the waters of the Gulf of Mexico, combined with other factors, sets up a whirlpool-like current or drift in a clockwise direction. This eddy tends to carry some of the sediments from the rivers northward and eastward toward the Florida coast. A brief study of these currents will aid the beachcomber in following the travels of his "treasure" from its far-away source. Leipper (1954, p. 125) writes that

a description of the currents of the Gulf of Mexico is provided in the United States Coast Pilot (1949).

Under normal conditions, at all seasons of the year the great volume of water passing northward through Yucatán Channel into the Gulf of Mexico, spreads out in various directions. Surface flows set westward across Campeche Bank, the Gulf of Campeche, and the Sigsbee Deep; northwestward toward Galveston and Port Arthur; north-northwestward toward the Mississippi Passes; and eastward into the Straits of Florida.

A straight line drawn from Buenavista Key, Western Cuba, to the Mississippi Passes forms an approximate boundary between movements having different directions. West of this line the drift is generally northward or westward, while east of it the drift is eastward or southeastward toward the Straits of Florida.

There are northward flows along the west side of the Gulf between Tampico and Corpus Christi, in the vicinity of the 100 fathom and 1,000 fathom curves, north of the Sigsbee Deep between the 2,000 fathom and 100 fathom curves, and along the west coast of Florida.

In general, the surface circulation is the same at all seasons. There is, however, some seasonal change in velocity, the flow being generally stronger in spring and summer than in autumn and winter.

The current near the Florida Keys is variable and uncertain.

In other words, there is much to be learned about the currents off the shores of Texas. However, it is known that there is only a slight and slow exchange of water between the Gulf and its neighboring bodies of water (Atlantic and Caribbean). This rate of exchange is a contributing factor to the higher overall salinity of the Gulf of Mexico as compared to that of the Atlantic and Caribbean.

TIDES. Tides are one of the factors that keep the ocean in motion. They may be simply explained as twice-daily waves that move like the hour hand of a clock about a central point mid-ocean. These waves run twelve hours and twenty-five minutes apart; their crests are high tides and the troughs are low tides. They are the response of the waters of the ocean to the gravitational pull of the moon and the more distant sun. These heavenly bodies set the seas in motion, but the local nature of tides is determined by the underwater topography of the area—the width of an inlet's entrance, the depth of the channel, or the slope of the bottom. In the open ocean underwater structures are hardly significant, but closer to the shore the continental shelf acts as a wedge against the wave to produce the kind of tide that man notices.

The moon, closer to the earth, has a stronger pull than the sun. When sun and moon are in line with the earth and pull together, the high tide is highest and the low tide is lowest. The extreme tides that accompany a full moon are known as *spring tides*, and the moderate tides on the quarters of the moon are called the *neap tides*. The Gulf of Mexico has only one- or two-foot tides; elsewhere they may be four to eight feet or more. Compare this with Juneau, Alaska, where the tide is more than twenty-three feet. What the Gulf of Mexico tides lack in impressiveness, they make up for in complexity.

There are three types of tides to be found in the world, and the Gulf of Mexico has all three. They are daily, semidaily, and mixed. Marmer (1954, p. 110) describes them by saying:

The semidaily type of tide includes all those in which the tidal cycle is completed in half a day; that is, there are two high and two low waters a day with only little difference between corresponding morning and afternoon tides.

The daily type of tide includes tides in which only one high and one low water occur in a day. The mixed type of tide includes those tides which feature two high and two low waters a day but with considerable difference between the two high waters and/or between the two low waters.

Due to the revolution of the moon about the earth in twenty-eight days, the times of either high or low tides are about fifty minutes later each day. Several factors affect the clocklike regularity of the tides, but the declination of the moon is the most important factor affecting the Texas tides. As the moon revolves around the earth from east to west, it also has a north-south movement. The declination is the measurement in degrees of latitude that the moon is north or south of the equator. The plane of the moon's orbit is not in the same plane as the equator; therefore, the declination of the moon is constantly changing. Another factor that affects tides is the shape of the basin. In the Gulf of Mexico this shape is such that it, combined with the moon's declination, responds better to daily forces than to semidaily forces. Meteorological conditions, such as winds and pressure, as well as the shape of the coastline, affect this regularity. On open coasts like those of Texas, the tide is usually less than eight feet, but in funnel-shaped bays, like the Bay of Fundy in Nova Scotia, the tides may rise thirty to forty feet.

It has been pointed out that there are two kinds of daily tides, and that they are affected by the declination of the moon. In the moon's fortnightly change from maximum northerly to maximum southerly declination, the differences between morning and afternoon tides are greatest near the times the moon is over the equator. Those at maximum declination are known as *tropical tides* and those at minimum declination as *equatorial tides.* The water level variation coincident with the cycle of tropical and equatorial tides is probably the most important so far as the lagoons and bays of Texas are concerned. It is these variations that are responsible for the largest regular exchange of water between bay and Gulf. This exchange causes the tidal currents in the narrow inlets that affect the migrations of organisms through the passes. Molluscan larvae are transported primarily by these tidal currents.

Sessile mollusks, such as mussels, oysters, and pen shells, owe their existence to the flow of the tides, which brings them the food they are unable to go in search of. The breeding rhythms of certain mollusks coincide in some unknown way with the phases of the moon and the stages of the tide.

In the bays the periodic tide is negligible and follows the Gulf tide about an hour later. The water level of these landlocked bodies is dependent principally on the wind. The north winds that push the water out through the passes often cause extremely low tides in the winter months. There are five natural passes along this coast that are kept open by this scouring action of wind and water. Annual migration of marine life in this area between the bays and the Gulf is dependent on the tidal current in these passes.

If a norther occurs at the time of very low, or minus, tide, the collector has a field day in the bays. To determine when a minus tide is due check the annual book of tide tables published by the United States Coast and Geodetic Survey. The daily paper in the area where you wish to collect also will give the hour of each high and low tide.

WAVES AND SURF. When one thinks of the ocean he thinks of waves. They may appear low and regular or high and frightening. Waves are not actually regular, but differ in size and form. The inconsistency of the wind has much to do with the irregularity of the waves. Wave motion is largely a surface phenomenon, the wave motion decreasing rapidly with increasing depth. As the wave nears the shore, the friction of the bottom causes it to rise higher until it dips forward and breaks. The breaker rushes to the beach until its energy is exhausted. Wave action can also set up currents and undertows along the beaches.

The size of the unbroken water area plus the velocity and direction of the wind determine the size of waves. Obviously waves in the Gulf of Mexico could not compare in size to those of the Pacific coast. In closed bays and lagoons, there is almost no surf. A gently sloping beach and broad continental shelf such as ours do not produce crashing surf. They deprive the wave of a considerable amount of its energy because bottom friction is increased so slowly that the wave remains stable most of its way to the shore. By then its force is largely spent. This is the type of wave that surfers love to ride. Ours is a wide surf zone with a series of sand bars and troughs that have big breakers on the outer bar, and smaller ones on each succeeding bar toward shore.

The action of the waves on a sandy beach tends to move the sand, making it a more inhospitable habitat than a rock shore. Only those animals which can burrow rapidly in the sand and can remain buried for some time are to be found on a sandy beach. Where the surf is an important factor only those animals which have developed structures that will minimize the impact of the breakers against their bodies, or the means of holding securely to rocks and seaweed, have been able to live.

The shell collector may wonder why on some days he finds greater numbers of the right valves of clams and on other days left valves. This variation is a result of wave action; the shape of the shell is the determining factor. Waves approaching from the right as you face the sea tend to deposit right valves.

TURBIDITY. As a factor governing marine life, turbidity has had little study on the Texas coast. Turbidity is expressed by the percentage of light transmission through a sample of water. Dirty or disturbed water affects the amount of light transmitted. The shallow-

Fig. 8. Rocky shores: *A*, typical Texas jetty, south jetty at Port Aransas; *B*, close-up of jetty rock with *Littorina lineolata* and *Siphonaria pectinata*, dwellers of the splash zone; *C*, rock fill at Cline's Point, Port Aransas, uncovered by low tide; *D*, close-up of fill rock with *Thais* snails and false limpets.

ness of the Texas bays permits turbulence to the bottoms by a minimum amount of wave action. Particles suspended in the water cut out light as well as put a strain on animals that must filter water for their food. Insufficient light discourages plant and algal growth, and animals dependent on these sun-loving forms will disappear when the plants die. Silting of the rivers from erosion in the watersheds, industrial wastes, and drilling in the bays contributes to the turbidity of Texas bay waters, but the most important factor is the minute size of the particles that make up the bottom sediments.

SUBSTRATUM. Two major faunal divisions exist in Texas waters, one in open-shelf, unprotected waters and the other in the protected waters of the bays and lagoons. Within these two divisions the substratum becomes one of the important factors in limiting the distribution of mollusks. The inshore area has a sandy bottom, the water is usually turbulent, and the temperature reflects that of the air. An unprotected sandy beach is sparsely populated by a few forms of burrowing animals and is not the place one can expect to find more than a few species of live mollusks, except after a storm or a hard freeze when inhabitants of the open shelf are stranded on the beach.

Fig. 9. A variety of mollusks are shown in this highly generalized diagram. The burrowing animal is shown under the sand, while those that live on the surface of the bottom or attached to a substrate are in their relative positions. *Left to right*: *A*, rock shell; *B*, false limpet; *C*, pecten; *D*, whelk; *E*, angel wing; *F*, quahog; *G*, *Janthina*; *H*, *Mactra*; *I*, olive; *J*, tellin; *K*, *Cardium*; *L*, *Crepidula* on old shell; *M*, oyster on shell and piling; *N*, *Littorina*; *O*, *Teredo* in cross-section of piling; *P*, razor clam.

On the outer shores it is not always easy to group marine animals into communities, but in the bays and estuaries with their mud or sand bottoms the animals are characteristically grouped. Many areas have rocky shores, but on the Texas coast the man-made jetties at the major inlets furnish the only rocky habitat (fig. 8). On the jetties the universal problems of rocky shores exist—wave shock and the development of powerful attachment devices. The jetties have been constructed by covering a fill of rubble and small stones with large quarry blocks, forming a long wall

that is triangular in cross section. These structures are about three to five feet above mean low water, and the slopes are covered with green algae. During storms and in the spring, when waves break over them, the jetties become slick and dangerous to walk on. Because they project hundreds of yards into the open Gulf, they are favorite spots for anglers. Life on the jetties is sparse, since they have been available for colonization less than one hundred years. The shifting sand base, wave shock, sudden variations in salinity, and low winter temperatures of the region limit the colonization to

A

B

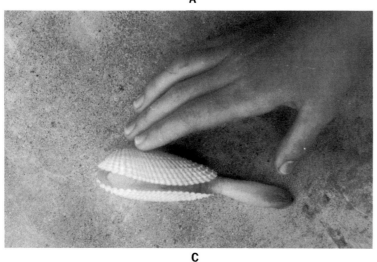

C

Fig. 10. Angel wing in the mud flats: *A*, a collector digging in the mire for the beautiful angel wing clam; *B*, next to the open valves of *Tagelus plebeius* can be seen the exhalent and inhalent siphons of *Cyrtopleura costata*, the angel wing, who resides about eighteen inches below the surface in the black mud; *C*, an angel wing after having been removed from its burrow with the siphon contracted. The siphon cannot be drawn into the shell.

very hardy forms from various sources. The jettied passes are highways by which fish, crabs, and shrimp migrate from the bays to the Gulf to spawn.

The type of substratum has a great deal of influence on the molluscan populations (fig. 9). Most of these animals are incapable of rapid movement and might be kept from an area by the size of the grains of sand on the bottom. Many shells that can live in sand cannot exist in mud. Much of the area near the mouths of rivers is mud with a surface layer of fine unconsolidated material that is easily stirred by waves. Most mollusks are unable to keep silt from clogging their gills and as a consequence are unable to maintain life in these regions; however, certain clams like the delicate angel wing and some snails have adapted to the mud flat (fig. 10). The lack of attachment sites in mud

Fig. 11. A typical sand flat with a group of gulls in winter plumage.

Fig. 12. Marine grass flats: *A, Ruppia maritima* Linné grass at low tide with empty *Busycon; B,* to the left are small snail trails in the sand, while a gelatinous mass of mollusk eggs may be seen on the grass just above and to the right of the *Busycon; C, Aequipecten amplicostatus* in feeding posture, a typical grass flat dweller (photo by Carl Young); *D,* the spotted sea hare, *Aplysia willcoxi,* swimming quietly in the shallows of the grass flat.

flats also makes for a rather specialized fauna. Some snails crawl around on top of the mud at low tide, but most of the mollusks burrow to rather unusual depths. Food getting becomes a problem and the lack of oxygen is another hazard to which the animal must adapt.

Sand flats in quiet waters with no shoal grass are ideal locations for many marine animals (fig. 11).

Here the substratum is soft enough so that feeble burrowing powers will suffice and yet not so soft as to require special adaptations to avoid suffocation. There is no attachment problem, for a small amount of burrowing is sufficient to assure that the animal will not be washed out, and a little more burrowing protects it from the drying winds and bright sunlight. Overpop-

Fig. 13. Pilings and wood. *A*, a protected, barnacle-covered piling with stranded *Mercenaria* clam and cabbagehead jelly-fish; *B*, wood after being drilled by *Teredo*, the little bivalve known as the ship-worm; *C*, wood that has been broken apart to expose living *Martesia*, who make their homes in burrows in the wood.

A

B

C

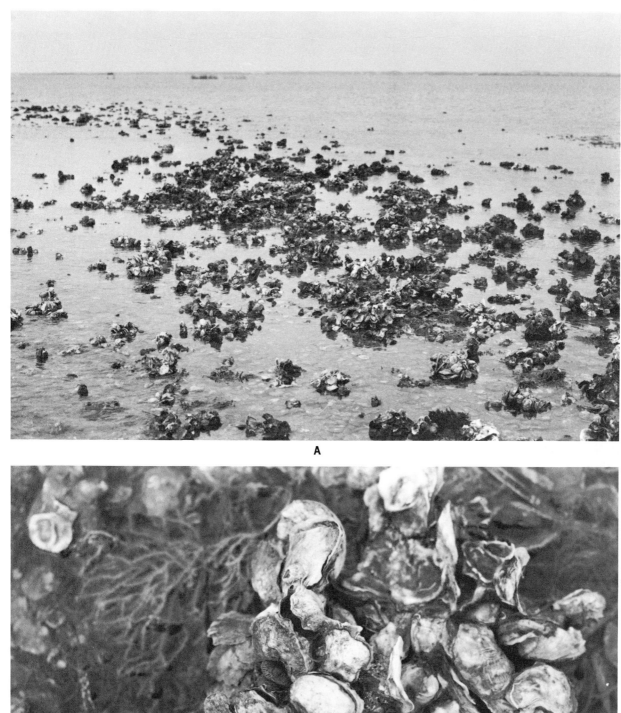

Fig. 14. Oyster reef: *A*, an extreme low tide exposes the oyster *Crassostrea virginica* in Aransas Bay; *B*, the edible oyster provides a home for many plants and animals, such as the mussel *Brachidontes recurvus*, which can be seen in the left center of the large clump of oysters.

ulation is the main problem that develops in this area.

Marine grass flats can be on either sand or pure muck in warm, clear, shallow water (fig. 12). Eastern Texas coastal waters have sparse marine plant growth, but the lower coastal lagoons are rich with many algae and grasses. The establishment of a bed of marine grass is an important step in the conversion of an ocean region into wet meadow and then into dry land. Turtle grass (*Thalassia testudinum* Koenig & Sims) and eelgrass (*Diplanthera wrightii* Aschers) support a group of mollusks that live on the blades, about the bases, and among the roots. Summer sunlight or extremes in temperature cause the animals that live among grasses to move downward for protection against exposure; as a result the flats seem quite barren at certain times.

These tidal flats are a meeting ground between the sea and the land, a tension zone (Ricketts & Calvin, 1960, p. 315) where the reaction of animals and their environment can be observed as nowhere else. The communities, or groupings, found in the flats are the result of the complicated interplay of the physical factors of environment—influx of fresh water, tidal action and currents, salinity and temperature—and of the biological factors imposed by the animals on their environment. A change in any one of these factors will affect the various communities. Consequently, in an area like Redfish Bay, the communities will change from year to year.

Here on the tidal flats, a change in the factors could wipe out the species as it did following the influx of fresh water from Hurricane Beulah in 1967, but the territory is rapidly restored when the balance is reestablished.

Man-made wharf pilings may be considered a distinct type of substratum and are generally considered under two headings: (*a*) exposed and (*b*) protected (fig. 13). All wharves and pilings are built in relatively protected positions, but the outer pilings experience more pounding during storms. On the outer pilings marine growth will be rather sparse and almost flush with the wood. The animal most associated with wooden pilings is the woodboring *Teredo* or shipworm. As mentioned earlier, this small clam is the bane of shipping throughout the world. It is completely helpless outside the protected life in timber, but can reduce an untreated pile to the collapsing point in six months. The life of a pile may be prolonged to three or four years by chemical treatment; however, nothing permanently protects pilings from the shipworm, which, unseen, works on the inside of a piece of wood.

Oyster reefs occur in shallow water, many of them aligned perpendicular to the prevailing currents (fig. 14). The conditions that promote the formation of an oyster reef are salinities ranging from about 12 to 25 $^0/_{00}$, a firm bottom, and temperature cool enough in winter to permit a resting period in their reproductive activities. The living as well as the dead oyster reefs provide shelter and food for many kinds of organisms.

2. General Features of the Mollusk

The shelled mollusk was one of the earliest forms of life on this planet. Very highly evolved snails crept about the earth 600 million years ago in the Cambrian period when the earliest widespread fossils occurred. Today, the fossil remains of their shells are important to the study of the earth's geologic past.

The phylum Mollusca contains more than 150,000 living species and is out-numbered only by the Arthropoda, which includes the crustaceans and insects. For many years several theories about the origin of the mollusks have been advanced. Some biologists believe they were derived from the segmented annelid worms. In 1957 a dramatic discovery was made off the northwestern coast of Central America by Dr. Henning Lemche of the Danish Galathea Expedition. From a depth of five thousand meters (3.1 miles) he recovered several living specimens of a limpetlike shell, *Neopilina*. Previously, these members of the class Monoplacaphora were known only as fossils from the Cambrian to the Devonian periods (600 million to 345 million years ago). Dr. Lemche's remarkable discovery of this segmented mollusk tipped the scales in favor of the annelid relationship.

The usual concept of a mollusk is that of an animal with a soft, slippery body and a hard shell, but there are many exceptions, for some, such as the squid and sea slug, have no visible shell. However, they all have a soft, fleshy body that is referred to in the name Mollusca, "soft bodied."

The various classes of molluscan animals include chitons that cling to rock; clams that burrow, anchor, or skip; snails that float, dig, or climb; the tooth shell that burrows its wide end into the sand; and the jet-propelled octopus that zips through the water leaving a smoke screen of ink in its wake. They may be found in any habitat: deserts, trees, rivers, lakes, mud, gardens, coral reefs, the depths of the sea, inside or attached to other animals, and buried in the ground.

BASIC MOLLUSCAN PATTERN

The unsegmented molluscan body generally contains a variation of the following: (*a*) head, with tentacles; (*b*) ventral muscular foot; and (*c*) conical visceral hump covered by the mantle (fig. 15). The visceral mass remains permanently within the confines of the shell, but the head and foot may protrude when the animal is active (fig. 16). Most mollusks have a head with tentacles at the anterior end of the body. This head generally contains the "mouth" or buccal cavity; however, in bivalves the head is not present. Most of the special sense organs are found in the head region.

The foot is common to all mollusks in one form or another. This organ of locomotion is located on the

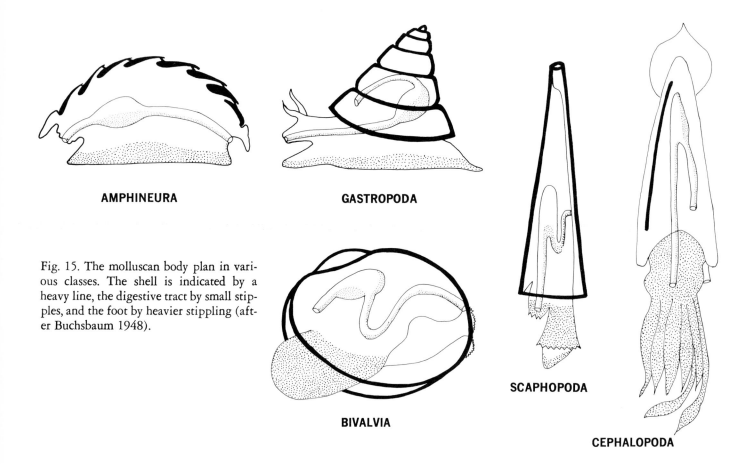

AMPHINEURA

GASTROPODA

BIVALVIA

SCAPHOPODA

CEPHALOPODA

Fig. 15. The molluscan body plan in various classes. The shell is indicated by a heavy line, the digestive tract by small stipples, and the foot by heavier stippling (after Buchsbaum 1948).

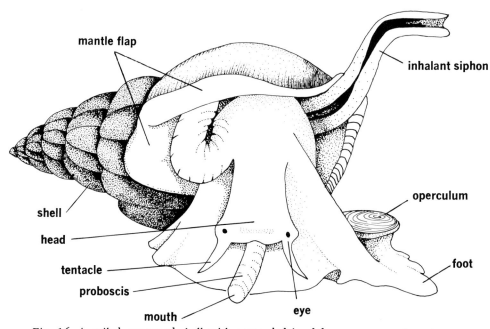

mantle flap

inhalant siphon

operculum

foot

shell

head

tentacle

proboscis

mouth

eye

Fig. 16. A coiled gastropod shell with protruded head-foot mass, anterior view (after Cox 1960; courtesy of The Geological Society of America and The University of Kansas).

ventral, or apertural, side of the animal and always takes the form of a highly muscular, fleshy projection. The operculum, when present, is attached to the rear of this appendage. The foot is typically a tough, broad, muscular structure containing special glands for secreting the operculum and, in certain mollusks, the byssus threads for attachment. In digging forms, the foot is highly developed. Gastropods creep by a succession of muscular waves or contractions, which move forward along the flattened foot. The expanded parts of the foot stay fixed to the surface with mucus while the contracted area is in forward motion. Many bivalves move over the bottom of the sea by extending the foot and then contracting the muscles, violently producing a jumping motion. Other clams dig or creep by extending the foot ahead and then drawing the body after it. In certain pelagic forms, as the pteropods, the foot is much modified into fan-shaped, finlike structures (parapodia), which aid in swimming.

The visceral hump is covered by a mantle, or pallium. The folds of the body wall that constitute the mantle envelop a cavity that contains the gill-like breathing organs (ctenidia), digestive tract, and blood and nervous systems. The exterior surface and edges of the mantle secrete the shell. The siphon is a prolongation or protrusion of the mantle edges. In some snails like the cowries and olives, the mantle flaps completely envelop the shell. In clams, Abbott (1958a) likens the mantle to the flyleaves of a book. The two equal portions of the mantle, flyleaves, enclose the visceral mass, or body of the book, and secrete the two shells, or book covers, that they line. The mantle edges are joined where the valves of the shell are joined and are generally free along the other edges; however, some are fused together along all or most of the lower margins, with openings for the foot and siphon.

The shell secreted by the mantle is a protective, calcareous covering that has evolved into many varied forms. The shape of the shell is a poor guide to natural relationships, for it may be external, partly external and partly internal, or completely absent. The mantle lays down a coating of carbonate of lime and magnesium on a framework of the protein conchyolin or keratin. The shell is made in three or more highly variable layers: the outer, periostracum; the middle, prismatic; and the inner, nacreous. Sculpturing or ornamentation of the shell is only produced at the edge of the mantle. Irregularities in the mantle edge produce corresponding sculpture on the shell, such as spines and flutings. At the time of formation, all shell ornamentation followed the edge of the outer lip. Successive stages of growth (additions to the shell at ventral or apertual edge) remain marked on the surface by distinct growth lines. From a study of the growth lines it is possible to reconstruct the shape of the outer lip of the shell even when the lip has been damaged (Cox 1960). When the shell is not protected by the mantle, the outer coat, or periostracum, is a horny, rough, sometimes hairy covering that prohibits marine growths and acids from damaging the shell. The periostracum soon disappears after the animal dies.

In snails the animal is held in its shell by a columellar muscle that originates on the axial columella, follows the right side of the body into the foot, and attaches to the under side of the operculum. This muscle retracts the animal into the shell. The head precedes the foot and if there is an operculum on the rear of the foot it partially or completely closes the aperture. Bivalves are held in their shells by strong adductor muscles that act to open and close the valves. In scallops and oysters there is only one adductor muscle. In mussels there is one large muscle near the posterior end and a small one near the anterior. In clams there are two adductor muscles, one near the anterior end and one near the posterior.

The shell's color, which is only on the surface of most shells, is influenced by light and temperature. The secretion of color pigment is actually a part of the disposal of the waste products of metabolism. Thus many gastropods harmonize with the sea weeds or corals on which they live and feed (e.g., *Neosimnia uniplicata* and *Gorgonia*). Color patterns are not necessarily protective, because many of the most elaborately colored shells are obscured with a coating of periostracum in actual life. Species from clear tropical waters are more highly colored than those from colder climates.

The shell grows rapidly the first year or two if the food supply is abundant. In most cases it continues to grow during the lifetime of the animal. The largest gastropod, the horse conch of the Gulf of Mexico, attains a length of two feet; the giant squid, a cephalopod, reaches fifty-five feet; and the *Tridacna* clam may grow to four or five feet and weigh several hundred pounds. The lifetime of a mollusk may range from one to twenty-five years depending on the species; however, the *Tridacna* clam is thought to live one hundred years. The majority of the mollusks die at an early age, one year or less, due to changes in salinity or temperature, exhaustion from egg laying, or falling prey to

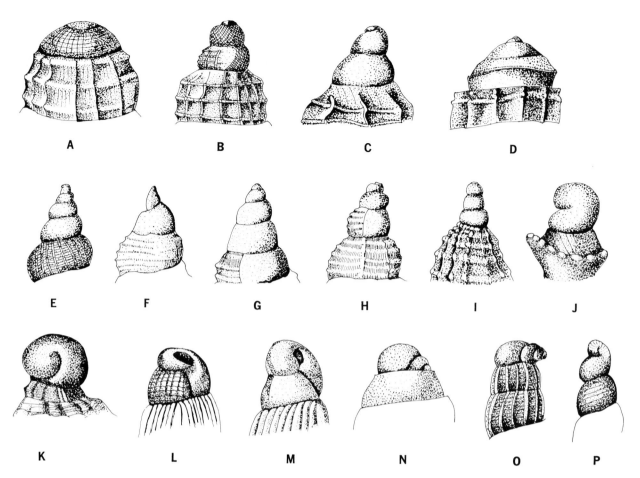

Fig. 17. Various gastropod protoconchs: *A*, domelike, paucispiral, with reticulate ornament, *Melatoma*; *B*, with decussate ornament, *Daphnella*; *C*, with papillose ornament, *Murex*; *D*, obtusely conical, *Nassarius*; *E*, conical, *Scala*; *F*, disjunct with erect tip, *Charonia*; *G*, conical, multispiral, *Cymatium*; *H*, mammillated, *Clavilithes*; *I*, mammillated, *Cerithiopsis*; *J*, deviated, paucispiral, *Columbarium*; *K*, deviated, paucispiral, *Pterospira*; *L*, heteroscopic and submerged, *Partulida*; *M*, heteroscopic, *Turbonilla*; *N*, heteroscopic, *Odostomia*; *O*, heteroscopic, *Odostomia*; *P*, heteroscopic, *Eulimella* (after Cox 1960; courtesy of The Geological Society of America and The University of Kansas).

another organism. Many shells form a thickened lip when adulthood is reached; others form varix or transverse axial ridges along the edges of the shell between periods of growth and rest. Ninety percent of the life of a shell is spent in the varix stage because additional growth takes place in less than two days.

A rudimentary shell or nucleus (protoconch) is present when the mollusk leaves the egg. The protoconch of marine shells usually consists of two to three smooth whorls that are clearly delimited from the adult shell (fig. 17). In some families, such as the Pyramidellidae, the protoconch is coiled in the opposite direction of the remainder of the shell and is described as *heterostrophic*. The immature or juvenile shells are usually thin, with thin lips, and are often so different

from the adult that they are mistaken for different species.

Life goes on within the protection of the shell. When the shell is nonexistent the animal may have protective coloration or ink sacs that expel a "smoke screen" when danger approaches. Some species develop long, sharp spines on the shell to keep other animals from swallowing them. Any injury to the soft mantle will result in a deformity in the shell.

The digestive system consists of a tube leading from the mouth to the anus. In carnivorous forms the mouth is on a snout or proboscis on the head. At the anterior end it is enlarged to form the buccal cavity, which contains the mandibles, or jaws; the radula, or rasping tongue; the salivary glands; and the opening to the esophagus, leading into the stomach. The radula is a ribbonlike organ with rows of rasplike teeth that tear food and draw it into the mouth. In some species it contains a device to poison the victim, thus quieting it to make feeding easier. In other species the teeth and rasping motion of the radula are used to drill holes in bivalves through which the soft parts of clams are eaten. The type of hole can be used to determine the species of gastropod that drilled it. When present, the radula is located in the mouth of the mollusk. The arrangement of the minute teeth on this ribbonlike organ is so distinctive that it may be used as a reliable means of identification of a species and the present arrangement of the prosobranch gastropod families is based largely upon the radula.

The pulleylike motion of the radula aids the mandibles in grinding the food. The food then passes through the esophagus by means of muscular contractions into the stomach, which is followed by an intestine. In the more complex mollusks, the intestine forms loops and folds. The intestine ends in the anus through which wastes are expelled. In snails the intestine has twisted forward so that the anus is located anteriorly near the head.

Many snails are carnivorous, feeding on other animals and mollusks, others are vegetarians, and still others are parasites. Clams have only a vestigial mouth and use their gills as the main organ of feeding. They take minute particles of food from the water in one of two ways: suspension, or filter feeding (pumping water through the mantle cavity), or by deposit, or detritus feeding (sucking up food from the muddy bottom with the siphon). Detritus is a scum of decaying plant and animal material and bacteria that forms on the surface of mud flats, estuaries, and ocean bottoms.

The circulatory system contains a heart that receives aerated blood from the breathing organs and propels it to every part of the body. Arteries carry the blood from the heart to the other parts, and the veins return the blood to the breathing organs after passing through the kidneys where waste materials are extracted. The blood is usually colorless or faintly blue, but a few species have red blood.

The respiratory apparatus consists of branchiae, or ctenidia (gills), which aerate the blood on its way to the heart. Water is pumped in through the incurrent siphon, carried over the gills by cilia, and then forced out the excurrent siphon. In some clams these two siphons are joined, while in others they may be long and separate. Many snails have a tubelike projection of the shell (siphonal canal) to protect the siphon. The gills are so distinctive in position and character that the divisions of the Gastropoda are named for these characteristics. The nature of the gill may be taken as indicating different degrees of removal from the primitive form of bivalve.

The nervous system, which is less complicated in bivalves, consists of a nerve ring with three swellings, or concentrations of ganglia: (a) the cerebral ganglia controlling mouth, mantle, ears, and eyes, (b) the visceral ganglia controlling the adductor muscles and visceral mass, and (c) the pedal ganglia controlling the foot.

The sense of touch is well developed in only a few species. The foot is the most sensitive area, then the edges of the mantle, the tentacles, and the skin surface. Taste is present to some extent in all head-bearing mollusks.

Sight is a curious sense in many mollusks. The range in development is great—from the highly developed eye of the octopus down to the simple light-sensitive spots in some clams. It is believed that only the cephalopods use their eyes to observe. In head-bearing mollusks the "eyes" are generally in the region of the tentacles but not on the tips. Usually pelagic, burrowing, and deep-sea mollusks have no eyes. The tidal species have the best developed sense of sight, which they use for protection against man and diving birds. Bivalves may have light-sensitive spots scattered over the surface of the mantle, concentrated around the siphon, or as in the scallop highly developed ocelli on the edge of the mantle.

Perhaps one of the most important senses to the mollusk is that of smell. The osphradium, or smell organ, is situated intimately with the breathing organs. The carnivorous gastropods have the most highly developed sense of smell, while the bivalves apparently do not have this sense. Very little is known about the hearing sense of the mollusk.

In Mollusca, reproduction is invariably from an egg, which is developed in the ovary of the female and fertilized by spermatozoa of the male. The eggs undergo subsequent development apart from the parent, except in a few cases where the parent hatches the young before expelling them. This latter form of giving birth to young alive is known as viviparity (ovoviviparity). The number of eggs is highest among the bivalves and in the case of *Ostrea edulis* can be as many as 60 million. The mother sea hare, a gastropod, deposits 41,000 eggs per minute, or half of a million in one seventeen-

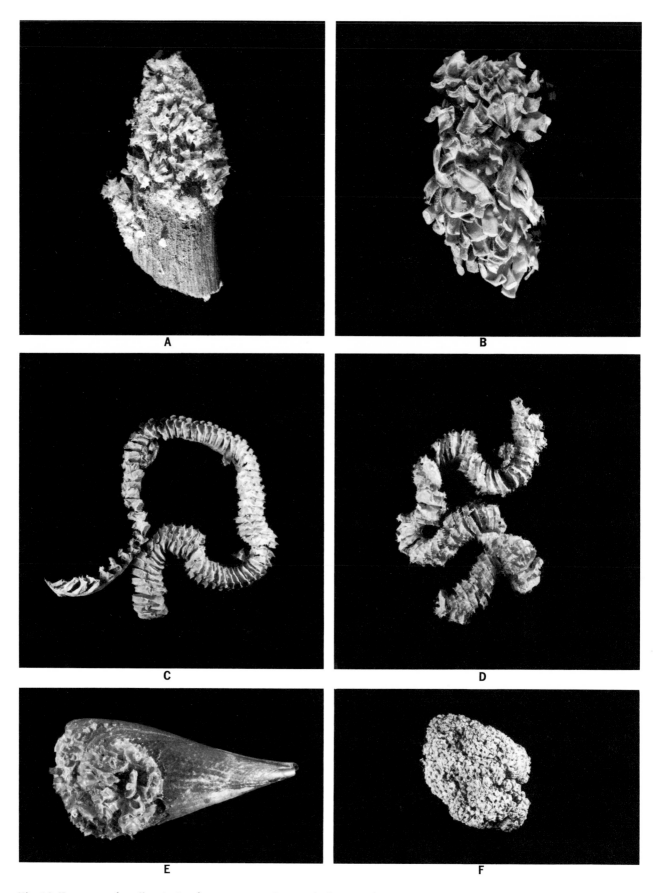

Fig. 18. Egg cases of snails: *A, Cantharus* egg capsules attached to wood; *B,* the purplish egg cases of a *Thais; C,* a rope of horny egg capsules constructed by the female *Busycon contrarium; D, Busycon spiratum plagosus* constructs her capsules with serrated edges; *E,* the tulip shell, *Fasciolaria,* has attached its vase-shaped egg capsules to a pen shell; *F,* a spongelike mass of egg capsules produced by *Murex pomum.*

week laying period; on the other hand, the abalone lays only one egg at a time.

The female snail may lay the eggs in a variety of places that include inside strings of horny egg cases, enclosed in bladdery capsules glued with grains of sand into sand collars, in jellylike masses or strings on sea weed, attached to the underside of floats, or attached to the outside of their own shells (fig. 18). The foot is often used to shape complicated, horny strings of egg cases like those of the whelk. The clams usually release the unfertilized eggs into the water where they come in contact with the liberated sperm. Some species keep the eggs in the mantle fold and suck sperm into them through the siphon.

A great number of the male gastropods have external copulatory organs, but none of the bivalves have copulatory organs or external sexual characteristics. Some univalves and most bivalves are of separate sexes, but frequently snails have dual sexuality (hermaphroditism); some mollusks exhibit sex reversal. In most cases, the gonocoele, which houses the gonads, passes the gametes through the kidneys or renal organs before expelling them. The kidneys function as a part of the genital duct and, in some, one of the kidneys has lost its renal function entirely.

The winter months appear to be the usual time for depositing eggs. The time and length of breeding differ depending on geographic locality, phases of the moon, water temperature, and inherent characteristics of the species. Some breed once a year for a few weeks, while others spend half of the year breeding. In those mollusks which shed their eggs and sperm into the plankton—floating or drifting animal or plant life of the sea—great waste might be expected from failure of fertilization. Many species have adjusted their spawning habits to permit economy of gametes. Some will not shed their sexual products unless they are close to another of the opposite sex; egg laying in others is induced by the presence of sperm in the water; in still other forms there is "epidemic spawning," controlled by lunar phase, food abundance, or temperature.

The fertilized egg develops into one of four types of larvae: (a) planktotrophic larvae with a long larval life of up to three months, (b) planktotrophic larvae with a short swimming life of never more than a week, (c) lecithotrophic larvae that take no food in the plankton but hatch from yolky eggs and swim little, or (d) direct development exhibited by many gastropods that develop completely in the egg capsule and are nonpelagic. It is the function of the larvae to locate a settling spot before metamorphosis takes place. This embryo shell, the veliger, may differ completely from the adult shell. The size and number of embryonic whorls differ according to the length of time the larva passed in the plankton. The more whorls, the longer the life in the planktonic stage.

Only a small number of the eggs develop into adult mollusks; some serve as nurse eggs for the few cannibalistic survivors that hatch at the crawling stage and others fall prey to plankton feeders. The adult also has many enemies. Birds, including gulls and the other aquatic birds, crows, and vultures are some of the main enemies. Gulls have been seen flying into the air with a clam and letting it drop to the ground, repeating the process until the shell has broken so that the bird can have its meal. Fish, such as the bottom-feeding catfish, drum, and flounder, find mollusks a staple of diet. Parasitic worms cause great havoc in some shells, principally the fresh-water varieties. Snails are one of the greatest enemies of the bivalves. Carnivorous species, such as *Polinices* and *Thais*, drill holes in bivalves and eat the flesh. The *Thais*, or oyster drill, is one of the plagues of the oyster reef. Man as an enemy also must not be forgotten.

To overcome these enemies, the mollusk has developed many passive modes of resistance. When danger approaches, the clams snap their valves shut and the snail draws into his shell, sealing the opening with the operculum. Some hide in a nest they have constructed of marine refuse held together with byssus threads. The *Xenophora*, or carrier shell, glues bits of shell and rocks to the top of his shell and is thus difficult to locate. To keep fish from swallowing them, many have developed a variety of spines. The vulnerable sea slug, or nudibranch, produces an exudate that makes it unpalatable to its predators. The cones are able to inject poison into their prey or enemies, while other mollusks can only rely on protective coloration. A few mimic other shells; for example, the juvenile *Strombus* takes on the appearance of the poisonous cone. Some become active and leap about with their strong foot, while the shell-less sea hares and octopuses resort to a cloudy screen of ink to shield their getaway.

The phylum Mollusca is, indeed, quite distinct from other modern groups.

IDENTIFYING THE SHELL

One of the most rewarding features of an interest in the field of natural history, whether it be insects, wildflowers, or sea shells, is the pleasure to be had from identifying one's collection. Once the collector is able to name his shell he can compare the specimen with others of its class as well as be able to compare his observations of the habits of the animal with the studies of zoologists.

Identification of a shell is not always a simple task. A look at illustrations may spot the shell or a near relative so that a reference to the text will point out the real identity. Although the shell of the mollusk may be unsatisfactory as a basis of classification, it is convenient for identification because one seldom has the live animal.

Within a species there might be great variation in color or shape because marine mollusks are extremely responsive to varying environmental conditions. Conversely, some species may be so similar to others that only an anatomical study of the animal will prevent misnomer. Quite often the differences caused by environment are difficult to distinguish from the inherent characteristics of a species.

For the ability to find his way around at all, within the multiplicity of forms, the collector is indebted to the great Swedish naturalist, Carolus Linnaeus, or Karl von Linné (1707–1778). It was he who established the first useful system of classification for the whole of the plant and animal kingdoms. His remarkable powers of observation permitted him to discover the distinguishing characteristics that establish the species to which any given animal or plant belongs.

Standardized names, which can be recognized by students throughout the world, are used to discuss the many kinds of mollusks. Because these names are in international use, Latin or Latinized forms are employed for nomenclature. No one, not even professional malacologists, can remember all of them, but it adds to the pleasure of collecting to be able to recall a few of the more common species.

Like people, a mollusk is given a name of two parts—a binominal. The generic name, always capitalized, is used like the surname, such as Jones or Smith —for example, *Tellina*; the trivial, or specific, name, which is not capitalized and which follows the generic name, corresponds to the first name, such as Jane or George—for example, *texana*. Both of these names are italicized. The name of a man, printed in Roman type, and a date, follows the scientific name—for example, *Tellina texana* Dall, 1900—indicating the person who first described the species and the year in which it was described. This information is helpful in tracing the original description. If the author's name is in parentheses it indicates that someone has since studied that particular species and placed it in a different genus from that decided upon by the original writer. Thus, the Common Rangia, first described as *Gnathodon cuneata* Gray, 1831, is now written as *Rangia cuneata* (Gray, 1831). Once an animal has been named, the specific name becomes unchangeable unless it is a homonym within that genus. It may be transferred to another genus but the specific name remains the same.

If there is a subgeneric name it is capitalized and placed in parentheses following the generic name. If there is a subspecific name it follows the specific name without separation by sign or symbol. Thus, "*Genus (Subgenus) species subspecies* Author, date."

The generic name is a noun, while the specific name takes the form of an adjective that agrees in gender with the name of the genus. The name may describe some characteristic of the mollusk, may indicate geographic localities in which it is found, or may be either the name of a person who had made an important contribution to this field of science or the name of a mythological person. Some names are without definite meaning, but usually the name selected is a store of information about the animal and is seldom a nonsense name.

The names of orders and families (collective classification) always end in plurals of Latinized nouns. Families have the ending *idae* and subfamilies have *inae* added to the stem of their type genus. These names originate in three ways:

1. Adoption directly with appropriate modifications in spelling from Greek, Latin, and other languages
2. Composition by compounding and affixation
3. Outright or arbitrary creation without use of evident antecedent root or stem material.

The International Commission for Zoological Nomenclature devises the rules by which animals are named. The starting point of zoological nomenclature and the Law of Priority is accepted to be the date of January 1, 1758, when Linné's *Systema Naturae* was published. According to Article 25 of the Rules of Zoological Nomenclature the valid name of a genus or species may be only the name by which it was first designated. As our knowledge of mollusks increases and old papers are reinvestigated it often becomes necessary to change a name to an older, little used name. Because in some cases this has caused confusion, the commission ruled (Article 11, 1961) to make exceptions when a name published in synonymy has been

TABLE 1: TYPE TERMS

TYPES OF FAMILIES
>   *Type-genus*: the genus upon which a family is based

TYPES OF GENERIC CATEGORIES
>   *Genosyntype*: one of several species included within a genus at the time of its proposal, if none was designated as genotype
>   *Genotype*: the single species upon which a genus is based

TYPES OF SPECIFIC CATEGORIES
1. Primary types (chief purpose is the elevation of specific names):
   >   *Syntype*: one of several specimens of equal rank upon which a species is based
   >   *Lectotype*: a specimen, selected from a syntypic series, upon which a revised species is based
   >   *Holotype*: the single specimen taken as "the type" by the original author of a species
   >   *Paratype*: a specimen, other than the holotype, used as a basis of a new species
   >   *Neotype*: a specimen selected to replace the holotype, in case all type material of a species is lost or destroyed
2. Secondary types (comparative material used to check published identifications and to extend knowledge of the species):
   >   *Hypotype*: a described, figured, or listed specimen
   >   *Topotype*: a specimen from the type locality of a species
   >   *Homoeotype*: a specimen compared with the type and found to be conspecific with it
3. Reproductions (for distribution of facsimiles of rare specimens):
   >   *Plastotype*: a cast of a type

accepted through the years by authors prior to 1961, thus maintaining a widely accepted name which then becomes a *nomen conservandum.*

In spite of the care taken by this august body, the seemingly constant switching of scientific names is a perplexing problem for the amateur. In 1905 David Starr Jordan said in his *Guide to the Study of Fishes*: "In taxonomy it is not nearly so important that a name be pertinent or even well chosen as that it be stable. In changing his own established names, the father of classification, Linnaeus, set a bad example to his successors, one which they did not fail to follow."

When it has been determined that a new genus or species has been found, a specimen of the shell is placed under lock and key in a type collection, such as those located at the U.S. National Museum and other major study centers. The various type categories are shown in Table 1.

Pronunciation of the genus and species names is a personal matter but it is generally based on the Latin and Greek pronunciations. Words of two syllables are accented on the first; those of more than two syllables are accented on the penult (next to last). The purpose of pronunciation is to be understood. If the name is a patronymic (formed by adding a prefix or suffix to a proper name) the pronunciation of the area or the person's name is retained regardless of how it is syllabicated. A boon to those interested in correct pronunciation is a long-playing recording of the pronunciation of twelve hundred scientific species by Dr. R. Tucker Abbott. All the names and terms used in *Seashells of North America* have been recorded on this record, which has been produced and published by Dr. Abbott.

Many shells have common, or popular, names. These names may vary for the same shell according to region, but in most cases they are names in use by local collectors and fishermen. Abbott (1954) has listed eleven hundred popular names of American sea shells and in so doing has gone a long way toward their standardization. However, many small shells are too little known to have been given a common name.

The development of a species is a gradual and continuous process. Ernst Mayr in his *Systematics on Origin of Species* defines species as groups of actually or potentially interbreeding natural populations, which are reproductively isolated from other such groups by geographical, physiological, or ecological barriers.

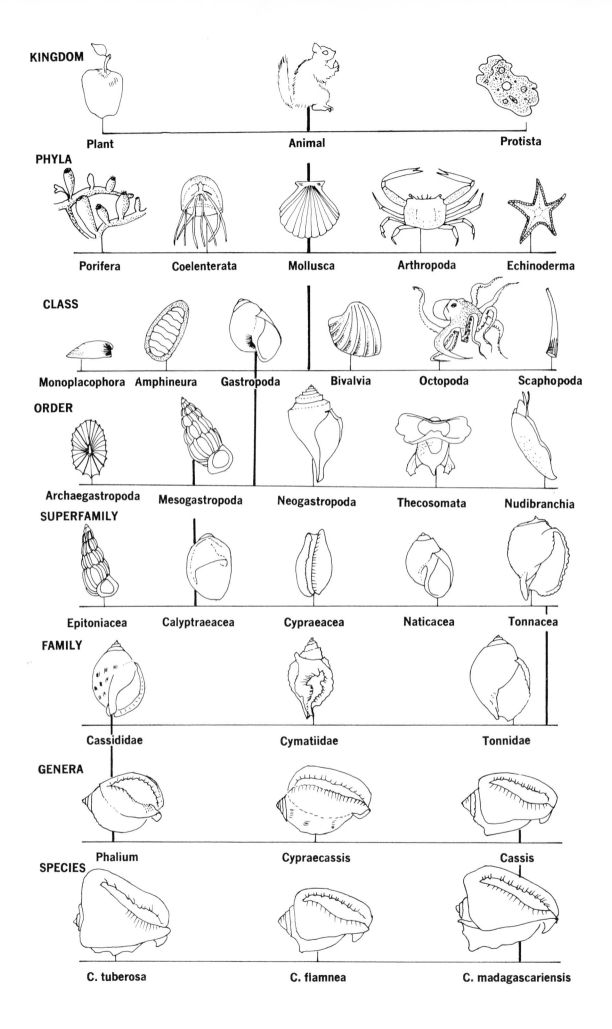

KINGDOM

Plant　　　　　　　　　　Animal　　　　　　　　　　Protista

PHYLA

Porifera　　Coelenterata　　Mollusca　　Arthropoda　　Echinoderma

CLASS

Monoplacophora　Amphineura　Gastropoda　Bivalvia　Octopoda　Scaphopoda

ORDER

Archaegastropoda　Mesogastropoda　Neogastropoda　Thecosomata　Nudibranchia

SUPERFAMILY

Epitoniacea　Calyptraeacea　Cypraeacea　Naticacea　Tonnacea

FAMILY

Cassididae　　　　　Cymatiidae　　　　　Tonnidae

GENERA

Phalium　　　　　Cypraecassis　　　　　Cassis

SPECIES

C. tuberosa　　　　C. flamnea　　　　C. madagascariensis

A genus is merely an arbitrary and convenient grouping of related species. This is also true of the higher categories, subfamily and family, which are groupings of related genera. Whatever the faults of the system, it is extremely useful. With the generic system we may arrange the species in our collections in a biological, evolutionary, and logical sequence. The usual order is kingdom, phylum, class, order, family, genus, species, and subspecies. The chart in Figure 19 illustrates the classification of mollusks.

The differences in characteristics of a mollusk may be morphological (form and structure), physiological (pertaining to function), or genetic (hereditary characteristics). The morphological features of a mollusk are the characteristics most commonly used to identify the shell so that it may be assigned its proper category. Such characteristics as spines, color, and number of whorls are used to distinguish species; therefore, it will be necessary to become acquainted with the names of these features in order to identify one's shells. The use of technical words is unavoidable because many of these names have no counterpart in our daily language. Familiarization with such terms as operculum, periostracum, umbone, and beaks is gained easily with a few trial identifications using the illustrations and glossary provided in this book.

The amateur should understand taxonomy (the study of the rules of classification) and should try to identify his collection because it adds to the value of the collection, but taxonomy should not be his primary interest. Today we are bogged down in a tangle of names—one shell may have many synonyms; even the experts are often stumped. Without an extensive library or a large collection from many locations the amateur is greatly handicapped.

This does not mean that he cannot make a significant contribution to the sciences of malacology (generally considered to be the study of the soft parts of the mollusk) or conchology (the study of the hard shell). The amateur may have more opportunity to observe the living mollusks than does the busy professional malacologist who usually teaches or directs a museum. Much information is needed as to the type of bottom that the animal lives in, his spawning habits, how he feeds, and so on. Consider the mollusk as a living animal, record your observations, publish them if you wish. The professional is usually only too happy to help you because your observation often helps him. But without data the most prolonged study of an animal's habits is worthless. Record your observations.

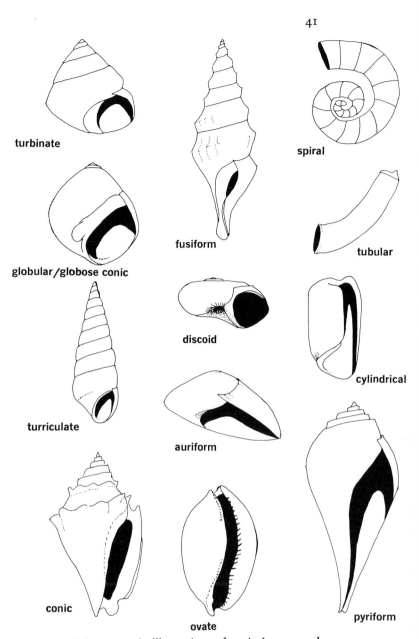

Fig. 20. Diagrammatic illustrations of typical gastropod shapes.

The Snail

The name Gastropoda means "stomach-footed ones" (L. *gaster* stomach; Gk. *pod* foot). The gastropod shell, which consists of one unit, is called a univalve. This univalve shell is distinct from the other orders of mollusks due to the phenomenon called torsion. In the first few hours of the snail's life, the larva twists so that the front end revolves halfway around. The organs that began bilaterally symmetrical now cease developing

Fig. 19. Classification chart for mollusks.

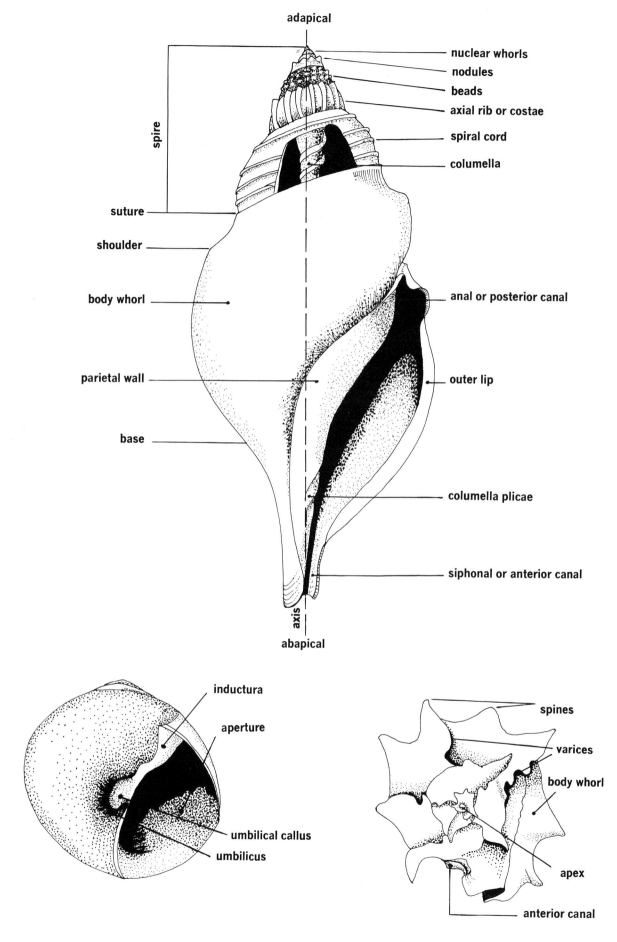

adapical

nuclear whorls

nodules

beads

axial rib or costae

spiral cord

columella

spire

suture

shoulder

body whorl

parietal wall

base

anal or posterior canal

outer lip

columella plicae

siphonal or anterior canal

axis

abapical

inductura

aperture

umbilical callus

umbilicus

spines

varices

body whorl

apex

anterior canal

Fig. 21. Parts of the gastropod shell. In this composite shell the columella is seen through a cutaway section.

in this manner, becoming asymmetrical and spirally coiled.

The living mollusk carries its shell with the aperture down and the apex toward the rear, but to examine the shell you should hold it in a position with the apex uppermost or away from you, the aperture up, and the axis vertical. The shape of the shell is so varied that some of the typical shapes and the names applied to them are illustrated in Figure 20.

The external parts of the snail shell have been given various names that are useful in describing each type. In the following discussion, some of the terms used frequently are given in italics and are illustrated in Figure 21. The *apex* (at the *posterior*, or *adapical*, end) of the shell is where the *nuclear whorls*, or *protoconch*, are located. In many cases the apex is the nucleus, or *embryonic shell*, with which the young mollusk came from the egg. The nuclear whorls are often quite different from the other whorls, or turns, which are below the apex. The last and largest whorl is known as the *body whorl*. These successive whorls are added as the gastropod grows. These whorls are separated by indentations called *sutures*. The sutures may be shallow or deeply cut, wavy or regular, marking the juncture of each whorl. The whorls may vary greatly in sculpture and shape. Some species bear pointed spines, ridges, and bright colors, while others are flat and smooth. The body whorl ends with the aperture.

All the whorls together with the exception of the body whorl form the spire. To count the whorls, place the aperture of the shell downward and count each whorl up to but not including the *nuclear*, or *apical*, *whorl*.

The *aperture* of the shell is an opening through which the animal can extend or withdraw its body. When the mollusk is crawling it has protruded through this aperture in the body whorl and carries this end of the shell in front. Hence, the aperture becomes the *anterior*, or *abapical*, end of the shell. The apex is of course located on the opposite end. Thus, the apex that would ordinarily appear to be the front, or uppermost part, of the shell becomes the posterior, or back end, of the shell. When the posterior, or adapical, side of a whorl or spine is mentioned we mean the side away from the aperture and nearest the apex of the shell. The length, or height, of the shell is the total difference between the two ends of the shell.

The aperture varies greatly in shape, being round, oval, or a long narrow slit as found in the olive shell. The edge of the body whorl adjoining the aperture is

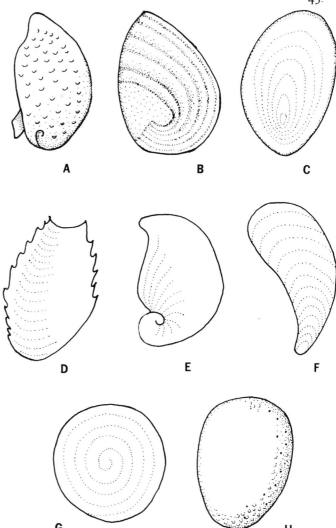

Fig. 22. Gastropod opercula: *A*, calcareous, paucispiral with apothesis; *B*, calcareous, paucispiral; *C*, horny, concentric; *D*, horny, lamellar with serrated edge; *E*, horny, paucispiral; *F*, horny, unguiculate; *G*, horny, multispiral; *H*, calcareous.

the *outer lip*. It too may take a variety of shapes. At the end of each growth period it may become thickened to form a *varix* and at adulthood it might become extremely flared or ornate. At each stage in the growth of the shell a varix, or rib, might be produced and the position and total number of these ribs are used in the identification of the shell. However, some snails dissolve the thickened outer lip before making an addition to the shell.

The side of the aperture opposite the outer lip is known as the *inner lip* and consists of two parts, the

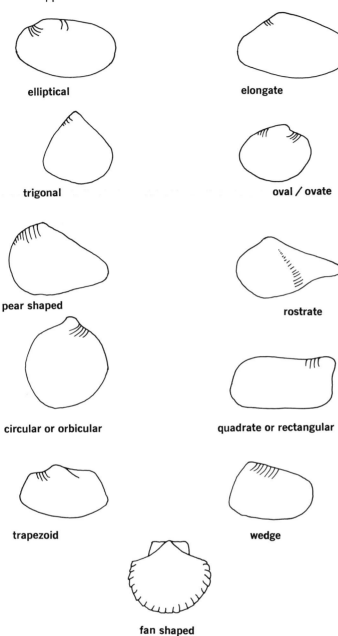

elliptical

elongate

trigonal

oval / ovate

pear shaped

rostrate

circular or orbicular

quadrate or rectangular

trapezoid

wedge

fan shaped

Fig. 23. Diagrammatic illustrations of typical bivalve shapes.

columellar lip and the parietal lip. It may be smooth or it may bear teeth. The inner lip, a continuation of the columella, or axis, of the shell, often extends into a tubelike *siphonal canal.* Most gastropods that have canaliculated apertures are known to be carnivorous in habit.

The *operculum* is a special hornlike, or calcareous,

development on the foot (fig. 22). It serves to close the aperture against enemies, drying, or pollution, as well as aiding in movement and in protecting the soft foot. When the operculum is too small to close the aperture, it serves a secondary function as an aid to locomotion. It is only present in gastropods but not in all families. Even when it is absent in the adult snail, the operculum was usually present in the larval or juvenile stage. The characteristic shape and color of the operculum are often important factors in identification, for it has growth rings that correspond with those of the shell. When the animal dies the noncalcareous operculum disintegrates and is lost.

The sculpture found on the exterior of the shell —spines, ribs, costae, indentations, nodules, cords, threads, and colored bands—may be put into two basic groups: (*a*) the *spiral sculpture*, which is spirally arranged in the direction of the suture going around the whorls; and (*b*) the *transverse*, or *axial, sculpture*, that is, ribs, lines, or nodules that run across the whorl from suture to suture parallel with the axis. Growth lines, varices, and the outer lip are transverse features. *Growth lines* refer to irregularities in the shell that mark the place where the growth of the shell was arrested for a relatively long period. At that time these features were the outer lip of the shell but when additional growth took place they were left as growth lines.

When the whorls are not wound closely at their anterior end a depression is left in the shell adjoining the base of the columella, forming a feature known as the *umbilicus.* Only about a fourth of the gastropods have this characteristic, but it can be quite a prominent feature in those that are umbilicated.

Often the exterior of the shell of a mollusk is covered with a brownish, horny coating known as the *periostracum.* This protective coating is secreted by a lobe of the mantle separate from the one making the shell and may take various forms. In some cases it is thin and transparent like a coating of varnish; in others it might be thick and hairy; and in some groups like the olives it is entirely absent.

The Clam

The Bivalvia (Pelecypoda), or "two-valved ones" (L. *bi* two, *valvae* folding doors), as a class contains fewer genera and species than the Gastropoda, but in number of individuals it outweighs all other classes of Mollusca, and it is the most important economically. On the Texas coast, at the areas known as Big Shell and

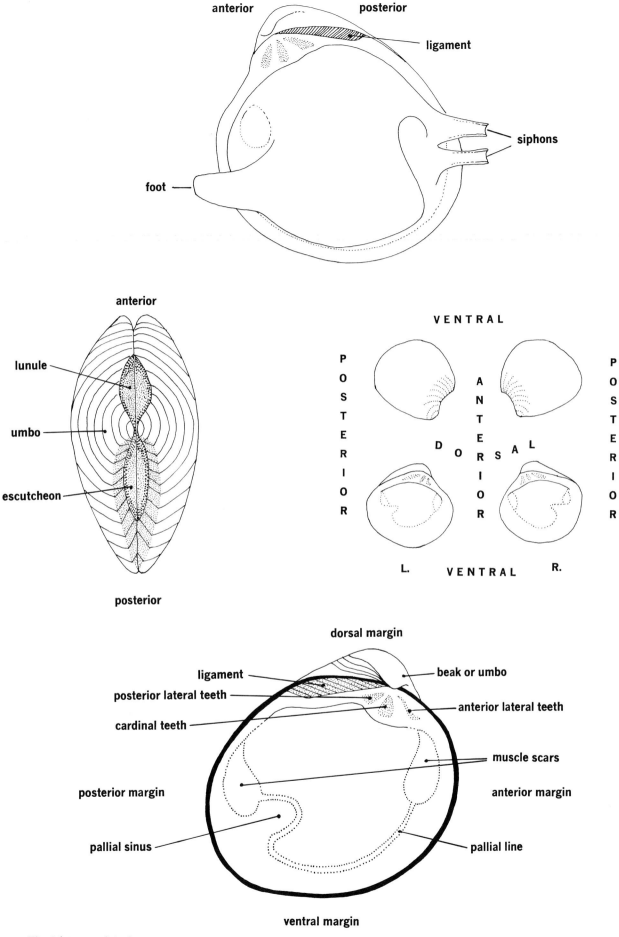

Fig. 24. Parts of the bivalve shell.

Little Shell on Padre Island, the wind and waves have banked the shore with incredible numbers of the shells of the coquina, the arks, and others of this class.

The recent discovery of two-valved opisthobranch gastropods has confirmed the theory of the derivation of a two-valved form from a group of cap-shaped univalve mollusks (Cox, Nuttal, & Turner 1969, p. 107). Unlike their ancestor, the gastropod, the Bivalvia can live only in water. The shape (fig. 23) and the structure of the bivalve shell have been adapted through the ages to survive in various aqueous environments and modes of life. In general, there are three habitat categories:

1. *Epifaunal* bivalves, which normally live exposed above the substrate (sea floor) surface; they may be with or without attachment.
2. *Semi-infaunal* bivalves, which live partially within the subtrate and partially exposed.
3. *Infaunal* bivalves, which are sessile or mobile mollusks that spend part or all of their lives buried beneath the sea floor sediment.

The bivalves invariably consist of two shells on either side of the body. Typically the two valves are of equal convexity (*equivalve*), but in some forms this symmetry has been lost and the valves will differ in size to a varying degree (*inequivalve*). These valves are held together dorsally by a leathery, brown, elastic *ligament*. Some of the characteristics of the bivalve shell are illustrated in Figure 24 and are described in the following text.

The two valves are joined at their *dorsal margin* but are open along their *anterior, posterior,* and *ventral* margins. The ventral margin is opposite the dorsal margin. The valves are closed by action of the *adductor* muscles attached to the inner face of each. When the two valves do not meet at the ventral margin when closed, they are said to *gape*.

Along the dorsal margin will be found the *beak,* or *umbo* (plural, *umbones*), which is the point where growth of the shell began. If the adult shell has a beak that occupies a position close to the middle of the length of the shell, the shell is said to be *equilateral*, but if the beak lies closer to one end or the other it is said to be *inequilateral*. The *umbonal cavity* is that part of the interior of the valve which lies within the umbones.

In the umbonal cavity are found the *hinge teeth*. These teeth function to prevent any rocking or shearing movements of the valves. Infaunal bivalves, with their more sheltered lives, do not require the stronger,

interlocking teeth of epifaunal genera, such as *Spondylus. Cardinal teeth* are those immediately below the umbones; the *anterior* and *posterior laterals* are those on the respective sides of the cardinals. Often in the case of inequilateral bivalves the teeth have become so distorted that they are difficult to distinguish. The dentition is the most satisfactory basis of classification within the class Bivalvia.

Two characteristic features of the umbonal area to be found in infaunal bivalves are the *lunule* and the *escutcheon*. The lunule is a heart-shaped impression anterior to the beaks, while the escutcheon is a similar depression posterior to the beaks. The function of these two features has not yet been adequately interpreted.

The interior of the shell bears an impression, or *pallial line*. This line running parallel and close to the shell margin is a band, or series, of pits made by the insertion of muscles attaching the mantle to its shell. This line may be *continuous*, as in the *Mercenaria*, or *discontinuous* and farther within the shell, as in many epifaunal bivalves. The latter location is a result of a need for longer muscle fibers in order to rapidly withdraw the mantle far into the shell of the more vulnerable epifaunal species.

An obvious feature of the pallial line is a U-shaped embayment, or indentation, the *pallial sinus*, which is always located posteriorly at the position of the siphon. The sinus will only be found in infaunal bivalves, although the very short siphons of some infaunal clams do not necessarily produce a sinus. If the pallial line is without a pallial sinus it is described as a *simple* pallial line.

The primary function of the bivalve shell is protective, and any accessory structure of the two valves shares this function (Kauffman 1969, n. 142). The forms in which the shell does not completely enclose the soft parts have compensated for the loss of protection by such adaptations as deep burial, secretion of tubes, building of nests, or through other structural and ecologic modifications of the usual bivalve pattern.

The shape and convexity of infaunal bivalves are closely correlative with the depth and rate of burrowing. In general the more obese bivalves will be found nearer the surface than the more streamlined shell shapes. In either case the depth of burial is relative to the length of the shell, not to the distance below the sea floor.

The sculpture of the exterior of the shell reflects the environmental demands and is strikingly similar in even distantly related genera that occupy the same type of environment. Epifaunal groups are more variable than those of the infauna due to the greater variety in their modes of life and habitats. Ornamentation, such as spines, costae, lamellae, and flutes, functions to strengthen the shell, to provide support and anchorage in the substrate, to keep the feeding margin elevated above the bottom, to support sensory extensions of the mantle margin, and to discourage predators. The exterior sculpture may be put into two basic groups: (*a*) the concentric sculpture, which is arranged parallel to the margins of the valves, generally indicating former growth and resting stages; and (*b*) the radial sculpture, which runs fanlike across the shell from the umbones to the ventral margins of the valves. When both concentric and radial sculpture occur a cancellate sculpture is formed. In most bivalves the exterior is covered with a protective periostracum.

To identify the bivalve shell, it is often necessary to determine which valve is the right valve and which is the left. This identification is more difficult in the siphonless epifaunal forms than in the infaunal. Hold the paired shells of a bivalve with the ventral margins down, the dorsal region (beaks) up, and the anterior end (usually the flatter end) away from you; the beaks will usually be pointed toward the anterior end, and the ligament will be found posterior to the beaks in most cases. Holding the shell in this manner, the right valve is on your right and the left valve is on your left. A method for distinguishing the right valve from the left of infaunal bivalves (*Tellin, Mactra*) is to lay the open shell on the table with the beaks uppermost, or away from you, and the interior of the shell up. In this position it is possible to locate the U-shaped pallial sinus. If the sinus is indented or open toward the right, it is a right valve, and if to the left, a left valve. The indentation is always nearest the posterior end where the siphons are located. This margin is nearest the surface when the siphons are extended for feeding.

To determine the length of a bivalve shell, a straight line from the extreme anterior margin to the extreme posterior margin is measured; the height is measured on a line from the umbones to a point directly opposite on the ventral margin. The greatest distance between the sides of the closed valves determines the thickness.

Other Mollusks

The classes of Amphineura, Scaphopoda, and Cephalopoda have such few known species along the Texas coast that a brief discussion of their identifying features will suffice.

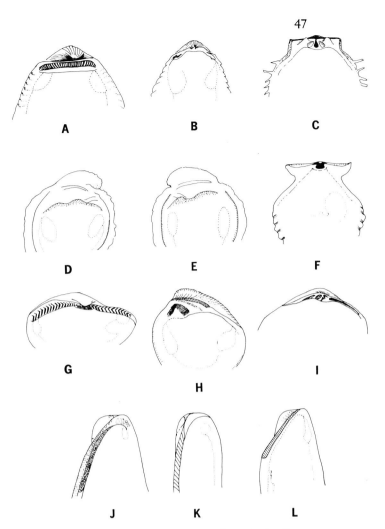

Fig. 25. Various bivalve hinges: *A, Anadara; B, Cardium; C, Spondylus; D, Chama; E, Pseudochama; F, Pecten; G, Nuculana; H, Chione; I, Mactra; J, Brachidontes; K, Lithophaga; L, Modiolus.*

The name Amphineura (Gk. *amphi* both, in both sides, *neuron* nerve) refers to the bilateral symmetry of the nervous system. This class contains two subclasses, the Aplacophora, which are shell-less and considered by some to be the lowest form of molluscan life, and the Polyplacophora, which bear an eight-plated shell.

The Polyplacophora, or chitons, have a many-plated shell similar in appearance to that of the garden pill bug. The chiton has a broad, suckerlike foot on the ventral side that it uses to creep about rocks and shell in search of the vegetation on which it feeds. Each of the eight plates of its shell is greatly arched and overlaps the valve next behind it. These valves are held together by a leathery tissue, or girdle, that surrounds the entire periphery of the lapped valves. This flexible girdle permits the chiton to roll up into a ball for protection as the armadillo does. Often the interior of the valves is beautifully colored pink, blue, or green, while the exterior is camouflaged so that detection is difficult.

The Scaphopoda (Gk. *scapha* boat, *pod* foot) burrow with a foot shaped like the prow of a boat. Fewest in number of the Mollusca, this class was long thought to be marine worms with calcareous tubes. The shells are tubular, open at both ends, nonspiral, and generally tapered toward the posterior end (tusk shaped). All growth takes place at the wider anterior end. The foot protrudes through the anterior end of the shell and is used to bury this deposit feeder just below the surface.

The Cephalopoda (Gk. *cephalos* head, *pod* foot), or "head-footed ones," are the most highly developed of the phylum Mollusca. These strictly marine animals are found in all the oceans of the world. Today there are about 650 known species in the world, but fossil evidence shows that the class abounded in prehistoric seas. These primeval cephalopods bore coiled and many-chambered shells that clearly showed their relationship to mollusks. Only the nautilus of the southwestern Pacific and the female argonaut produce this kind of shell today, and the shell of the argonaut is an egg case, not an exoskeleton. More species will probably be found to exist in the unexplored areas of the oceans.

The Nautiloidea deviate from the usual pattern of the class in other ways than just the shell; they have up to ninety arms and four gills. The other subclass, Coleoidea, contains the orders Vampyromorpha, Octopoda, and Decapoda—all with two gills, two kidneys, and three hearts with blue blood.

The Vampyromorpha are strange creatures long thought to be extinct but now known by several surviving members. One species, a jet black animal with ten arms, *Vampyroteuthis infernalis* Chun, was recovered from the Gulf of Mexico by the *Atlantis* expedition. The first specimen was found off the Cameron-Congo River in Africa and the order was monographed by G. E. Pickford (1949). It is an inhabitant of deep water.

The Octopoda, or eight footed, range in size from two inches to thirty-two feet. However, the body of the "giant" octopus is only eighteen inches long. Octopuses have a globular body that narrows slightly into a "neck." Eight arms arise from the head and are united at their base by a membranous web. These arms are often incorrectly referred to as tentacles (Lane 1960). The octopuses are the best known of all cephalopods, and the most studied species of octopus are inhabitants of shallow water, where they live in natural holes or lairs of their own construction. The octopus lies in wait for its favorite food, the crab, in its lair where it is camouflaged by its ability to change color and appear a part of the surroundings. It captures its prey by snaring the victim with its sucker-clad arms and enveloping it in the membranous web while holding on securely with the other arms. It actively stalks prey at night and uses its beak to open clams, another principal item of its diet. By some unknown method, the octopus quiets his victim with a nerve poison or venom secreted from the salivary glands; then he breaks it apart and picks the shell clean.

The Decapoda, ten footed, include the squid and the cuttlefish. At present there are no living cuttlefish known from American waters (see Harry 1969c), but the squid is familiar to all fishermen who find it a choice bait. This animal has a cigar-shaped or fusiform body with one end or two side fins at the posterior end. The head bears eight arms and two tentacles used to seize prey. The mantle covers a chitinous shell called a pen. Most squid live in deep water by day and rise to the surface to feed at night.

The deep-water dweller *Spirula spirula* differs from this pattern by bearing a spirally coiled internal shell. This shell, which is enclosed by the mantle at the posterior end of the animal, is often found by the thousands on the beaches. The animal that formed this beautiful, white coil was long a mystery. Prior to 1920 only 13 specimens had ever been recovered. A Danish expedition in search of the breeding place of the American and European river eel in the Sargasso Sea netted 188 *Spirula* about a quarter of a mile below the surface of the sea. The shell is constructed to resist great pressure, enabling it to rise from the depths and float unbroken to distant shores.

The squid feed on plankton, crustaceans, and fish. They reach out and catch their prey with their two long tentacles, hold it with the suckers on their arms, and then bite it with their sharp beaks.

Although Aristotle described the reproduction of the cephalopods, it was not studied again until the middle of the nineteenth century and it is still imperfectly known except in a few species. In most members of the class the male, which is smaller, has one or more of the arms modified into a sex organ, the hectocotylus, which he uses to transfer the fertilized sperm to the female.

The female usually lays her eggs inshore in clusters attached to rocks or debris and guards them until they hatch. Squid lay their eggs in gelatinous strings, which they often attach to other squid egg masses or seaweed. The female does not brood her eggs but awaits death in a very weakened state after she has performed her biological function.

Cephalopods display luminescence by a variety of means. This ability is more developed in deep-sea squid, such as the *Spirula*, than in shallow-water species. This bioluminescence, light without heat, may be produced by any one of the following three means:

1. Intracellular: intrinsic, depends on the animal's own biochemical processes and takes place within luminescent cells or photocytes
2. Extracellular: intrinsic, from the discharge of luminous secretions
3. Bacterial: from symbiotic luminous bacteria

Bioluminescence is one of the most striking characteristics of these strange creatures.

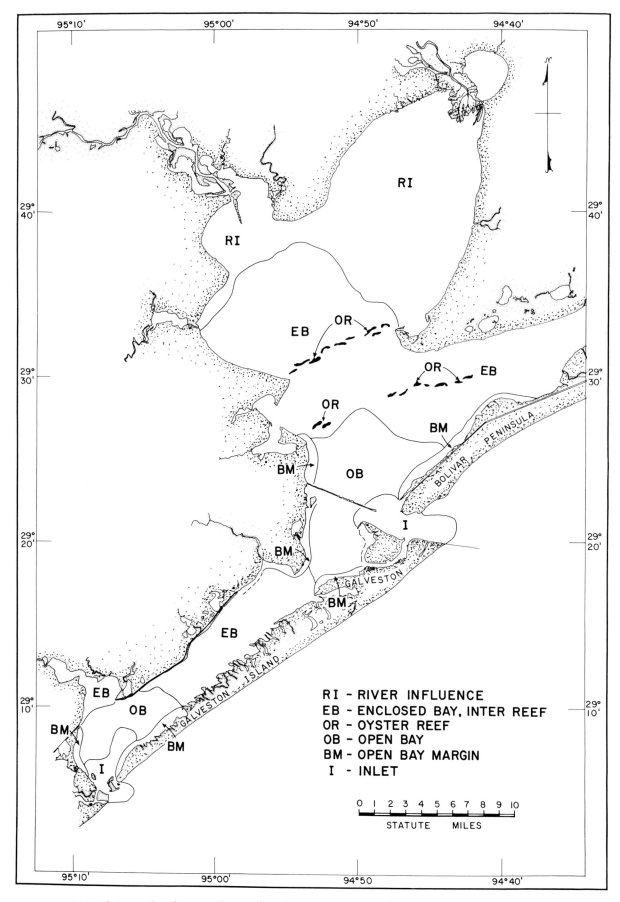

Fig. 26. Areal distribution of macrofaunal assemblages in Galveston Bay. Note the large extent of river-influenced and enclosed-bay assemblages, as well as the alignment of oyster reefs (Parker 1960).

3. Systematic Descriptions

The mollusks of the Gulf of Mexico are only imperfectly known. The first publication that listed them was by Ferdinand von Roemer in his work on Texas in 1849. Dall cataloged them in 1889 and Singley (1893) gave a listing of 342 species. Other listings have been published by Mitchell (1894), C. W. Johnson (1934), Stenzel (1940), Clench and Turner (1951–1952), Hedgpeth (1950), Whitten et al. (1950), Ladd (1951), Pulley (1952), Abbott (1954), Hildebrand (1954), and Parker (1956, 1959, and 1960).

The basis and background for the present list is a summary of the author's knowledge of Texas molluscan life derived from a long period of field work and continued study of the accumulated information available. It is not a complete list of the shallow-water mollusks of Texas, but it is intended as an aid in identifying those shells which may be collected on the beach. Many other small shells are to be found on the beach, and the offshore waters are the home of still more, both large and small, that are under study.

Because the classification of mollusks is still a matter of considerable debate, the order of arrangement is a compromise between the principal systems being used today. Within this larger framework the order of species under genus is alphabetical. Grassé (1968), Keen (1960), Clench (1954–1959), and Taylor and Sohl (1962–1964) have been the basis for the order of arrangement of the gastropods, with emphasis on Taylor and Sohl. The pelecypod order is based on the more recent study of Vokes (1967).

Some of the abbreviations inserted in the scientific names are cf. (*confer*), compare; s.l. (*sensu lato*), in the broad sense; s.s. (*sensu stricto*), in the strict sense; ex. gr. (*ex grupo*), of the group of; in litt. (*in litteris*), in correspondence; syn. (synonym), a scientific name superseded or discarded; and emend., emended, or corrected, by.

Dimensions are given in the metric scale if the shell is under one inch; larger shells are measured in inches.

Range is a compilation from published literature and some unpublished records of collectors known to the author.

Relative abundance, or occurrence, is given according to the occurrence on the beach and does not reflect the quantity to be found where the mollusks live offshore. *Rare, uncommon, fairly common*, and *common* are the categories used. These are subjective estimates, for there are wide variations, and the use of these categories will serve only as a general guide.

Locality is usually designated as *east, central, south*, or *entire. East* refers to that section of the coast from Sabine Lake to the Colorado River, where humid and subhumid climates prevail. *Central* is from the Colorado to Padre Island and represents the subhumid river-bay complexes. Padre Island and the Laguna Madre comprise the semiarid *south* region. Some localities overlap more than one region and are therefore described more explicitly; for example, *southern half of coast* indicates from Port O'Conner, south. After studying the various maps in this book and noting the habitat of the species, the collector will learn that estuarine or river-influenced fauna do not occur in the southern range, because no rivers empty into this area, that offshore marine species will not occur in the enclosed bays, and so on (fig. 26). Continued study of the coast is broadening these ranges. With few exceptions, the shells described inhabit near-shore, shallow-water environments at depths of less than one hundred feet.

In 1899 Charles Hedley remarked that "by some strange unwritten law these conchologists have invariably maintained a proportion between the size of a shell and its illustration. Thus, a large shell, however simple in structure, demanded a large figure; and a small shell, however complex its details, a small drawing. Had this school encountered *Pachyderms* or *Foraminifera* one or both would surely have fallen beyond the focus of their vision." Heeding Hedley, no attempt has been made to maintain a proportion between the size of the shell and its illustration; as a result a specimen only 7 mm. in length may be shown the same size as one whose length is 7 inches. The actual size is the first category in the accompanying descriptions.

This book describes more than 300 shells. You might wonder, as I did, who first described each species. It is interesting to note that only about a dozen people were responsible for naming more than half of these shells, yet about 125 authors' names appear in the descriptions. Listed in order of frequency, those who first described three or more Texas beach shells are:

Thomas Say (1787–1834), American malacologist, Philadelphia Academy of Natural Science.

Timothy Abbott Conrad (1803–1877), American paleontologist, Philadelphia Academy of Natural Science.

Carolus Linnaeus (Karl von Linné) (1707–1778), Swedish naturalist and botanist; originated biological systematics.

Charles Baker Adams (1814–1853), American naturalist, Amherst, Harvard Museum.

William H. Dall (1845–1927), American naturalist, U.S. National Museum and Harvard; prolific writer in many fields.

Jean Baptist P. A. de M. Lamarck (1774–1829), French naturalist; revised the Linnean system, proposing many new genera.

Johann F. Gmelin (1748–1804), German naturalist, student of Linné; published thirteenth edition of Linné's *Systema Naturae.*

Alcide Dessalines d'Orbigny (1802–1857), French biologist and geologist; conducted expedition to South America.

Paul L. McGinty (1877–1957), American collector and writer.

Henry Augustus Pilsbry (1862–1957), American, Philadelphia Academy of Natural Science; prolific writer, supplied 5,600 new names.

R. A. Philippi (1808–1904), German conchologist and writer.

Paul Bartsch (1871–1960), American malacologist, U.S. National Museum.

Peter Freidrich Röding (1767–1846), German; catalogued the Bolton collection.

Katherine J. Bush (1855–1937), American malacologist, U.S. National Museum.

P. C. A. Sander L. Rang (1784–1859), wrote *Histoire Naturelle de Mollusques-Pterepodes.*

J. G. Bruiguière (1750-1798), French writer on invertebrates; physician.

Lowell Augustus Reeve (1814–1865), British; wrote *Conchologia Iconica,* first large set of illustrated volumes on world-wide sea shells.

William Stimpson (1832–1872), American malacologist; collection and manuscript destroyed in Chicago fire of 1871.

Henri M. Ducrotay de Blainville (1777–1850), French zoologist.

Francis Holmes (1815–1882), American zoologist.

Harold A. Rehder (1907–), curator of mollusks, U.S. National Museum.

C. A. Lesueur (1778–1846), French naturalist from voyages of *Géographe.*

Sowerby Family. British writers, illustrators, and publishers who published books in parts and sold them separately over a period of years. Thus, it is difficult to determine dates for an article's composition or to say which Sowerby wrote the article, and it is equally hard to find the true date of publication. The Sowerbys were:

James (1757–1822), the father.

James de Carle (1787–1871), son.

George Brettingham I (1788–1854), son.

George Brettingham II (1812–1884), son of G. B. I.

George Brettingham III (1843–1921), son of G. B. II.

Class AMPHINEURA Von Ihering, 1876

Order NEOLURICATA Bergenhayn, 1955

Suborder ISCHNOCHITONINA Bergenhayn, 1930

Family ISCHNOCHITONIDAE Dall, 1889

Genus *Ischnochiton* Gray, 1847

MESH-PITTED CHITON
Ischnochiton papillosus (C. B. Adams, 1845), *Proc. Boston Soc. Natur. Hist.* 2:9.
Gk. *ischnos* slender, *chiton* a girdle; L. *papilla* a nipple.
SIZE: Length 8 to 12 mm.
COLOR: Whitish, mottled with olive green. Girdle alternately white and olive.
SHAPE: Oval.
ORNAMENT OR SCULPTURE: Upper surfaces of valves sculptured with microscopic pittings. End valves with concentric rows of fine, low beads. Lateral areas with fine, sinuous, longitudinal lines. Posterior slope of posterior valve is concave with 9 slits. Girdle narrow.
REMARKS: Several other chitons have been reported on the Texas coast, but this is the only species found with any frequency.
HABITAT: High-salinity lagoons on shell and reefs.
LOCALITIES: Entire.
OCCURRENCE: Fairly common.
RANGE: Tampa to the Lower Keys, West Indies and Texas.

Ischnochiton papillosus

Class GASTROPODA Cuvier, 1797

Subclass STREPTONEURA Spengel, 1881 =
PROSOBRANCHIA Milne Edwards, 1848

Order ARCHAEOGASTROPODA Thiele, 1925

Suborder PLEUROTOMARIOIDEA Swainson, 1840

Superfamily FISSURELLACEA Fleming, 1822

Family FISSURELLIDAE Fleming, 1822

Subfamily DIODORINAE Odhner, 1932

Genus *Diodora* Gray, 1821

CAYENNE KEYHOLE LIMPET
Diodora cayenensis (Lamarck, 1822) *Hist. Natur. Anim. sans Vert.* 6:12.
Gk. *dia* through, *dora* hide or fur; Cayenne, the capital of French Guiana.
SIZE: Diameter 1 to 2 in.
COLOR: Variable from pinkish, whitish, to dark gray. Interior white or blue gray, polished.
SHAPE: Conical, oblong oval, rather thick shell. Like a coolie hat.

Diodora cayenensis

ORNAMENT OR SCULPTURE: Orifice keyhole shaped, just in front of and slightly lower than the apex. Every fourth rib is noticeably larger. Ribs are crossed by concentric, lamellar ridges.
APERTURE: Oval with heavy callus. Truncate behind with a depression immediately posterior. Margins are irregularly and finely crenulated. Interior smooth.
OPERCULUM: None.
PERIOSTRACUM: None visible.
REMARKS: Grazes on vegetation. Moves about freely but returns to its home base. Animal has pair of tentacles with eyes on small projections near their base. Snout short, oval, and disc shaped with mouth in middle of disc. Unfertilized eggs and genital products are expelled through apical hole into the sea. Exterior of shell usually covered with algae, making live specimens difficult for the untrained eye to see.
HABITAT: Intertidal to moderately deep water. Mainly on rocks, jetties.
LOCALITIES: Entire.
OCCURRENCE: Uncommon in east, becoming common south to Mexico.
RANGE: Virginia to southern half of Florida, Texas and Brazil.
GEOLOGIC RANGE: Recent with related species to Miocene.

Genus *Lucapinella* Pilsbry, 1890

FILE FLESHY LIMPET
Lucapinella limatula (Reeve, 1850) *Conch. Iconogr.* 6:15.
L. *lucanus* a beetle, dim. of *lima* file.
SIZE: Length 15 mm.
COLOR: Dull white with brownish mottlings.
SHAPE: Oblong conical with almost central apical hole.
ORNAMENT OR SCULPTURE: Radial ribs are alternate-

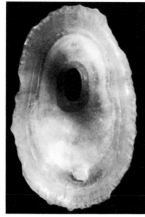

Lucapinella limatula

ly large and small. Growth lines form concentric lamellations or scales as they cross ribs.

APERTURE: Large, interior smooth, white, porcelaneous. Callus rounded, smooth, with crenulate margins.

OPERCULUM: None.

PERIOSTRACUM: None visible.

REMARKS: Grazes on minute algae, which it scrapes with radula. Lives completely underwater. Apical hole is the former marginal notch in lip of aperture, which closes to a hole as shell grows. Uses foot to hold itself to rock. Sexes separate. Figured specimen worn.

HABITAT: On rocks and jetties, fairly deep.

LOCALITIES: Port Aransas, south.

OCCURRENCE: Uncommon.

RANGE: North Carolina to southern half of Florida, and West Indies, Texas.

GEOLOGIC RANGE: Pliocene to Recent.

Superfamily TROCHACEA Rafinesque, 1815

Family TROCHIDAE Rafinesque, 1815

Subfamily CALLIOSTOMATINAE Thiele, 1924

Genus *Calliostoma* Swainson, 1840

Subgenus *Kombologion* Clench & Turner, 1960

SCULPTURED TOP-SHELL
Calliostoma (Kombologion) euglyptum (A. Adams, 1854) *Proc. Zool. Soc. London* 22:38.

Gk. *kalos* beauty, *stoma* a mouth; L. *eu* well, *glyphein* to carve.

SIZE: Height 25 mm.

COLOR: Pinkish brown, often mottled with white.

SHAPE: Turbinate, or conical; toplike.

ORNAMENT OR SCULPTURE: Seven and one-half whorls, slightly convex, with a rounded keel. Periphery of whorls rounded. Spire extended. No umbilicus. Six large, beaded, spiral cords with smaller, intermediate, weakly beaded cords on each whorl.

APERTURE: Subquadrate, outer lip simple and produced at an angle from the base. Interior nacreous. Columella white, arched near base.

OPERCULUM: Thin, corneous, circular, golden brown, multispiral.

PERIOSTRACUM: None visible.

REMARKS: Grazes on algae or plant detritus. Produces eggs in gelatinous ribbons.

HABITAT: Low water to 32 fathoms.

LOCALITIES: On offshore oil-rig platforms and 7½ fathom reef off Padre Island.

OCCURRENCE: Rare.

RANGE: North Carolina to Florida and Texas.

GEOLOGIC RANGE: Cretaceous to Recent.

Calliostoma (Kombologion) euglyptum

Subfamily MONODONTINAE Cossmann, 1916

Genus *Tegula* Lesson, 1832

SMOOTH ATLANTIC TEGULA
Tegula fasciata (Born, 1778) *Index Mus. Caes. Vind.*,
    p. 337.
L. *tegula* tile, *fasci* bundle.
SIZE: Width 12 to 18 mm.
COLOR: Background light or a reddish brown. Mottled
    with reds, browns, blacks, white. Callus and um-
    bilicus white.
SHAPE: Conical, or turban shaped, with rounded shoul-
    ders on whorls.
ORNAMENT OR SCULPTURE: Twelve postnuclear
    whorls sculptured with 3 or more fine spiral threads.
    Umbilicus deep, round, and smooth.
APERTURE: Rounded with 2 teeth at base of columella.
OPERCULUM: Corneous and multispiral.
PERIOSTRACUM: None visible.
REMARKS: To date, this shell has not been reported
    alive on the Texas coast, although it is abundant in
    the drift around Port Isabel and Port Aransas in bay
    areas. Figured specimen worn.
HABITAT: On grass blades and under rocks at low tide
    when living.
LOCALITIES: Entire, more to south.
OCCURRENCE: Common.
RANGE: Southeastern Florida, West Indies and Texas.
GEOLOGIC RANGE: Pliocene to Recent.

Tegula fasciata

Family PHASIANELLIDAE Swainson, 1840

Genus *Tricolia* Risso, 1826

CHECKERED PHEASANT
Tricolia affinis cruenta Robertson, 1958, *Johnsonia*
    3(37):267.
L. *tri* three, *colis* stalk, *affinis* adjacent, *cruenta* stain
    with blood.
SIZE: Length 6.2 mm.
COLOR: Background white or pale orange patterned
    with spiral rows of squarish, dark red spots.
SHAPE: Inflated, conical.
ORNAMENT OR SCULPTURE: Four and one-half
    smooth, rounded whorls. Apex rounded. Suture im-
    pressed. Umbilicus only a chink.
APERTURE: Elongate ovate. Columella with a thick,
    white callus. Outer lip smooth.
OPERCULUM: Calcified, white with dark olive green
    at the margin.
PERIOSTRACUM: None visible.
REMARKS: Herbivorous, may be dependent on specific
    plant substrates. Sexes separate.
HABITAT: Shallow water.
LOCALITIES: St. Joseph Island, south.
OCCURRENCE: Fairly common.
RANGE: Caribbean coast of Central America and
    northern South America. Sporadic along western
    Gulf of Mexico.
GEOLOGIC RANGE: Lower Miocene to Recent.

Tricolia affinis cruenta

Nerita fulgurans

Superfamily NERITACEA Rafinesque, 1815

Family NERITIDAE Rafinesque, 1815

Genus *Nerita* Linné, 1758

ANTILLEAN NERITE

Nerita fulgurans Gmelin, 1791, *Syst. Natur.*, 13th ed.,
    p. 3685.
Gk. *nerites* pertaining to coastline or a sea shell; L.
    fulgar lightning.
SIZE: Length 18 to 25 mm.
COLOR: Dark gray to black with occasional blurred
    markings.

SHAPE: Conical globose, body whorl expanded, spire
    low.
ORNAMENT OR SCULPTURE: Strong spiral cords.
    Heavy shell.
APERTURE: Large, rounded with 2 prominent teeth on
    inside of outer lip. Inner lip is yellowish white,
    toothed and with a decklike callus. Porcelaneous.
OPERCULUM: Calcareous, light gray to yellowish gray.
    It has an odd shape with a clawlike projection, or
    apophysis, that fits under the edge of the columella,
    permitting a tight fit when the shell is closed.
PERIOSTRACUM: None visible.
REMARKS: The tightly closed shell is adaptable to
    changes in environment. Can live from 12 to 77
    days out of water. Places its eggs upon its own shell
    and those of other mollusks. Feeds by grazing.
HABITAT: Salt to brackish water. Jetties and pilings
    near high-tide mark.
LOCALITIES: Port Aransas, south.
OCCURRENCE: Uncommon.
RANGE: Southeastern Florida and West Indies, and
    Texas.

Genus *Neritina* Lamarck, 1816

Subgenus *Vitta* Mörch, 1852

OLIVE NERITE

Neritina (Vitta) reclivata (Say, 1822) *J. Acad. Natur.
    Sci. Phila.* 2:244.
Gk. *nerites* pertaining to coastline or a sea shell; L.
    reclinis leaning back, sloping.
SIZE: Length 12 mm.
COLOR: Brownish green or brownish yellow back-
    ground with numerous transverse lines of brown,
    purple, or black.
SHAPE: Globular with expanded body whorl.
ORNAMENT OR SCULPTURE: Smooth and polished.
    Spire usually eroded away.
APERTURE: Semilunar. Parietal area smooth, white to
    yellowish with a variable number of small, irregular
    teeth on the columellar edge.
OPERCULUM: Black to slightly brownish. Calcareous.
    Has a projection that fits into a corresponding in-
    dentation on the inner lip.
PERIOSTRACUM: None visible.
REMARKS: A grazer. Places eggs on own shell or other
    mollusk shells.
HABITAT: Brackish water. At times, fresh water.
LOCALITIES: Entire.

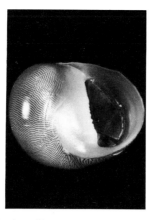

Neritina (Vitta) reclivata

OCCURRENCE: Uncommon to rare.
RANGE: Florida to Texas and the West Indies.
GEOLOGIC RANGE: Pleistocene to Recent.

VIRGIN NERITE
Neritina (Vitta) virginea (Linné, 1758) *Syst. Natur.*,
    10th ed., p. 778.
Gk. *nerites* pertaining to coastline or a sea shell; L.
    virgineus maidenly.
SIZE: Length 4 to 12 mm.
COLOR: Variable. Background may be olive, white,
    gray, red, yellow, purple, black with black and/or
    white waves, stripes, dots, lines, or mottlings.
SHAPE: Globular with expanded body whorl.
ORNAMENT OR SCULPTURE: Smooth and polished.
APERTURE: Semilunar. Parietal area smooth, convex,
    white to yellow, with a variable number of small,
    irregular teeth.
OPERCULUM: Calcified, usually black, smooth.
PERIOSTRACUM: None visible.
REMARKS: During winter months it lays eggs on own
    shell or shell of other mollusks. The eggs are in
    clustered, yellowish, gelatinous capsules. More easily
    found at night or on cloudy days. Patterns so varied
    one might think they were several different species.
    Is more colorful in brackish water than in more
    saline water.
HABITAT: Bay margins and grass flats.
LOCALITIES: Entire.
OCCURRENCE: Common. More common in south.
RANGE: Florida to Texas, the West Indies and Ber-
    muda.
GEOLOGIC RANGE: Pleistocene to Recent.

Genus *Smaragdia* Issel, 1869

EMERALD NERITE
Smaragdia viridis viridemaris (Maury, 1917) *Bull.
    Amer. Paleontol.* 5:316.
Gk. *smaragdo* emerald; L. *viridis* green.
SIZE: Length 6 to 16 mm.
COLOR: Bright pea green with a few fine, broken white
    lines near the apex; these are outlined in maroon
    color in some specimens.
SHAPE: Obliquely oval, subglobular, low spire.
ORNAMENT OR SCULPTURE: Glossy, smooth. Body
    whorl is swollen or expanded.
APERTURE: Simple, semilunar, outer lip thin and
    sharp. Parietal area is green with 7 to 9 minute
    teeth.
OPERCULUM: Calcareous, green, and smooth except
    for microscopic lines. Apophysis on columellar edge.
PERIOSTRACUM: None visible.
REMARKS: Animal is green with dark eyes at the end
    of pointed tentacles. More abundant when cloudy
    or in late afternoon. It will climb above the water
    level when placed in a bucket.

Neritina (Vitta) virginea

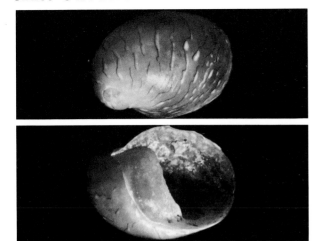

Smaragdia viridis viridemaris

HABITAT: Grass flats in shallow bays.
LOCALITIES: Port Aransas, south.
OCCURRENCE: Uncommon.
RANGE: Southeastern Florida, West Indies and Bermuda and Texas.

ORNAMENT OR SCULPTURE: About 6 convex whorls; body whorl about one-half the height; slightly channeled sutures. Spiral sculpture of irregular, inequidistant incised lines. Anomphalous.
APERTURE: Rounded oval; outer lip smooth, does not flare, sharp and nearly horizontal at the body whorl. Inner lip with a thin deposit over body whorl. Columellar area smooth, moderately wide, with central groove in lower portion.
OPERCULUM: Corneous, light brown, thin, and flexible.
PERIOSTRACUM: None visible.
REMARKS: Usually is attached to leaves, roots, or trunks of the mangrove tree. Climbs high into the tree and can spend much of the day out of water. At times on pilings and rocks. It probably lives on the black mangrove *Avicennia nitida*, which is subject to freeze in this area.
HABITAT: Mangrove thickets. Brackish water.
LOCALITIES: Extreme south and on rocks at Corpus Christi Naval Air Station.
OCCURRENCE: Uncommon.
RANGE: Southern half of Florida, West Indies and Bermuda.

Order MESOGASTROPODA Thiele, 1925 = CAENOGASTROPODA Cox, 1959

Superfamily LITTORINACEA Gray, 1840

Family LITTORINIDAE Gray, 1840

Genus *Littorina* Férussac, 1821

Subgenus *Littoraria* Gray, 1834

ANGULATE PERIWINKLE
Littorina (Littoraria) angulifera (Lamarck, 1822), *Hist. Natur. Anim. sans Vert.* 7:54.
L. *littus* the seashore, *angulatus* angular.
SIZE: Height 25 to 30 mm.
COLOR: Background color varies: bluish white, orange yellow, dull yellow, reddish brown, or grayish brown. Darker marks of elongated spots on the ribs; often fused to form oblique stripes on the body whorl. Early whorls have regularly spaced, vertical white spots below the suture.
SHAPE: Conical with elongate spire. Higher than wide.

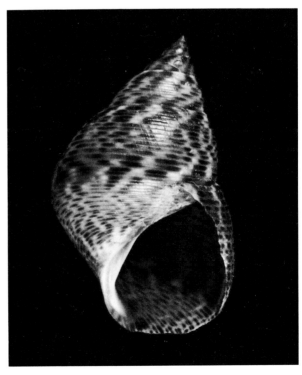

Littorina (Littoraria) angulifera

Littorina (Littoraria) irrorata

MARSH PERIWINKLE

Littorina (Littoraria) irrorata (Say, 1822), *J. Acad. Natur. Sci. Phila.* 1(2):239.

L. *littus* the seashore, *irror* bedew, dotted.

SIZE: Length 1 in.

COLOR: Grayish white with tiny, short streaks of reddish brown on the spiral ridges. Opaque, dull.

SHAPE: Elongate conical. Longer than wide.

ORNAMENT OR SCULPTURE: Eight to 10 gradually increasing, flat whorls. Suture weak. Body whorl about one-half total height. Numerous, regular shaped spiral grooves. Shell thick.

APERTURE: Oval, whitish to yellowish. Columella and callus usually pale reddish brown. Outer lip stout, sharp, and with tiny, regular grooves on the inside edge.

OPERCULUM: Corneous, dark brown.

PERIOSTRACUM: None visible.

REMARKS: Herbivorous on algae. When tide is out, animal goes into shell and may remain dry and exposed to sun for many hours. It is often seen completely out of water on stems of marsh grass where it leaves a trail of mucus.

HABITAT: On march grass in low-salinity bays.

LOCALITIES: East, central.

OCCURRENCE: Common.

RANGE: New Jersey to central Florida to Texas.

GEOLOGIC RANGE: Upper Miocene to Recent.

Subgenus *Melarhaphe* Menke, 1828

ZEBRA PERIWINKLE

Littorina (Melarhaphe) lineolata d'Orbigny, 1842, *Voy. Amer. Morid.* 5(3):392.

L. *littus* the seashore, *linea* lined.

SIZE: Length 12 to 25 mm.; females larger than males.

COLOR: Background gray with oblique zigzag lines of dark brown. Apex reddish brown.

SHAPE: Elongate conical.

ORNAMENT OR SCULPTURE: Six to 8 gradually increasing convex whorls. Body whorl more than one-half total height, more convex than other whorls. Suture well marked. Male more strongly sculptured with spiral grooves. Shell moderately thick.

APERTURE: Pear shaped. Outer lip not flaring, edge sharp and thin. Meets body whorl at sharp angle. Columellar area long and wide, slanting inward, smooth. No umbilicus. Purple to mahogany brown.

OPERCULUM: Corneous, dark brown.

PERIOSTRACUM: None visible.

REMARKS: Marked sexual dimorphism. Reproduction unknown. Herbivorous on algae. Can stay alive for hours out of water. This species was formerly known

Littorina (Melarhaphe) lineolata

as *L. zic zac* in Texas. *L. lineolata* is darker in color, is smaller, and has a wider apical angle than *L. zic zac*.

HABITAT: Intertidal on rocks, jetties, pilings. Often in large colonies in crevices.

LOCALITIES: Entire.

OCCURRENCE: Common.

RANGE: Southern half of Florida to Texas, Bermuda to Brazil.

GEOLOGIC RANGE: Upper Miocene to Recent.

CLOUDY PERIWINKLE

Littorina (Melarhaphe) nebulosa (Lamarck, 1822), *Hist Natur. Anim. sans Vert.* 7:54.

L. *littus* the seashore, *nebulosus* misty, cloudy.

SIZE: Length 15 mm.

COLOR: Background bone yellow or white with a bluish tinge; early whorls have spotting of white and reddish brown, which stops abruptly on the sixth or seventh whorl. Variable in juveniles.

SHAPE: Elongate conical. Higher than wide.

ORNAMENT OR SCULPTURE: Seven to 9 gradually in-creasing convex whorls. Body whorl about two-thirds of total height. Nuclear whorls usually eroded in adults. Suture well marked, smooth or slightly crenulate behind the outer lip. Spiral engraved lines become evident on fourth whorl, becoming numerous and more regularly spaced on body whorl. Strong transverse striations cross the spiral lines at angles.

APERTURE: Subcircular; outer lip not flaring; edge thin, sharp, smooth, meeting body whorl at sharp angle. Slight callus over body whorl. Columella long, wide, smooth, and without umbilicus. Mouth yellowish brown to pale purplish within; outer lip whitish. Columella white.

OPERCULUM: Corneous, pale mahogany brown.

PERIOSTRACUM: None visible.

REMARKS: Herbivorous on algae. Sexes separate. Prefers wooden jetties, wreckage, and logs. Avoids rocks exposed to heavy surf. Seldom survives Texas winters. Greatly reduced since heavy freeze of 1962.

HABITAT: Pilings, jetties in more saline waters.

LOCALITIES: Entire.

OCCURRENCE: Fairly common.

RANGE: West Indies and Texas, Caribbean coast of Central America and northern South America.

GEOLOGIC RANGE: ? to Recent.

Littorina (Melarhaphe) nebulosa

Subgenus *Neritrema* Recluz, 1869

SPOTTED PERIWINKLE

Littorina (Neritrema) meleagris (Potiez & Michaud, 1838), *Gal. Moll. Doval* 1:311.

L. *littus* the seashore; Gk. Meleager, mythological son of Oeneus, or a guinea fowl.

SIZE: Height 7.5 mm., width 4.5 mm.

COLOR: Light to dark brown with more-or-less spiral rows of white spots giving a checkered appearance. Spots may be faded on body whorl.

SHAPE: Elongate conical. (Ovate turriculate.)

ORNAMENT OR SCULPTURE: Five to 6 rapidly increasing convex whorls, nuclear whorls not eroded; suture well marked, smooth. Body whorl about two-thirds of total height and more convex than other whorls. Surface smooth or with faint spiral striations; transverse striations very weak. A crescent-shaped umbilical slit.

APERTURE: Pear shaped; outer lip not flaring, edge smooth, sharp, and thin but slightly thickened within; meets body whorl at a sharp angle; inner lip has

Littorina (Neritrema) meleagris

Littordina (Texadina) sphinctostoma

slight callus over body whorl. Reddish brown within. Columellar area long and moderately wide, smooth, usually flat; inner edge long and nearly straight, but not ending abruptly below; a very sharp outer ridge, less sharp where it merges evenly with the bowlike basal lip. Inner lip whitish.

OPERCULUM: Corneous, dark mahogany brown.

PERIOSTRACUM: Thin, often worn.

REMARKS: Herbivorous. Sexes separate.

HABITAT: Intertidal zone on rocks and pilings in quiet water.

LOCALITIES: Port Aransas, south.

OCCURRENCE: Uncommon to rare.

RANGE: Southern Florida and West Indies.

Superfamily RISSOACEA Gray, 1847

Family HYDROBIIDAE Troschel, 1857

Subfamily LITTORIDINAE Thiele, 1929

Genus *Littoridina* Eydoux & Souleyet, 1852

Subgenus *Texadina* Abbott & Ladd, 1951

Littoridina (Texadina) sphinctostoma Abbott & Ladd, 1951, *Nautilus* 41(10):335.

L. *littus* the seashore; Gk. *sphinkter* binder, *stoma* mouth.

SIZE: Length 2 to 3 mm.

COLOR: Translucent gray alive, opaque white dead.

SHAPE: Ovate conic.

ORNAMENT OR SCULPTURE: About 5 convex whorls, thin but strong. Whorls increase regularly in size until the last third of body whorl, which becomes constricted and in some specimens detached. Surface smooth except for minute growth lines. Suture fine, moderately impressed. Umbilicated. Juvenile specimens may be keeled at base of body whorl.

APERTURE: Oval to round. It may be free from parietal wall.

OPERCULUM: Thin, corneous, transparent yellowish, paucispiral. Size and shape of aperture.

PERIOSTRACUM: Thin, smooth, translucent, grayish to yellowish.

REMARKS: Lives associated with *Rangia cuneata* in brackish water, originally described from the Texas coast.

HABITAT: Soft mud in river-influenced areas.
LOCALITIES: Entire.
OCCURRENCE: Common.
RANGE: Texas and Louisiana.
GEOLOGIC RANGE: Pleistocene to Recent.

Family TRUNCATELLIDAE Gray, 1840

Subfamily TRUNCATELLINAE Gray, 1840

Genus *Truncatella* Risso, 1826

BEAUTIFUL TRUNCATELLA
Truncatella pulchella Pfeiffer, 1839, *Wiegm. Arch.*
 1:356.
L. *truncatus* cut off, *pulchella* very pretty.
SIZE: Length 5.5 to 7.5 mm.
COLOR: White to pale amber, shiny.
SHAPE: Tall conical.
ORNAMENT OR SCULPTURE: Four to 4½ convex
 whorls. Deeply impressed suture. Seventeen or more
 poorly developed transverse costae; some specimens
 may be nearly smooth. Costae appear only on upper
part of whorl. Mature specimens appear very different from young due to loss of early whorls.
APERTURE: Ovate and somewhat flaring; entire. Angled above and rounded below. Outer lip simple,
 thin; thickened at its union with body whorl. Columella not apparent. Anomphalous.
OPERCULUM: Corneous and paucispiral with thin
 calcareous plate on the outer surface.
PERIOSTRACUM: None visible.
REMARKS: Can exist for a time submerged in salt water
 but usually found at high-tide line or slightly above
 under some protective material, such as boards,
 rocks, or seaweed. Probably distributed by flotsam,
 as the distribution on a beach is very spotty. Near
 maturity, the middle of a whorl is plugged by a
 septum and the unoccupied whorls are broken off
 by environmental stresses. (Figure on right is juvenile specimen.)
HABITAT: Above high tide and inlet areas.
LOCALITIES: Entire.
OCCURRENCE: Fairly common.
RANGE: Southeastern United States and West Indies.

Truncatella pulchella

Family RISSOIDAE Gray, 1847

Subfamily RISSOINAE Gray, 1847

Genus *Rissoina* d'Orbigny, 1840

Subgenus *Schwartziella* Nevil, 1884

CATE'S RISSO
Rissoina (*Schwartziella*) *catesbyana* d'Orbigny, 1842,
 [La Sagra, *Hist. d'Ile de Cuba*], *Moll.* 2:24.
Dim. of *Rissoa*, a genus dedicated to the French collec-
 tor A. Risso (1777–1845); dedicated to Catesby, a
 pre-Linnean malacologist.
SIZE: Length 3 to 5 mm.
COLOR: Shiny white.
SHAPE: Conical, elongated.
ORNAMENT OR SCULPTURE: Eight whorls, slightly
 rounded, suture fairly deep. About 14 strong ribs to
 each whorl.
APERTURE: Oval, slightly oblique. Strong tooth on
 inner side of outer lip.
OPERCULUM: Corneous, light brown.
PERIOSTRACUM: Not known. Probably thin as in other
 members of genus.

Rissoina (Schwartziella) catesbyana

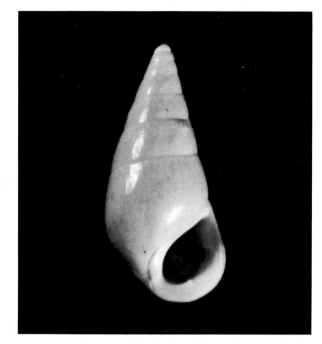

Rissoina (Cibedezebina) browniana

REMARKS: All *Rissoina* are minute in size. The word *Rissoina* is pronounced "riss-o-ee-na." Syn. *R. chesneli* Michaud.

HABITAT: Inhabits seaweeds in shallow bays and inlets.

LOCALITIES: Entire.

OCCURRENCE: Fairly common.

RANGE: North Carolina, Florida, West Indies and Texas.

GEOLOGIC RANGE: Pliocene to Recent.

Subgenus *Cibedezebina* Woodring, 1928

SMOOTH RISSO

Rissoina (Cibedezebina) browniana d'Orbigny, 1842, *Contr. Conch.*, p. 114.

Dedicated to French collector A. Risso (1777–1845) and Captain Thomas Brown (1785–1862), conchologist.

SIZE: Length 4 mm.

COLOR: Shiny white or pale amber.

SHAPE: Conical, elongated.

ORNAMENT OR SCULPTURE: Nine to 10 slightly convex whorls, smooth. Suture shallow. Apex small, prominent.

APERTURE: Oval, entire. Outer lip thickened.

OPERCULUM: Corneous, light amber color.

PERIOSTRACUM: Thin, transparent.

REMARKS: Syn. *R. laevigata* (C. B. Adams, 1850). Consult *Journalede Conchylogic* 89 (4): 205.

HABITAT: Grassy bottom beyond littoral zone.

LOCALITIES: Entire.

OCCURRENCE: Common.

RANGE: North Carolina to Gulf of Mexico and West Indies.

GEOLOGIC RANGE: Lower Miocene to Recent.

Family VITRINELLIDAE Busch, 1897

Genus *Anticlimax* Pilsbry & McGinty, 1946

Anticlimax pilsbryi (McGinty, 1945), *Nautilus* 54(4): 142.

Gk. *anti* against, *klimax* ladder, staircase; dedicated to Henry A. Pilsbry (1862–1959), American malacologist.

SIZE: Diameter 3.4 mm.

Anticlimax pilsbryi

Cochliolepis parasitica

COLOR: White.

SHAPE: Discoid. Spire slightly dome shaped.

ORNAMENT OR SCULPTURE: Two and one-half post-nuclear whorls sculptured with numerous, fine, zig-zag spiral grooves. More grooves on dorsal side than on the base. The periphery just above the base is marked with a strong keel. Umbilicus partly filled with a heavy callus. May be entirely filled.

APERTURE: Oblique. Lip slightly thickened in adults. Parietal callus is thick and on preceding whorl.

OPERCULUM: Not known.

PERIOSTRACUM: None visible.

REMARKS: The callus-filled umbilicus often causes confusion with the *Teniostoma*. Look for the zigzag grooving.

HABITAT: Inlets and along shore.

LOCALITIES: Entire.

OCCURRENCE: Fairly common.

RANGE: Southern Florida, Texas and Mexico.

Genus *Cochliolepis* Stimpson, 1858

Cochliolepis parasitica Stimpson, 1858, *Proc. Boston Soc. Natur. Hist.* 6:308.

L. *cochlea* snail shell, spiral; Gk. *lepis* scale; L. *parasitus* parasite.

SIZE: Diameter 3.55 mm.

COLOR: White, glassy.

SHAPE: Discoid.

ORNAMENT OR SCULPTURE: Two postnuclear whorls are smooth except for an occasional growth line.

Spire is flat and partially covered with a thin callus of shell from each successive whorl. Umbilicus is shallow and broadly open.

APERTURE: Oblique. Lip incomplete. Parietal callus thin.

OPERCULUM: Thin and flexible.

PERIOSTRACUM: None visible.

REMARKS: Smaller than *C. striata*. Heavy growth lines give nautiloid appearance to shell. Syn. *C. nautiliformis* (Holmes, 1860).

HABITAT: Shallow coastal bays, probably living on a worm.

LOCALITIES: Entire, more to south.

OCCURRENCE: Uncommon.

RANGE: Western coast of Florida and Texas.

Cochliolepis striata Dall, 1889, *Bull. M.C.Z.* 18:360.

L. *cochlea* snail shell, spiral; Gk. *lepis* scale; L. *striae* line.

SIZE: Diameter 6.5 mm.

COLOR: White.

SHAPE: Discoid.

ORNAMENT OR SCULPTURE: Thin shell has 4 whorls sculptured with numerous spiral striations. Base is smooth, but umbilicus is striated. First whorl of nuclear whorls is slightly projecting, while the remainder are covered by a thin callus from the succeeding whorls. Umbilicus is slightly constricted.

APERTURE: Oblique. Lip thin, incomplete. No parietal callus.

OPERCULUM: Thin, flexible.

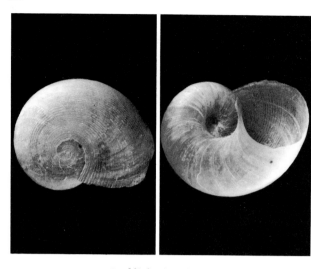

Cochliolepis striata

PERIOSTRACUM: None visible.

REMARKS: This is the largest of the Texas Vitrinellidae.

HABITAT: Shallow coastal bays, probably commensal with a worm.

LOCALITIES: Entire.

OCCURRENCE: Uncommon.

RANGE: North Carolina to Caribbean, western coast of Florida and Texas.

Cyclostremella humilis

Genus *Cyclostremella* Bush, 1897

Cyclostremella humilis Bush, 1897, *Trans. Conn. Acad. Sci.* 10(1):140.

Gk. *kyklos* 'circle, *trema* aperture; L. *humilis* on the ground.

SIZE: Diameter 1.7 mm.

COLOR: White.

SHAPE: Spiral, slightly depressed.

ORNAMENT OR SCULPTURE: Two postnuclear whorls with several weak spiral cords on the periphery and numerous fine, wavy growth lines. Suture is deep, forming a channel around the spire. Beyond the suture, the whorl is flattened and slopes at a 45-degree angle to the periphery. Base of shell is rounded. Umbilicus widely opened; apex can be seen.

APERTURE: Trigonal. Lip is deeply notched at the upper inner angle.

OPERCULUM: Paucispiral, nucleus subcentral.

PERIOSTRACUM: None visible when dead, light tan when living.

REMARKS: The nuclear whorls turn opposite (heterostrophic) to the body whorls. Animal is translucent. Moore (1964) suggests that, due to operculum and protoconch, this genus probably belongs in another family.

HABITAT: Along shore on sandy bottom, probably commensal with a worm.

LOCALITIES: Entire.

OCCURRENCE: Common.

RANGE: North Carolina, northwestern Florida and Texas.

Genus *Cyclostremiscus* Pilsbry & Olsson, 1945

Subgenus *Ponocyclus* Pilsbry, 1953

JEANNE'S VITRINELLA

Cyclostremiscus (Ponocyclus) jeannae Pilsbry & McGinty, 1946, *Nautilus* 59(3):82.

Gk. *kyklos* circle, *trema* hole or aperture; dedicated to Dr. Jeanne S. Schwengel, Florida malacologist.

SIZE: Diameter 1.8 mm.

COLOR: Translucent white.

SHAPE: Discoid.

ORNAMENT OR SCULPTURE: There are 3½ whorls. Penultimate whorl has a medial keel extending on to the first part of the last whorl. Last whorl has

Cyclostremiscus (Ponocyclus) jeannae

strong upper and lower carina with submedian keels or 4 keels. Umbilical region is funnel shaped with a strongly keeled base.

APERTURE: Rounded pentagonal, slightly oblique.

OPERCULUM: Not known, probably circular.

PERIOSTRACUM: None visible.

REMARKS: This species is found offshore in deeper waters and is common on the 7½ fathom reef off Port Mansfield. It is included here for those who collect in offshore waters.

HABITAT: Offshore.

LOCALITIES: South.

OCCURRENCE: Not yet found on beach.

RANGE: Both sides of Florida and Texas.

TRILIX VITRINELLA

Cyclostremiscus (Ponocyclus) pentagonus (Gabb, 1873), *Proc. Acad. Natur. Sci. Phila.* 24:243.

Gk. *kyklos* circle, *trema* hole or aperture, *penta* five, *gonia* angle.

SIZE: Diameter 2.1 mm.

COLOR: White, glassy.

SHAPE: Discoid, nuclear whorls projecting.

ORNAMENT OR SCULPTURE: Postnuclear whorls almost flat on top and sculptured with 3 keel-like ridges (tricarinate). Base is flattened. Umbilicus open, deep, and bordered with a spiral ridge. Faint spiral lines within the umbilicus.

APERTURE: Oblique. Lip incomplete.

OPERCULUM: Circular, multispiral.

PERIOSTRACUM: None visible.

REMARKS: Syn. *C. trilix* Bush.

HABITAT: Along shore in surf zone.

LOCALITIES: Entire.

OCCURRENCE: Common.

RANGE: Southeastern United States to the West Indies and Texas.

Cyclostremiscus (Ponocyclus) suppressus (Dall, 1889), *Nautilus* 60:82.

Gk. *kyklos* circle, *trema* hole or aperture; L. *supressus* low or sunken.

SIZE: Diameter 1.85 mm.

COLOR: White.

SHAPE: Discoid.

ORNAMENT OR SCULPTURE: There are 1¾ postnuclear whorls sculptured with 3 narrow, sharp keels on the periphery. The central keel is higher than the other 2. Near the suture is a less conspicuous keel. At the suture, which is covered by a thin callus of shell, is another ridge, which is the peripheral keel

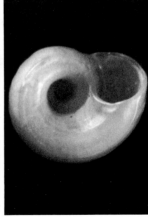

Cyclostremiscus (Ponocyclus) pentagonus

of the prior whorl. Base is flat with a raised ridge around the umbilicus. Umbilicus is narrow with a flat wall. The last half of the body whorl is twisted so that the umbilical ridge joins the aperture in the middle of the base.

APERTURE: Oblique. Parietal callus heavy with a groove at the upper inner angle.

OPERCULUM: Thin, multispiral.

PERIOSTRACUM: None visible.

REMARKS: Look in the drift along the Aransas ship channel and Cline's Point.

HABITAT: Bays and inlets.

LOCALITIES: Entire.

OCCURRENCE: Fairly common.

RANGE: Lower eastern coast and entire western coast of Florida and Texas.

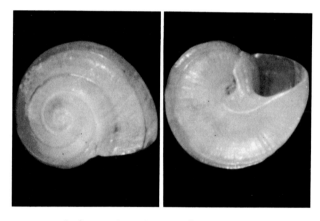

Cyclostremiscus (Ponocyclus) suppressus

Genus *Episcynia* Mörch, 1875

HAIRY VITRINELLA

Episcynia inornata (d'Orbigny, 1842), [La Sagra, *Hist. l'Ile de Cuba*], *Moll.* 2:67.

Gk. *epi*- upon; L. *scynium* the skin above the eyes, *in* without, *orno* decoration.

SIZE: Diameter 3.4 mm.

COLOR: White, glassy.

SHAPE: Trochiform, top shaped.

ORNAMENT OR SCULPTURE: Four and one-half postnuclear whorls sculptured with a narrow peripheral keel that is fringed with minute projecting teeth. Umbilicus is narrow, flat sided, and deep with a stepped appearance.

APERTURE: Flattened oval. Lip incomplete.

Episcynia inornata

OPERCULUM: Not known.

PERIOSTRACUM: Thin, yellowish brown, extending beyond the periphery in a curious fringed arrangement.

REMARKS: An extremely beautiful species due to peripheral sculpture.

HABITAT: Offshore and along shore.

LOCALITIES: Port Aransas, south.

OCCURRENCE: Uncommon.

RANGE: Greater Antilles and Texas.

Genus *Macromphalina* Cossmann, 1888

Macromphalina palmalitoris Pilsbry & McGinty, 1950, *Nautilus* 63(3):86.

Gk. *macro* long, *omphalos* umbilicus; L. *palma* palm, *littus* seashore (palm-lined shore).

SIZE: Diameter 1.7 mm.

COLOR: White.

SHAPE: Auriform. Nuclear whorls projecting.

ORNAMENT OR SCULPTURE: One and three-fourths postnuclear whorls, which are sculptured with recurved radial riblets that continue over the periphery into the base. Numerous fine, spiral grooves on entire shell. Base rounded. Umbilicus shallow and wide.

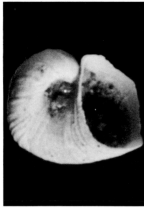

Macromphalina palmalitoris

APERTURE: Oblique oval.
OPERCULUM: Not known.
PERIOSTRACUM: None visible.
REMARKS: Feeds on algae. Figured specimen has sand in umbilicus.
HABITAT: On algae, in deeper water offshore.
LOCALITIES: Port Aransas, south.
OCCURANCE: Rare.
RANGE: Southeastern Florida, Northwestern Florida, and Texas.

Parviturboides interruptus

Genus *Parviturboides* Pilsbry & McGinty, 1950

Parviturboides interruptus (C. B. Adams, 1850), *Monogr. Vitrinella*, p. 6.
L. *parvus* small, *turbo* a top, *interruptus* broken apart.
SIZE: Diameter 1.43 mm.
COLOR: White.
SHAPE: Trochiform.
ORNAMENT OR SCULPTURE: One and three-fourths postnuclear whorls sculptured with 8 or 9 distinct spiral ridges and several indistinct ones. Low thread-like transverse sculpture between the ridges. Suture not impressed. Nuclear whorls large and prominent. Base rounded. Umbilicus, a narrow chink bordered by a ridge joining the base of the columella.
APERTURE: Circular, oblique, slightly angled. Lip incomplete.
OPERCULUM: Thin, multispiral, and circular.
PERIOSTRACUM: None visible.
REMARKS: The tentacled animal is light shy.
HABITAT: Shallow water in rocks and crevices.
LOCALITIES: Port Aransas, south.
OCCURRENCE: Fairly common.
RANGE: South Carolina, Florida, Texas, Mexico, Panama, Puerto Rico, Haiti, and Jamaica.

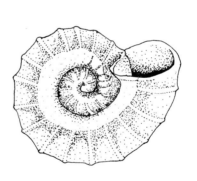

Pleuromalaxis balesi
(after Moore 1964, p. 221)

Genus *Pleuromalaxis* Pilsbry & McGinty, 1945

BALES' FALSE DIAL
Pleuromalaxis balesi Pilsbry & McGinty, 1945, *Nautilus* 59(10):10.
Gk. *pleure* side, *homalos* even, smooth; dedicated to Dr. Blenn R. Bales (1876–1946).

SIZE: Diameter 1.6 mm.

COLOR: White.

SHAPE: Turbinate. Depressed.

ORNAMENT OR SCULPTURE: One and three-fourths postnuclear whorls. Suture slightly impressed. Base flattened. Two spiral keels on the periphery with a concave space between. Transverse riblets on both top and bottom of the whorls terminate at the peripheral keels, giving the keels a nodulose appearance. Umbilicus shallow and widely open.

APERTURE: Circular and oblique.

OPERCULUM: Circular.

PERIOSTRACUM: None visible.

HABITAT: Inlet areas, under rocks, and offshore.

LOCALITIES: Port Aransas, south.

OCCURRENCE: Uncommon.

RANGE: Florida, Cuba, Puerto Rico and Texas.

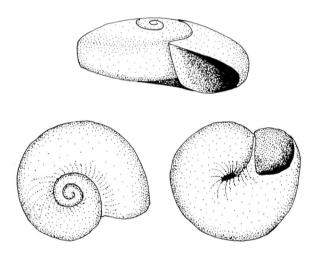

Solariorbis (Solariorbis) blakei
(after Moore 1964, p. 122)

Genus *Solariorbis* Conrad, 1865

Subgenus *Solariorbis* s.s.

Solariorbis (Solariorbis) blakei Rehder, 1944, *Nautilus* 57 (3):97.

L. *solariorbis* in the sun's orbit; *Blake*, a ship used for the United States coast survey in 1877–1878.

SIZE: Diameter 1.45 mm.

COLOR: White.

SHAPE: Discoid.

ORNAMENT OR SCULPTURE: One and one-half post-

nuclear whorls sculptured with microscopic spiral grooves and brief transverse wrinkles fanning from the suture. Suture only slightly impressed. Nuclear whorls project slightly. Base is broad, smooth, and evenly rounded. Umbilicus may be nearly or entirely closed; however, there is always a narrow chink and it never looks like the *Teinostoma*.

APERTURE: Oblique, rounded. Parietal callus extends out beyond aperture.

OPERCULUM: Multispiral, round.

PERIOSTRACUM: None visible.

REMARKS: The thickening around the umbilicus is the most distinctive characteristic of this genus.

HABITAT: Inlets and along the shore.

LOCALITIES: Entire.

OCCURRENCE: Fairly common.

RANGE: Eastern coast of United States, Florida, Gulf states, Mexico.

Solariorbis (Solariorbis) infracarinata Gabb, 1881, *J. Acad. Natur. Sci. Phila.* 8(2):365.

L. *solariorbis* in the sun's orbit, *infra-* below, *carina* keeled underneath.

SIZE: Diameter 2.0 mm.

COLOR: White.

SHAPE: Discoid.

ORNAMENT OR SCULPTURE: One and two-thirds postnuclear whorls, which are carinate on the periphery and sculptured with low radial waves on the first whorl. Nuclear whorls project slightly. Below the peripheral keel are three spiral ridges. The largest is next to the periphery and the smallest nearest the umbilicus. The largest is usually beaded on its inner surface with the beads disappearing on the last half

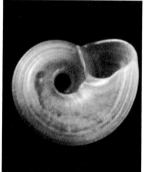

Solariorbis (Solariorbis) infracarinata

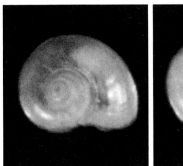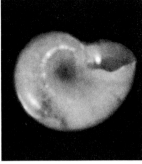

Solariorbis (Solariorbis) mooreana

whorl. Umbilicus is constricted with a strong ridge occupying the lower part of the umbilical wall.

APERTURE: Oblique, parietal callus fairly heavy.

OPERCULUM: Multispiral, circular.

PERIOSTRACUM: None visible.

REMARKS: Spiral sculpture of this microscopic shell is not visible from above. Syn. *S. euzonus.*

HABITAT: Bays and inlets.

LOCALITIES: Entire.

OCCURRENCE: Fairly common.

RANGE: Both sides of Florida, Texas Campeche Bank, Mexico, and Panama and Guatemala.

Solariorbis (Solariorbis) mooreana Vanatta, 1903, *Proc. Acad. Natur. Sci. Phila.* 55:758.

L. *solariorbis* in the sun's orbit; dedicated to Clarence B. Moore, Florida archaeologist.

SIZE: Diameter 2.75 mm.

COLOR: White.

SHAPE: Discoid.

ORNAMENT OR SCULPTURE: Two and one-half post-nuclear whorls, which are sculptured with about 8 spiral ridges on upper half of shell crossed with fine axial threads strongest near the suture. Base is smooth and rounded. Umbilicus narrow, deep.

APERTURE: Oblique. Parietal callus heavy. Columella thick.

OPERCULUM: Multispiral, circular, thin, and light tan in color.

PERIOSTRACUM: None visible.

REMARKS: The animal is unknown.

HABITAT: Near shore on sandy bottom.

LOCALITIES: Entire.

OCCURRENCE: Uncommon to rare.

RANGE: Cape San Blas, Florida, to Aransas Bay, Texas.

Genus *Teinostoma* H. & A. Adams, 1854

Teinostoma biscaynense Pilsbry & McGinty, 1945, *Nautilus* 59:5.

Gk. *teine* to stretch, *stoma* mouth; from Biscayne Bay south of Miami, Florida.

SIZE: Diameter 1.9 mm.

COLOR: White.

SHAPE: Discoid.

ORNAMENT OR SCULPTURE: Three and one-half whorls in adult shell are very smooth. Both spire and umbilicus are covered with a shelly callus. Umbilical callus is convex and fills entire cavity.

APERTURE: Oblique. Parietal callus thickened. Upper inner angle of lip has a small groove.

OPERCULUM: Circular, multispiral.

PERIOSTRACUM: None visible.

REMARKS: Occasional specimens are minutely striate.

HABITAT: Inlet areas and surf zone.

LOCALITIES: Entire.

OCCURRENCE: Fairly common.

RANGE: Southern Florida, western Florida and Texas.

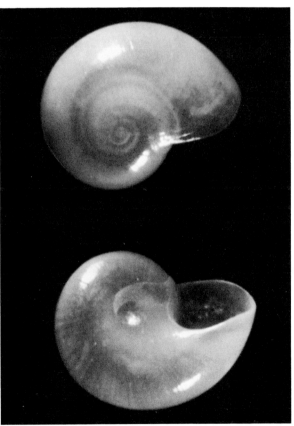

Teinostoma biscaynense

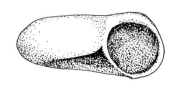

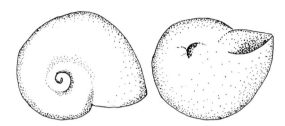

Teinostoma leremum
(after Moore 1964, p. 221)

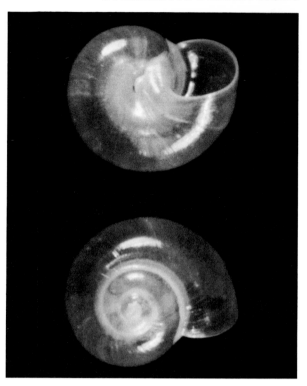

Teinostoma parvicallum

Teinostoma leremum Pilsbry & McGinty, 1945, *Nautilus* 59:6.
Gk. *teino* to stretch, *stoma* mouth, *lerema* a trifle.
SIZE: Diameter 1.12 mm.
COLOR: White.
SHAPE: Discoid.
ORNAMENT OR SCULPTURE: There are 2 nuclear whorls and but little more than 1 postnuclear whorl, which are smooth and glassy except for axial wrinkles adjoining the suture. Umbilicus is completely covered by a shelly callus.
APERTURE: Oblique, round. No groove at upper inner angle.
OPERCULUM: Multispiral, circular.
PERIOSTRACUM: None visible.
REMARKS: Very similar to *T. biscaynense* but smaller, and the umbilicus is not covered by a thin shelly callus.
HABITAT: Bays and inlets.
LOCALITIES: Entire, more to south.
OCCURRENCE: Fairly common.
RANGE: Florida Keys, northwestern Florida, Texas, Mexico and the Virgin Islands.

TINY-CALLOUSED TEINOSTOMA
Teinostoma parvicallum Pilsbry & McGinty, 1945, *Nautilus* 59:4.
Gk. *teino* to stretch, *stoma* mouth, *parvus* small; L. *callum* hard skin.
SIZE: Diameter 1.7 mm.
COLOR: White.

SHAPE: Slightly turbinate.
MARKINGS OR SCULPTURE: Two and one-half rounded postnuclear whorls, which are smooth and glassy. Nuclear whorls slightly elevated. Periphery and base both rounded. Chinklike umbilicus may not be covered with shelly callus until maturity is reached.
APERTURE: Oblique, rounded with a small groove at upper inner angle. Heavy parietal callus.
OPERCULUM: Multispiral, circular.
PERIOSTRACUM: None visible.
REMARKS: This minute shell is more rounded than *T. lereum* and *T. biscaynense*.
HABITAT: Along shore.
LOCALITIES: Entire.
OCCURRENCE: Rare.
RANGE: Florida, Puerto Rico and Texas.

Genus *Vitrinella* C. B. Adams, 1850

Vitrinella floridana Pilsbry & McGinty, 1946, *Nautilus* 60:16.
L. *vitrum* glass; from Florida.

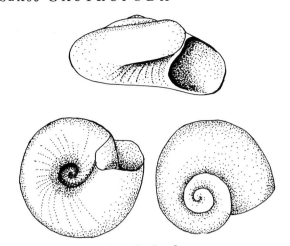

Vitrinella floridana
(after Moore 1964, p. 219)

SIZE: Diameter 1.8 mm.
COLOR: White.
SHAPE: Slightly turbinate.
ORNAMENT OR SCULPTURE: No sculpture. Umbilicus broadly open. Protoconch slightly elevated. Two postnuclear whorls. Prior whorls can be seen; walls convex.
APERTURE: Circular, oblique. Columella slightly thickened.
OPERCULUM: Circular, thin, flexible.
PERIOSTRACUM: None visible.
REMARKS: Differentiate this from *V. helicoidea* by the uncarinated umbilicus.
HABITAT: Inlet-influenced areas and near shore.
LOCALITIES: Entire.
OCCURRENCE: Common.
RANGE: Southern Florida and Texas to Campeche Bank, Mexico.

Vitrinella helicoidea

HELIX VITRINELLA
Vitrinella helicoidea C. B. Adams, 1850, *Monogr. Vitrinella*, p. 9.
L. *vitrum* glass; Gk. *helico* spiral, coil.
SIZE: Diameter 2.7 mm.
COLOR: White, thin.
SHAPE: Turbinate.
ORNAMENT OR SCULPTURE: Two and one-half whorls. A spiral cord next to the suture of the first whorl. There may be low spiral sculpture on the periphery to the first 1½ whorls. Umbilicus narrow, deep, and flat sided, angled where it meets the base.
APERTURE: Round and slightly oblique. Columella curved inward.
OPERCULUM: Circular, multispiral.
PERIOSTRACUM: None visible.
REMARKS: Sift the drift carefully for all members of this family.
HABITAT: Under rocks. Near shore.
LOCALITIES: Entire.
OCCURRENCE: Fairly common.
RANGE: Southeastern United States to the West Indies, Texas, Panama.

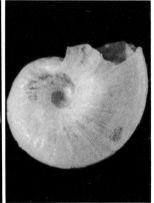

Vitrinella texana

Vitrinella texana Moore, 1964, Ph.D. dissertation, p. 66.
L. *vitrum* glass; from Texas.
SIZE: Diameter 1.72 mm.
COLOR: White.
SHAPE: Turbinate.
ORNAMENT OR SCULPTURE: One and one-fourth postnuclear whorls. The nuclear whorls are glassy and there is a narrow but distinct varix at their termina-

tion. Spire flat. Sculpture consists of many revolving dorsal grooves and low radiating ribs on the base. The base is flattened. Several grooves in umbilicus. Umbilicus narrow and deep.

APERTURE: Oblique. Parietal callus heavy with a groove at the upper inner angle.

OPERCULUM: Circular, multispiral.

PERIOSTRACUM: None visible.

REMARKS: This species has only been found near the inlet area to Aransas Bay. Originally described from Texas.

HABITAT: Inlet areas.

LOCALITIES: Port Aransas, south.

OCCURRENCE: Uncommon.

RANGE: South Texas and Florida.

Superfamily ARCHITECTONICACEA Gray, 1850

Family ARCHITECTONICIDAE Gray, 1850

Genus *Heliacus* d'Orbigny, 1842

ORBIGNY'S SUNDIAL

Heliacus bisulcata (d'Orbigny, 1845), [La Sagra, *Hist. l'Ile de Cuba*], *Moll.* 2:66.

Gk. *helico* spiral, coil; L. *bis* twice, *sulcare* to furrow.

SIZE: Width 6 to 12 mm.

COLOR: Tan to dull gray.

SHAPE: Broadly conical. Spire only slightly elevated.

ORNAMENT OR SCULPTURE: Whorls flat, more strongly sculptured than *Architectonica.* Spiral sculpture of revolving beaded cords; on the periphery

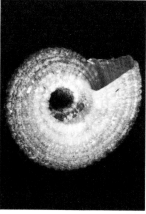

Heliacus bisulcata

there are two rows of stronger beads. The base is sculptured in a similar fashion. Umbilicus is deep and wide with crenulated edge.

APERTURE: Round; lip thin and irregularly crenulated.

OPERCULUM: Corneous, spiral with several turns and a fringed edge.

PERIOSTRACUM: None visible.

REMARKS: Has been found alive at Port Aransas at the base of sea whip coral living on the jetties.

HABITAT: Intertidal on rocks, jetties, probably on tunicates and soft corals.

LOCALITIES: Port Aransas, south.

OCCURRENCE: Rare.

RANGE: Southeastern United States, Texas and West Indies.

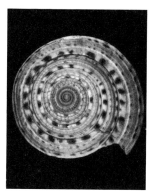

Architectonica nobilis

Genus *Architectonica* Röding, 1798

COMMON SUNDIAL

Architectonica nobilis (Röding, 1798), *Mus. Boltenianum,* p. 78.

L. *architecton* master builder, *nobilis* famous or noted.

SIZE: Diameter 1 to 2 in.

COLOR: Background cream with orange brown spots spirally arranged, more prominent next to suture. Porcelaneous.

SHAPE: Broadly conical, spire low.

ORNAMENT OR SCULPTURE: Early whorls marked with spiral cords that appear beaded due to growth lines of spire. Four or 5 prominent sulcations and traces of others on remaining whorls. Base flat, more strongly sculptured than topside. Umbilicus deep and wide and surrounded with a strongly crenulated spiral cord.

APERTURE: Subquadrate. Lip thin and irregularly crenulated. Interior white.

OPERCULUM: Brown and paucispiral. Corneous.

PERIOSTRACUM: None visible.

REMARKS: Lives offshore among sea pansies; found alive after storms and northers on outer beaches. The same species occurs on the Pacific coast. Formerly called *A. granulata*.

HABITAT: Offshore.

LOCALITIES: Entire.

OCCURRENCE: Common.

RANGE: Southeastern United States to the West Indies and Texas, also western Mexico and Peru.

GEOLOGIC RANGE: Miocene to Recent.

Superfamily CERITHIACEA Fleming, 1822

Family TURRITELLIDAE Woodward, 1851

Subfamily VERMICULARIINAE Kimoshita, 1932

Genus *Vermicularia* Lamarck, 1799

FARGO'S WORM SHELL

Vermicularia fargoi Olsson, 1951, *Nautilus* 65(1):6.

L. *vermis* worm; dedicated to William Fargo of Pass-a-Grill, Florida.

SIZE: Height 18 to 25 mm.

COLOR: Light brown, paler toward aperture.

SHAPE: Very elongate conical or turriculate.

ORNAMENT OR SCULPTURE: About 6 regular whorls closely coiled; later whorls are detached giving a wormlike appearance. About 3 spiral cords form keels at sutures of later whorls. Weak, irregular

transverse growth lines. Fine spiral threads between major cords.

APERTURE: Round, thin lipped.

OPERCULUM: Corneous, circular, closes aperture completely.

PERIOSTRACUM: None visible.

REMARKS: This species has not been found living on the Texas coast, but vast quantities of dead shells occur along the bay shores of the southern half. The living adult is sedentary, living in intertwined colonies or attached to rock or in sponge; also crawls on mud flats. Feeds by throwing mucus sheets to entangle fine food particles and drawing them into the mouth.

HABITAT: Intertidal, attached to rocks, in mud, in bays.

LOCALITIES: Entire, more to south.

OCCURRENCE: Common.

RANGE: Southeastern Florida, Texas and Caribbean.

Family CAECIDAE Gray, 1850

Genus *Caecum* Fleming, 1824

SMOOTH CAECUM

Caecum glabrum (Montagu, 1803), *Test. Brit.* 2:497.

L. *caecus* blind gut, *glaber* smooth.

SIZE: Height 2.5 mm.

COLOR: White.

SHAPE: Curved, tubular.

ORNAMENT OR SCULPTURE: Shell narrow, thin, smooth, subdiaphanous; aperture not contracted or tumid, white; septum without appendage (Tryon, 1885).

Vermicularia fargoi

Caecum glabrum

APERTURE: Circular, not constricted.

OPERCULUM: Externally convex, corneous.

PERIOSTRACUM: None visible.

REMARKS: First reported for Texas by Hulings (1955). Lays eggs in capsules attached to bottom.

HABITAT: In bays and oyster reef.

LOCALITIES: Entire.

OCCURRENCE: Common.

RANGE: Europe, North Carolina to Florida and Texas.

GEOLOGIC RANGE: Miocene to Recent.

COOPER'S CAECUM

Caecum imbricatum Carpenter, 1858, *Proc. Zool. Soc. London* 26:422.

L. *caecus* blind gut, *imbricatus* overlapping.

SIZE: Length 4 to 5 mm.

COLOR: Glossy, opaque, white.

SHAPE: Curved, tubular.

ORNAMENT OR SCULPTURE: Longitudinal sculpture consists of about 24 rounded ribs crossed by numerous rings. More prominent near aperture, giving a cancellate appearance. Apical plug has prong to one side. Fairly heavy shell.

APERTURE: Round.

OPERCULUM: Corneous, concave.

PERIOSTRACUM: None visible.

REMARKS: Syn. *C. cooperi* (S. Smith, 1870).

Caecum imbricatum

HABITAT: Offshore on sand and shell banks.

LOCALITIES: Port Aransas, south.

OCCURRENCE: Uncommon.

RANGE: South of Cape Cod to northern Florida and Texas.

Caecum (Micranellum) pulchellum

Protoconch of *Caecum* sp. plus part of teleconch

Subgenus *Micranellum* Bartsch, 1920

BEAUTIFUL LITTLE CAECUM

Caecum (Micranellum) pulchellum (Stimpson, 1851), *Proc. Boston Soc. Natur. Hist.* 4:112.

L. *caecus* blind gut, *pulcher* fair, beautiful.

SIZE: Length 2 mm.

COLOR: Translucent tan and shiny when living; mat white when dead.

SHAPE: Tubular, cucumberlike.

ORNAMENT OR SCULPTURE: Twenty-five to 30 fine, closely arranged axial ribs; plug rounded.

APERTURE: Round, thin lipped.

OPERCULUM: Corneous, concave with 8 whorls.

PERIOSTRACUM: None visible.

REMARKS: Develops in 3 stages. In first the shell is spiral; this part is discarded after formation of second tubular stage. The adolescent tube is discarded after the formation of a similar curved tube closed with a septum. Lays eggs in capsules attached to bottom.

HABITAT: In crevices and under stones, intertidal and offshore.

LOCALITIES: Entire.

OCCURRENCE: Common.

RANGE: Cape Cod south to North Carolina and Texas.

GEOLOGIC RANGE: Pleistocene to Recent.

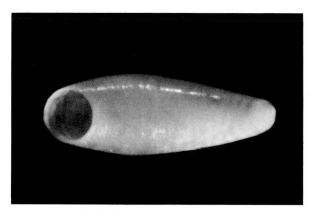

Meioceras nitidum

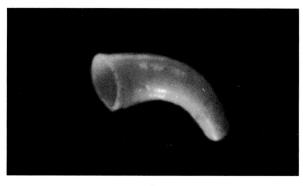

Second stage

Genus *Meioceras* Carpenter, 1858

LITTLE HORN CAECUM
Meioceras nitidum (Stimpson, 1851), *Proc. Boston Soc. Natur. Hist.* 4:112.
Gk. *meiōn* smaller, *keras* horn; L. *nitidus* glossy, sleek.
SIZE: Length 2.5 mm.
COLOR: Shiny white or brownish with irregular mottlings of opaque white.
SHAPE: Tubular with swollen center.

ORNAMENT OR SCULPTURE: Smooth, polished. Apical plug convexly rounded with tiny projection to one side.
APERTURE: Round, oblique, sharp edged. Larger than apical end.
OPERCULUM: Corneous.
PERIOSTRACUM: None visible.
REMARKS: Found on eel grass in Florida, so far only dead in Texas.
HABITAT: Intertidal.
LOCALITIES: South.
OCCURRENCE: Fairly common.
RANGE: Southern half of Florida, West Indies and Texas.
GEOLOGIC RANGE: Upper Miocene to Recent.

Family MODULIDAE Fischer, 1884

Genus *Modulus* Gray, 1842

ATLANTIC MODULUS
Modulus modulus (Linné, 1758), *Syst. Natur.*, 10th, ed., p. 757.
L. *modulus* a measure.
SIZE: Height 12mm.; diameter of base 17 mm.
COLOR: Grayish white spotted with brown.
SHAPE: Conical or turban shaped.
ORNAMENT OR SCULPTURE: About 3 small whorls in spire, body whorl large with sloping shoulders and a keeled base. Upper portion of shell marked with low spiral ridges and oblique, transverse growth lines. Five strong cords on base. Umbilicus deep, small.
APERTURE: Round with thin, slightly crenulated outer lip that is thickened and marked with low ridges within; white, porcelaneous. Columella short with distinctive single tooth near base.
OPERCULUM: Corneous, circular, reddish brown, multispiral and thin.
PERIOSTRACUM: Thin, brownish, deciduous.
REMARKS: No live shells have been reported in recent years. Abundant dead shells along Aransas ship channel. Animal is moss green with white dots, the head bears two thin, round tentacles with eyes at midpoint of each tentacle. Moss green animal blends with the green eel grass. Shells from this area well worn, dead specimens.
HABITAT: On grass. Intertidal.

Modulus modulus

LOCALITIES: Entire, more to south.
OCCURRENCE: Common.
RANGE: Florida Keys to the West Indies and Texas.
GEOLOGIC RANGE: Pliocene to Recent.

Family POTAMIDIDAE H. & A. Adams, 1854

Subfamily POTAMIDINAE H. & A. Adams, 1854

Genus *Cerithidea* Swainson, 1840

Subgenus *Cerithideopsis* Thiele, 1929

PLICATE HORN SHELL
Cerithidea (Cerithideopsis) pliculosa (Menke, 1829),
Verzeichm Conch. Samml. Ma'sburg, p. 27.

Cerithidea (Cerithideopsis) pliculosa

Gk. *keration* little horn; L. *plicare* to fold.
SIZE: Length 1 in.
COLOR: Dark brown or brownish black with bone yellow varices; sometimes a grayish yellow spiral band through middle of whorls.
SHAPE: High conical or turriculate with siphonal notch.
MARKINGS OR SCULPTURE: Eleven to 13 slightly convex whorls. Unevenly spaced transverse ribs, 18 to 25 per whorl. Many fine, uneven spiral striations. On the body whorl the ribs stop below the periphery at a strong cord; 6 to 9 spiral cords continue over base. Five to 8 prominent varices are the characteristic feature of the adult, usually beginning with the sixth whorl.
APERTURE: Subcircular; outer margin convex, columellar margin concave. Shallow indentation at base near columella for siphon. Outer lip greatly thickened, forming a raised, rounded varix.
OPERCULUM: Corneous, subcircular, spiral, closely coiled about a central nucleus.
PERIOSTRACUM: None visible.
REMARKS: Juvenile specimens without strongly developed varices are easily confused with other species. Before Hurricane Beulah in 1967, they could be seen crawling on the mud flats of Aransas Bay at low tide, but the huge influx of fresh water resulting from the storm appears to have decimated the population. A favorite food of water birds.
HABITAT: Mud flats. In bays and inlets.
LOCALITIES: Entire.
OCCURRENCE: Common.
RANGE: Texas, Louisiana and the West Indies, not Florida.

Family CERITHIIDAE Fleming, 1822

Subfamily CERITHIINAE Fleming, 1822

Genus *Cerithium* Bruguière, 1789

Subgenus *Thericium* ('Rochebrune') Monterosato, 1890

FLORIDA CERITH
Cerithium (Thericium) floridanum Mörch, 1876, *Mal. Blatt*, p. 23.
Gk. *keration* little horn; from Florida.
SIZE: Height 1 to 1½ in.

Cerithium (Thericium) floridanum

Cerithium (Thericium) variabile

COLOR: White with narrow, spiral brown bands.

SHAPE: Elongate conical or turriculate with siphonal canal.

ORNAMENT OR SCULPTURE: Eleven to 13 slightly convex whorls. Two to 3 white former varices on each whorl. Several rows of beaded spiral cords on each whorl with finer granulated cords separating them. Beads are fairly regular, giving a neat appearance.

APERTURE: Oval, oblique. Outer lip thickened into crenulated varix. Parietal area glossy white. Anterior siphonal canal short and upturned. Posterior canal is simple fold where lip joins body whorl.

OPERCULUM: Corneous, brown, thin, and paucispiral.

PERIOSTRACUM: None visible.

REMARKS: None have been found living on Texas coast as yet; dead shells are found in bay drift.

HABITAT: Littoral, on sea grasses.

LOCALITIES: Entire.

OCCURRENCE: Uncommon.

RANGE: North Carolina to southern half of Florida and Texas.

GEOLOGIC RANGE: Pliocene to Recent.

DWARF CERITH

Cerithium (Thericium) variabile (C. B. Adams, 1845), *Proc. Boston Soc. Natur. Hist.* 2:5.

Gk. *keration* little horn; L. *varius* diverse.

SIZE: Length 8 to 12 mm.

COLOR: Brown black or grayish white with mottlings of reddish brown. Nuclear whorls whitish.

SHAPE: Elongate conical, turriculate.

ORNAMENT OR SCULPTURE: Eight slightly convex whorls; sutures distinct; 7 to 8 beaded, spiral cords on body whorl interspaced with fine striations. Re-

peated but fewer in number on remaining whorls. One to 2 former varices on each whorl.

APERTURE: Oval; outer lip thin on edge with thickened varix behind edge. Parietal area white and glossy but thin. Columella short, edged in white with brief upturned siphonal canal. Only slight posterior canal.

OPERCULUM: Corneous, light brown.

PERIOSTRACUM: None visible.

REMARKS: Prior to Hurricane Beulah in 1967, this was the most common cerith on the Texas coast and could be found in the grasses and crawling along the mud flats of the Aransas Pass area in great numbers. Animal black with white mottlings, eyes behind base of tentacles. Herbivorous, sexes separate. Secretes mucus thread to suspend itself. Lays eggs in gelatinous strings.

HABITAT: Littoral, on marine grasses. In shallow bays.

LOCALITIES: Central, south.

OCCURRENCE: Common.

RANGE: Southern half of Florida to Texas and the West Indies.

Genus *Bittium* Leach, 1847

Subgenus *Bittium* s.s.

VARIABLE BITTIUM

Bittium (Bittium) varium (Pfeiffer, 1840), *Arch. für Natur.*, p. 256.

O.E. *bit* bite or particle, or from Bittion, a feminine proper name; L. *varius* diverse.

SIZE: Height 5 to 6 mm.

COLOR: Grayish brown.

SHAPE: Elongate conical, turriculate.

ORNAMENT OR SCULPTURE: Seven to 8 slightly con-

Bittium (Bittium) varium

vex whorls, sutures definite. Numerous, rounded, curved transverse ribs crossed by spiral grooves that give a nodulose appearance to sculpture. Base has spiral grooves but no ribs.

APERTURE: Oval, thin with varix adjacent. Anterior siphonal canal poorly developed.

OPERCULUM: Corneous, light brown.

PERIOSTRACUM: None visible.

REMARKS: During the warm months before Hurricane Beulah in 1967, this shell could be taken in vast numbers from blades of marine grass in Aransas Bay. Learning to identify this species early in the game will greatly facilitate future sorting of beach drift by eliminating its numbers, but old and faded specimens can be confusing. Food for drum.

HABITAT: On marine grass. In bays.

LOCALITIES: Entire.

OCCURRENCE: Common.

RANGE: Maryland to Florida, Texas and Mexico.

GEOLOGIC RANGE: Pliocene to Recent.

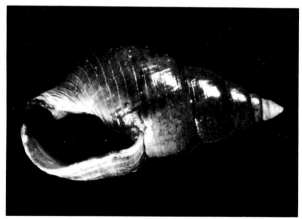

Litiopa melanostoma

Subfamily LITIOPINAE Gray, 1847

Genus *Litiopa* Rang, 1829

BROWN SARGASSUM SNAIL
Litiopa melanostoma Rang, 1829, *Bull. U.S. Bur. Fish.* 31:2.
Gk. *litos* simple, *melanos* black, *stoma* mouth.

SIZE: Height 3 to 6 mm.

COLOR: Translucent, dark brown.

SHAPE: Moderately elongate conical.

ORNAMENT OR SCULPTURE: Seven whorls; body whorl expanded and about half length of shell. Appears smooth but covered with numerous microscopic, spiral lines that are more strongly incised on base. Nuclear whorls have regularly spaced, vertical striations.

APERTURE: Semilunar with slightly convex columella. A strong ridge on inside edge of columella is characteristic.

OPERCULUM: Not known.

PERIOSTRACUM: None visible.

REMARKS: Lives in floating sargassum weed where it looks like the small round berries of this plant. Can be found by shaking sargassum or by sifting beach drift. Dead, faded specimens are cream colored with small brown spots on shoulder of whorl and a brown band bordering exterior of aperture.

HABITAT: Pelagic

LOCALITIES: Entire.

OCCURRENCE: Common.

RANGE: Pelagic in floating sargassum weed.

Family CERITHIOPSIDAE H. & A. Adams, 1853

Genus *Cerithiopsis* Forbes & Hanley, 1849

Subgenus *Laskeya* Iredale, 1918

AWL MINIATURE CERITH
Cerithiopsis (Laskeya) emersonii (C. B. Adams, 1838), *J. Boston Soc. Natur. Hist.* 2(2):284.
Gk. *keration* little horn, *opsis* aspect, appearance; dedicated to George B. Emerson of Boston.

SIZE: Length 12 to 18 mm.

COLOR: Light brown with suture sometimes darker.

SHAPE: Slender, elongated conical, or awl shaped.

Cerithiopsis (Laskeya) emersonii

ORNAMENT OR SCULPTURE: Ten to 14 flat-sided whorls, sutures distinct. Three strong spiral rows of raised, roundish beads, revolving thread between beaded rows. Faint axial ribs may connect the beads. Middle row of beads is less prominent than others. Base is concave with cordlike spiral ridges and fine transverse growth lines.

APERTURE: Oval, lip thin. Short, slightly flared anterior siphonal canal.

OPERCULUM: Corneous, brown.

PERIOSTRACUM: None visible.

REMARKS: Looks like an elongated *C. greeni* at first glance. Carnivorous. Screen beach drift for this one. Syn. *C. subulata* (Montagu, 1808).

HABITAT: Inlet, shelly sand along shore.

LOCALITIES: Central, south.

OCCURRENCE: Uncommon.

RANGE: Massachusetts to the West Indies and Texas.

GEOLOGIC RANGE: Upper Miocene to Recent.

Cerithiopsis (Laskeya) greeni

GREEN'S MINIATURE CERITH

Cerithiopsis (Laskeya) greeni (C. B. Adams, 1838), *J. Boston Soc. Natur. Hist.* 2(2):287.

Gk. *keration* little horn, *opsis* appearance; dedicated to Jacob Green (1790–1841), American naturalist.

SIZE: Length 3 mm.

COLOR: Glossy brown.

SHAPE: Elongate conical, turriculate.

ORNAMENT OR SCULPTURE: Nine to 12 convex whorls; nuclear whorls smooth, remaining whorls with 3 rows of glassy beads joined by spiral and transverse threads.

APERTURE: Oval, lip thin. Columella arched in juveniles but straight with a slight, flaring anterior notch in adults.

OPERCULUM: Corneous, brown.

PERIOSTRACUM: None visible.

REMARKS: Carnivorous, sexes separate. Lives on marine plants. Can be confused with *Bittium*, but whorls are more convex and siphonal canal better developed. To find, sift beach drift.

HABITAT: Inlet, shelly sand, and along shore.

LOCALITIES: Entire.

OCCURRENCE: Fairly common.

RANGE: Cape Cod to both sides of Florida and Texas.

GEOLOGIC RANGE: Pliocene to Recent.

Subgenus *Cerithiopsis* ss.

Cerithiopsis (Cerithiopsis) cf. *C. iota* C. B. Adams, 1845, *Proc. Boston Soc. Natur. Hist.* 2:5.

Gk. *keration* little horn, *opsis* appearance, *iota* as a numeral, 10.

SIZE: Length 12 mm.

COLOR: Reddish brown.

SHAPE: Conical, elongate.

ORNAMENT OR SCULPTURE: Ten slightly convex whorls. Each whorl has 3 nodulose, spiral ridges, except the last whorl, which has 4 ridges. Longitudinal ridges produce nodules at intersections. Sutures are distinct.

APERTURE: Oval with short canal.

OPERCULUM: Corneous.

PERIOSTRACUM: None visible.

REMARKS: Similar to *C. greeni*, but is longer and has a fourth row of nodules on the last whorl. This may be synonymous with *C. vinca* Olsson & Harbison, 1953.

Cerithiopsis (Cerithiopsis) cf. *C. iota*

HABITAT: Inlet, shelly sand.
LOCALITIES: Port Aransas.
OCCURRENCE: Rare in drift.
RANGE: Florida, West Indies and Texas.

Genus *Seila* A. Adams, 1861

ADAMS' MINIATURE CERITH
Seila adamsi (H. C. Lea, 1845), *Trans. Amer. Phil. Soc.* 2(9):42.
Gr. *seira?* cord, string; dedicated to A. Adams.
SIZE: Length 10 mm.
COLOR: Brown.
SHAPE: Elongate conical, nearly cylindrical.
ORNAMENT OR SCULPTURE: About 12 flat-sided whorls, regularly increasing in size. Three strong, flattened spiral ridges on each whorl. The concave spaces between them are marked with fine spiral

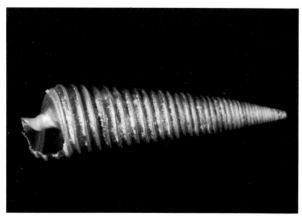

Seila adamsi

striations and delicate transverse lines that do not cross ridges. Base of shell is concave. Sutures are only wider spaces giving the initial impression of continuous ridges.
APERTURE: Oval, outer lip crenulated to correspond with external sculpture. Columella short, ending in centrally located siphonal canal, slightly recurved.
OPERCULUM: Corneous, brown, oval, paucispiral.
PERIOSTRACUM: None visible.
REMARKS: Can be found in Aransas Bay by turning broken shell and examining attached vegetation. Herbivorous. Sexes separate. Chalky and bleached when dead.
HABITAT: On shell in hypersaline bays and inlets.
LOCALITIES: Entire.
OCCURRENCE: Fairly common.
RANGE: Massachusetts to Florida, Texas and the West Indies.
GEOLOGIC RANGE: Lower Miocene to Recent.

Alaba incerta

Genus *Alaba* H. & A. Adams, 1853

UNCERTAIN MINIATURE CERITH
Alaba incerta (d'Orbigny, 1842), [La Sagra, *Hist. l'Ile de Cuba*], *Moll.* 1:218.
Gk. *alaba* coal dust or soot; L. *incertus* uncertain.
SIZE: Length 6 mm.
COLOR: Nucleus glossy brown, remainder whitish with light brown spots. Translucent.
SHAPE: Elongated conical slender.
ORNAMENT OR SCULPTURE: About 13 nicely rounded whorls, gradually increasing in size. Shell thin and

smooth except for numerous very fine spiral striations. Raised former varices are characteristic on whorls.

APERTURE: Oval, lip barely thickened, smooth inside. Siphonal canal not well developed.

OPERCULUM: Corneous.

PERIOSTRACUM: None visible.

REMARKS: Only reported recently, but at times can be found readily in beach drift near the jetties at Port Aransas. Herbivorous.

HABITAT: Inlets and offshore.

LOCALITIES: Central, south.

OCCURRENCE: Fairly common.

RANGE: Bermuda, Bahamas, southeastern Florida to Lesser Antilles and Central America, Brazil and Texas.

Family TRIPHORIDAE Gray, 1847

Genus *Triphora* Blainville, 1828

BLACK-LINED TRIPHORA

Triphora perversa nigrocinta (C. B. Adams, 1839), *Boston J. Natur. Hist.* 2:286.

Gk. *tri* thrice, *phoren* to bear, bearing 3 rows; L. *perversum* turn wrong way, *niger* black, *cincta* encircled.

SIZE: Height 5 to 10 mm.

COLOR: Dark brown.

SHAPE: Elongate conical, somewhat cylindrical.

ORNAMENT OR SCULPTURE: Ten to 12 slightly convex whorls tapering to an acute apex. Sutures slightly excavated. Three spiral rows of rather evenly

Triphora perversa nigrocinta

spaced, glossy beads that appear lighter under a lens. Keeled base has strong revolving cords.

APERTURE: Oval, oblique, brown with thin, crenulated outer lip. Well-developed posterior notch and short recurved anterior canal; in adults both of these become tubular. Sinistral.

OPERCULUM: Corneous, brown.

PERIOSTRACUM: None visible.

REMARKS: At a quick glance this shell is easily confused with *Cerithiopsis greeni*, but the "left-handed" aperture always sets it apart.

HABITAT: High-salinity reef, on algae.

LOCALITIES: Entire.

OCCURRENCE: Fairly common.

RANGE: Massachusetts to Florida, Texas, and the West Indies.

Superfamily EPITONIACEA Berry, 1910

Family EPITONIIDAE Berry, 1910

The members of this beautiful family, known as wentletraps or spiral-staircase shells, are a confusing lot. (*Wentletrap* is the Dutch equivalent of spiral staircase.) The Texas coast can boast a large number of these elegant little shells but most collectors will find them difficult to separate. Juveniles are virtually impossible to group.

The tendency of the characteristics of one species to merge with those of another compounds the problem. Any two species have extremes that are readily recognized with a hand lens, but in between the differences are not clear cut, especially if the distinguishing pattern is based on sculpture.

The wentletraps may be found by screening beach drift, or the lucky collector may happen on an occasional population washing in on the outer beaches. These populations have been of mixed species.

Most of these shells secrete a purple dye, an indication that they are carnivorous, in spite of the fact that they do not sport the siphonal canal that usually characterizes the carnivore. It is thought by some that they feed on sea anemones, worms, and foraminifera.

The Epitoniidae is one of the few Prosobranch families that are hermaphroditic. Eggs are produced in beadlike strings consisting of chitinous capsules made of agglutinized sand grains. Generally, the nuclear whorls are smooth and amber colored.

Clench and Turner have grouped the genera as to

those that have spiral sculpture and those that exhibit only the typical axial costae. On taxonomic grounds this is artificial but it is very convenient—and a great starting point for the beginner.

Genus *Amaea* H. & A. Adams, 1854

Subgenus *Amaea* s.s.

MITCHELL'S WENTLETRAP

Amaea (*Amaea*) *mitchelli* (Dall, 1896), *Nautilus* 9:112.

An invented word, may be Gk. *a* without; L. *maca* a crab; dedicated to J. D. Mitchell, early Texas naturalist.

SIZE: Length 1½ to 2½ in.

COLOR: Pale ivory with dark brownish band at the periphery and a solid brownish area below the basal ridge.

SHAPE: Elongate conic, turriculate.

ORNAMENT OR SCULPTURE: Fifteen strongly convex whorls gradually increasing in size from apex to base. Twenty-two irregular, low costae per whorl; spiral sculpture of fine threads produces a definite reticulated pattern. A thickened ridge and brownish color define the basal area.

APERTURE: Subcircular. Thickened but nonreflected lip. Columella short and arched.

OPERCULUM: Corneous, thin, dark brown.

PERIOSTRACUM: None visible.

REMARKS: One of the largest of the wentletraps and thought to be indigenous to this area. Collectors vie with each other for this one and even prize broken

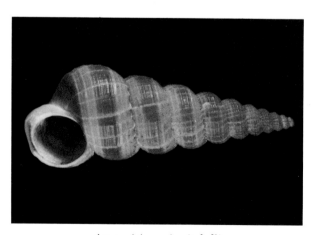

Amaea (*Amaea*) *mitchelli*

pieces. After a severe cold in early 1966, it was found in unusual numbers in the Port Aransas area. It is used as the symbol of the Coastal Bend Shell Club. Originally described from the Texas coast.

HABITAT: Offshore.

LOCALITIES: Entire.

OCCURRENCE: Broken pieces are not uncommon but good specimens are rare.

RANGE: Texas coast to Yucatán.

Epitonium (*Epitonium*) *albidum*

Genus *Epitonium* Röding, 1798

Subgenus *Epitonium* s.s.

WHITE WENTLETRAP

Epitonium (*Epitonium*) *albidum* (d'Orbigny, 1842), [La Sagra, *Hist. l'Ile de Cuba*], *Moll.* 2:17.

Gk. *epitonion* peg, turncock; L. *albidus* white.

SIZE: Height 20 mm.

COLOR: Shiny white.

SHAPE: Elongate conical, turriculate.

ORNAMENT OR SCULPTURE: Nine to 11 gradually increasing, moderately convex whorls, attached by costae only. Bladelike transverse costae are rather low and generally fused with the costae on the previous whorl. Body whorl contains 12 to 14 unangled costae. There may be microscopic spiral threads but not strong enough to be grouped with those considered to have spiral striations. Anomphalous.

APERTURE: Subcircular. Outer lip expanded and reflected, parietal area narrow, slightly thickened, and held away from the body whorl by costae. Nearly holostomatous.

OPERCULUM: Corneous, subcircular, thin, and paucispiral.

PERIOSTRACUM: None visible.

REMARKS: The low ribs of this rather "fat" little *Epitonium* appear to form continuous oblique lines from bottom to top. Sift for it.

HABITAT: Intertidal to 200 fathoms.

LOCALITIES: Entire.

OCCURRENCE: Fairly common.

RANGE: Texas, southern Florida, Bermuda, the West Indies and south to northern Argentina.

ANGULATE WENTLETRAP

Epitonium (Epitonium) angulatum (Say, 1831), *Amer. Conch.* 3:pl. 27.

Gk. *epitonion* peg, turncock; L. *angulatus* angled.

SIZE: Height 20 mm.

COLOR: Glossy white.

SHAPE: Elongate conical, turriculate.

ORNAMENT OR SCULPTURE: Eight moderately convex whorls gradually increasing in size from the apex. Sculpture on whorls consists of numerous reflected, bladelike costae. The costate form angles on the shoulder of each whorl that are stronger on the early whorls, a little less to almost absent on the later whorls. The costae are in line with those on the whorl above and are fused where they meet. Anomphalous.

APERTURE: Subcircular; outer lip thickened, expanded, and reflected. Parietal area slightly thickened and held away from the body whorls by costae. Entire columella not defined.

OPERCULUM: Corneous, brown, paucispiral.

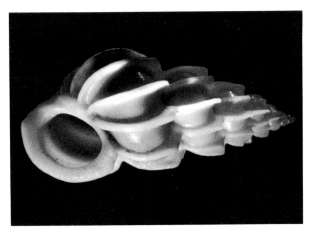

Epitonium (Epitonium) angulatum

PERIOSTRACUM: None visible.

REMARKS: This shell can fool you because some specimens are more slender than the typical form and can be confused with *E. humphreysi* (Kiener). The juvenile shells are more strongly angulated than the adults; as the shells reach maturity the costae become more thickened and rounded.

HABITAT: Intertidal to moderate depths with sea anemones.

LOCALITIES: Entire.

OCCURRENCE: Common.

RANGE: Eastern end of Long Island to Florida (excluding Lower Keys), and Texas.

Epitonium (Epitonium) cf. *E. humphreysi*

HUMPHREY'S WENTLETRAP

Epitonium (Epitonium) cf. *E. humphreysi* (Kiener, 1815), *Iconogr. Coq. Viv.* 10:15.

Gk. *epitonion* peg, turncock; dedicated to G. Humphrey, eighteenth-century London shell dealer.

SIZE: Height 20 mm.

COLOR: Flat white.

SHAPE: Elongate conical, turriculate.

ORNAMENT OR SCULPTURE: Nine to 10 strongly convex whorls. Numerous bladelike to rounded costae on whorls, also serve to connect whorls. Suture deep. The costae are more bladelike on early whorls, becoming thickened and rounded with maturity. They may be reflected backward and some are angled at the whorl shoulder, especially on early whorls. Body whorl has 8 to 9 costae. Anomphalous.

APERTURE: Subcircular; outer lip expanded and usually thickened. Parietal lip is thin and tightly pressed to body whorl. Columella short and arched.

OPERCULUM: Corneous, thin, dark brown, and paucispiral.

PERIOSTRACUM: None visible.

REMARKS: This shell is hard to distinguish from *E. angulatum* (Say). If typical it is more narrow, costae more rounded, with less developed shoulder angle but very variable. Less common than *E. angulatum*.

HABITAT: Inlet areas and along shore on sandy bottoms.

LOCALITIES: Entire.

OCCURRENCE: Fairly common.

RANGE: From Cape Cod, Massachusetts, south to Florida (not Lower Keys) and west to Texas.

GEOLOGIC RANGE: Miocene to Recent.

TOLLIN'S WENTLETRAP

Epitonium (Epitonium) cf. *E. tollini* Bartsch, 1938, *Nautilus* 52:34.

Gk. *epitonion* peg, turncock; dedicated to Oscar Tollin, who sent the specimen to Dall.

SIZE: Length 14 mm.

COLOR: White.

SHAPE: Elongate conical, turriculate.

ORNAMENT OR SCULPTURE: Nine to 10 convex whorls that can be appressed or separated and attached by costae only. Numerous bladelike costae, 11 or 16 on body whorl. The spacing of these costae is irregular and they do not always line up with those on the whorl above. They are even in height throughout with an occasional thick one, especially on the body whorl. They are not angled at the shoulder and there is no spiral sculpture. Suture deep. Anomphalous.

APERTURE: Subcircular; outer lip thickened and re-

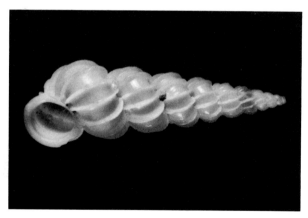

Epitonium (Epitonium) cf. E. tollini

flected. Parietal area thin and pressed closely to the body wall. Columella short and arched.

OPERCULUM: Corneous, yellowish brown, paucispiral.

PERIOSTRACUM: None visible.

REMARKS: Look for the mismatched costae and the occasional large one. This shell is the most common *Epitonium* in the northern part of range, less so to the south.

HABITAT: Probably intertidal.

LOCALITIES: Entire.

OCCURRENCE: Common.

RANGE: Texas, western Florida from Marco Island north to Gasparilla Island.

Epitonium (Gyroscala) rupicola

Subgenus *Gyroscala* de Boury, 1887

BROWN-BANDED WENTLETRAP

Epitonium (Gyroscala) rupicola (Kurtz, 1860), *Cat. Rec. Marine Moll. Portland Mus.*, p. 7.

Gk. *epitonion* peg or turncock; L. *rupes* rock, *incola* to inhabit.

SIZE: Height 20 mm.

COLOR: White to cream with two brownish, spiral bands at suture. Costae are white.

SHAPE: Elongate conical, turriculate.

ORNAMENT OR SCULPTURE: Eleven globose whorls attached at sutures. Suture deep. Transverse sculpture consists of numerous, low, bladelike costae, interspersed with rounded varices. The height and number of the costae are variable and they do not line up with costae on the adjoining whorl. Basal ridge well defined with a thin, threadlike line.

APERTURE: Subcircular; lip slightly thickened and reflected. Nearly entire.
OPERCULUM: Corneous, thin, yellowish, paucispiral.
PERIOSTRACUM: None visible.
REMARKS: Look for the brown bands of color. This shell is often large enough to see without screening drift. Olsson and Harbison (1953) have placed *E. rupicola* under the genus *Clathrus* and subgenus *Pitoscala*.
HABITAT: Below low water to 20 fathoms.
LOCALITIES: Entire.
OCCURRENCE: Fairly common.
RANGE: Massachusetts south to Florida and west to Texas.

Epitonium (Gyroscala) lamellosum

Epitonium (Gyroscala) lamellosum (Lamarck, 1822), *Hist. Natur. Anim. sans Vert.* 6:227.
Gk. *epitonion* peg or turncock; L. *lamina* thin plate or scale.
SIZE: Height 1½ in.
COLOR: White with brownish mottlings near suture. Costae white.
SHAPE: Elongate conical, turriculate.
ORNAMENT OR SCULPTURE: Eleven convex whorls, attached. Suture well defined. Numerous thin, high, bladelike axial costae that line up with those on the next whorl so that they look continuous from the base to the apex. Basal area set off by threadlike spiral line. Anomphalous.
APERTURE: Subcircular; outer lip thickened and reflected. There may be one varix on the body whorl. Holostomatous.
OPERCULUM: Corneous, thin, dark brown, paucispiral.

PERIOSTRACUM: None visible.
REMARKS: Much larger than any of the other *Epitonium* on the coast. The ribs are easily broken. It resembles a large *E. rupicola* but for the more bladelike ribs and spiral thread on base.
HABITAT: Offshore to 33 fathoms.
LOCALITIES: Mustang Island.
OCCURRENCE: Rare.
RANGE: Florida, Bermuda, West Indies, Texas and Lesser Antilles.

MULTIRIBBED WENTLETRAP
Epitonium (Gyroscala) multistriatum (Say, 1826), *J. Acad. Natur. Sci. Phila.* 5(1):208.
Gk. *epitonion* peg, turncock; L. *multi* many, *striatus* lined.
SIZE: Height 15 mm.
COLOR: White.
SHAPE: Elongate conical, turriculate.
ORNAMENT OR SCULPTURE: Eight to 10 strongly convex whorls, the later ones unattached. Suture deep. Transverse sculpture consists of very numerous cordlike to low bladelike costae not angled at the shoulder. Between these are many finely incised lines that do not cross ribs. The apical whorls are much smaller in proportion to the later ones and have more ribs. Anomphalous.
APERTURE: More oval than round with very narrowly expanded lip. Columella not defined, and parietal lip is tightly pressed to parietal area.
OPERCULUM: Probably typical.
PERIOSTRACUM: None visible.

Epitonium (Gyroscala) multistriatum

REMARKS: This shell is easy to separate because of the very numerous ribs and the spiral markings. Dainty in comparison to the others.

HABITAT: Offshore.

LOCALITIES: Entire.

OCCURRENCE: Common.

RANGE: Massachusetts south to Cape Canaveral, Florida, west to Texas.

NEW ENGLAND WENTLETRAP

Epitonium (Gyroscala) novangliae (Couthouy, 1835), *Boston J. Natur. Hist.* 2:96.

Gk. *epitonion* peg, turncock; L. *novus* new, Angliae, England.

SIZE: Length up to 14 mm.

COLOR: White, banded with light brown above and below the periphery of the whorl. Often light brown throughout.

SHAPE: Elongate conical, turriculate.

ORNAMENT OR SCULPTURE: Eight to 10 strongly convex whorls gradually increasing in size from apex, later whorls attached by costae only. Numerous bladelike to cordlike costae, the latter as a result of the costae being reflected backward and downward. The body whorl has 9 to 16 costae that are angled or hooked at the shoulder. Spiral sculpture has a reticulated pattern formed by numerous spiral threads crossed by finer transverse lines. Magnification may be needed to detect latter. No basal ridge.

APERTURE: Subcircular with narrow expanded lip. Parietal lip only moderately thickened and pressed to the body whorl above the umbilicus.

OPERCULUM: Corneous, dark brown, paucispiral.

PERIOSTRACUM: None visible.

REMARKS: Look for the brownish color and the spiral striations.' The axial ribs are not regular in width. This is the most common *Epitonium* on South Padre. Screen the beach drift.

HABITAT: Offshore.

LOCALITIES: Entire.

OCCURRENCE: Common.

RANGE: Massachusetts, Virginia south to Brazil and Texas.

GEOLOGIC RANGE: Pliocene to Recent.

Epitonium (Gyroscala) sericifilum (Dall, 1889), *Bull. M. C. Z.* 18:313.

Gk. *epitonion* peg, turncock; L. *sericum* silk, *filum* thread.

SIZE: Height 5.1 to 10 mm.

COLOR: Mat white.

SHAPE: Slender, elongate conical, turriculate.

ORNAMENT OR SCULPTURE: Ten convex whorls, gradually increasing from the apex. Suture moderately defined. Transverse costae are numerous, oblique, and low. Spiral threads are numerous but do not cross costae. No basal ridge. Anomphalous.

APERTURE: Subcircular; outer lip slightly thickened; no parietal shield.

OPERCULUM: Probably typical.

PERIOSTRACUM: None visible.

REMARKS: Little is known of this shell. It is easily recognized by the close, oblique ribs and angled periphery of whorls. Sift for it.

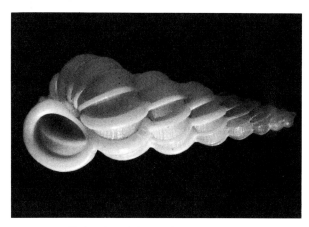

Epitonium (Gyroscala) novangliae

Epitonium (Gyroscala) sericifilum

HABITAT: Inlet areas.
LOCALITIES: Entire.
OCCURRENCE: Uncommon.
RANGE: Honduras and Texas.

Depressiscala nautlae

Genus *Depressiscala* de Boury, 1909

Depressiscala nautlae (Mörch, 1874), *Vidensk Med. Naturhist* 17: 265.
L. *depressus* press down, *nauta* sailor.
SIZE: Height 15 mm.
COLOR: Reddish brown, darker at early whorls with whitish ribs.
SHAPE: Slender, elongate conical, turriculate.
ORNAMENT OR SCULPTURE: Ten to 21 convex, attached whorls. Transverse costae are numerous, low, irregular width with slight angles on shoulder of whorls. No spiral sculpture; no basal ridge. Anomphalous.
APERTURE: Oval with slightly thickened lip. Parietal area smooth and tightly pressed to body whorl. Columella short and arched.
OPERCULUM: Corneous, thin, yellow brown, paucispiral.
PERIOSTRACUM: None visible.
REMARKS: The all-over brown color and white ribs make this shell easily recognizable, if one is lucky enough to find it in the beach drift.
HABITAT: Offshore.
LOCALITIES: Extreme south.
OCCURRENCE: Rare.
RANGE: North Carolina, Florida, Bahamas, Cuba, southern Mexico, and Texas (discontinuous).

Family JANTHINIDAE Leach, 1823

Genus *Janthina* Röding, 1798

Subgenus *Jodina* Mörch, 1860

DWARF PURPLE SEA SNAIL
Janthina (*Jodina*) *umbilicata* d'Orbigny, 1840, [La Sagra, *Hist. l'Ile de Cuba*], *Moll.* 2:85.
Gk. *janthinos* violet colored; L. *umbilicus* the navel.
SIZE: Diameter 6 to 19 mm.
COLOR: Deep violet.
SHAPE: Globose ovate.
ORNAMENT OR SCULPTURE: The thin shell is faintly striated, with the striae following the edge of the aperture. At the indentation of the latter they curve, forming a distinct keel on the last whorl.
APERTURE: The outer lip of the aperture is deeply indented at the mid-whorl point.

Janthina (*Jodina*) *umbilicata*

OPERCULUM: None.
PERIOSTRACUM: None visible.
REMARKS: Syn. *J. globosa* Blainville, 1825, and *J. megastoma* Adams, 1861. The float is comparatively small. Egg cases of this hermaphrodite are pear shaped and contain about 75 eggs.
HABITAT: Pelagic.
LOCALITIES: Entire.
OCCURRENCE: Uncommon.
RANGE: Worldwide.

Janthina (Janthina) janthina

Subgenus *Janthina* s.s.

COMMON PURPLE SEA SNAIL
Janthina (*Janthina*) *janthina* (Linné, 1798), *Syst. Natur.*, 10th ed., p. 772.
Gk. *janthinos,* violet colored.
SIZE: Diameter up to 1¾ in.

Janthina (Janthina) pallida

COLOR: Purple on the basal part of shell, lavender or white above. (Floats with the purple side up.)
SHAPE: Low conical, body whorl very large and gently angular.
ORNAMENT OR SCULPTURE: Smooth and very fragile.
APERTURE: More or less subquadrate with the columellar edge almost vertical. Outer lip very delicate, lower edge horizontal.
OPERCULUM: None.
PERIOSTRACUM: None visible.
REMARKS: The foot builds a float by trapping air bubbles in mucus. The animal exudes a purple stain when disturbed. Hermaphrodite. Viviparous. Carnivorous, feeding on Portuguese men-of-war, *Velella,* and other violet-colored coelenterates. It is "feast or famine" for the collector; a strong southeast wind in the spring will literally cover the beach with them; the next year they might not appear at all. Beware of the Portuguese men-of-war that wash in with them.
HABITAT: Pelagic in warm seas.
LOCALITIES: Entire.
OCCURRENCE: Seasonally common.
RANGE: Pelagic in most warm waters.

PALE PURPLE SEA SNAIL
Janthina (*Janthina*) *pallida* (Thompson, 1841), *Ann. Natur. Hist.* 5:96.
Gk. *janthinos* violet colored; L. *pallidus* pale.
SIZE: Diameter 12 to 18 mm.
COLOR: Purplish white with pale lavender on early whorls and along margin of outer lip.
SHAPE: Globose conical.
ORNAMENT OR SCULPTURE: Appears smooth but has fine, irregular axial growth lines. Body whorl large, rounded, and swollen.
APERTURE: Subquadrate, somewhat flaring at base. Thin outer margin of lip is slightly sinuate near the center where the float is attached. Interior is violet along columella and base, fainter over remainder.
OPERCULUM: None.
PERIOSTRACUM: None visible.
REMARKS: Very easily confused with *J. prolongata.* Only recently recognized as occurring in this area. Attaches eggs to bottom of float. Hermaphrodite.
HABITAT: Pelagic in warm seas.
LOCALITIES: Entire.
OCCURRENCE: Uncommon, seasonal.
RANGE: Pelagic in most warm waters.

Janthina (Violetta) prolongata

Subgenus *Violetta* Iredale, 1929

GLOBE PURPLE SEA SNAIL
Janthina (Violetta) prolongata (Blainville, 1822),
    Dict. Sci. Natur. 24:155.
Gk. *janthinos* violet colored; L. *prolongare* long.
SIZE: Diameter 12 to 18 mm.
COLOR: Deep violet throughout.
SHAPE: Globose with low spire, body whorl very large.
ORNAMENT OR SCULPTURE: Smooth and very fragile.
    Very fine growth lines.
APERTURE: Subquadrate and very large. Columella
    vertical. Outer lip has a sinuous margin; the indenta-
    tion marks the point of attachment of the float.
OPERCULUM: None.
PERIOSTRACUM: None visible.
REMARKS: Syn. *J. globosa* (Swainson, 1823). Attaches
    egg to float. This shell does not occur in the numbers
    that *J. janthina* does after a blow and is even more
    delicate than the others. Live shells are best cleaned
    by "watering out" (see section on cleaning, Chapter
    4). Left in the sun they will fade and get dull.
HABITAT: Pelagic in warm seas.
LOCALITIES: Entire.
OCCURRENCE: Fairly common.
RANGE: Pelagic in warm waters, both coasts of United
    States.

Genus *Recluzia* Petit, 1853

BROWN SEA SNAIL
Recluzia rollaniana Petit, 1853, *J. Conch.* [Paris] 5 (4).
Dedicated to C. A. Recluz (d. 1873); dedicated to
    Rolland du Roguan (1812–1863).
SIZE: Height up to 22 mm.; diameter 13 mm.
COLOR: White but periostracum makes it appear
    brown.
SHAPE: Moderately elongate conical.
ORNAMENT OR SCULPTURE: Body whorl large and
    globose. Spire somewhat extended. Appears smooth
    but has fine, oblique growth lines. Fragile.
APERTURE: Oval, large, lip very thin.
OPERCULUM: None.
PERIOSTRACUM: Thin, brown, completely coating
    shell, lacquerlike.
REMARKS: Looks like a large sargassum snail and more
    like a land snail than a marine snail. Uses its brown
    float as do *Janthina*. Feeds on *Minyas* sea anemones.
    Animal is yellow.
HABITAT: Pelagic in warm seas.
LOCALITIES: Port Aransas, south.
OCCURRENCE: Rare.
RANGE: Pelagic in most warm waters.

Recluzia rollaniana

Superfamily AGLOSSA Thiele, 1829

Family MELANELLIDAE = EULIMIDAE Risso, 1826

Genus *Niso* Risso, 1826

CUCUMBER MELANELLA

Niso interrupta Sowerby, 1834, *Proc. Zool. Soc. London* 2:70.

Gk. Nissus, king of Megara; L. *interruptus* interrupted.

SIZE: Length 18 mm.

COLOR: Pale brown background with occasional mottlings of darker brown. Reddish brown line near base of suture on each whorl. Highly polished.

SHAPE: Elongate conical. Apex acute.

ORNAMENT OR SCULPTURE: Ten to 11 flat-sided whorls. Angled at base. Base slightly convex. Transversely marked with fine reddish brown growth lines.

APERTURE: Tear shaped. Sharp outer lip. Outlined with fine reddish brown line. Deeply umbilicated by a conical depression.

OPERCULUM: Corneous, pale yellowish brown, closes aperture.

PERIOSTRACUM: None visible.

REMARKS: Animal is white and very active with ability to move forward and backward.

HABITAT: Shallow water in bay area, inlet influence.

LOCALITIES: Aransas Bay.

OCCURRENCE: Rare.

RANGE: Cape Hatteras, North Carolina, to Gulf of Mexico.

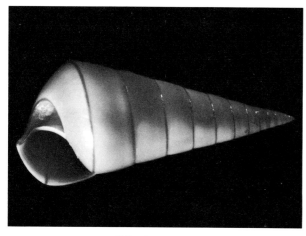

Niso interrupta

Strombiformis cf. *S. bilineata*

Genus *Strombiformis* Da Costa, 1778

TWO-LINED MELANELLA

Strombiformis cf. *S. bilineata* (Alder, 1848) *Moll. Northumberland & Durham,* p. 47 (Forbes and Hanley, *British Mollusca* 3:237).

Strombiformis in the form of a *strombus*; L. *bis* twice, *lineatus* lined.

SIZE: Height 8 mm.

COLOR: Whitish with two brownish lines on each whorl. Polished.

SHAPE: Elongate conic, very slender.

ORNAMENT OR SCULPTURE: Ten flattened whorls, sutures fairly distinct. Gradually tapers to a sharp apex that may be deflected.

APERTURE: Elongated pyriform. Entire. Outer lip sharp, columella concave.

OPERCULUM: Corneous.

PERIOSTRACUM: Thin, black.

REMARKS: A very elegant little shell that beach drift must be screened for. It has a long proboscis used to suck the juices of its prey. There is no gill or radula.

HABITAT: Thought to be ectoparasitic on holothurians.

LOCALITIES: Entire.

OCCURRENCE: Uncommon.

RANGE: North Carolina to West Indies, Texas and Europe.

Genus *Melanella* Bowdich, 1822

CURVED MELANELLA

Melanella cf. *M. arcuata* (C. B. Adams, 1850) *Contr. Conch.* (7):110.

Gk. *melanos* black; L. *arcus* curved.

SIZE: Length 4 mm.

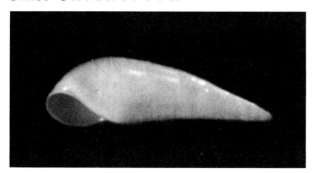

Melanella cf. *M. arcuata*

COLOR: Glossy white.

SHAPE: Ovate conic.

ORNAMENT OR SCULPTURE: Ten convex whorls with a lightly impressed suture. A fine impressed spiral line above the suture marks the smooth shell. Spire with the axis curved to an extraordinary degree in the upper whorls.

APERTURE: Rather long ovate.

OPERCULUM: Not known.

PERIOSTRACUM: None visible.

REMARKS: The type specimen of this tiny shell has been lost.

HABITAT: Ectoparasitic on sea cucumbers.

LOCALITIES: Central and south.

OCCURRENCE: Uncommon.

RANGE: Jamaica and Texas.

Subgenus *Polygireulima* Sacco, 1892

CONICAL MELANELLA

Melanella (Polygireulima) jamaicensis (C. B. Adams, 1845), *Proc. Boston Soc. Natur. Hist.* 2:6.

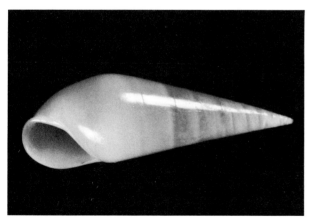

Melanella (Polygireulima) jamaicensis

Gk. *melanos* black; from Jamaica.

SIZE: Length 6 to 12 mm.

COLOR: Glossy white.

SHAPE: Elongate conic, slender.

ORNAMENT OR SCULPTURE: Twelve to 14 flat-sided whorls. The polished whorls taper to a sharp apex that is slightly bent. Sutures not well defined. Anomphalous.

APERTURE: Pyriform, slender; outer lip thin but slightly thickened at base. Columella concave.

OPERCULUM: Corneous.

PERIOSTRACUM: Thin, black.

REMARKS: Formerly *M. intermedia* Contraine and *Balcis conoidea* Kurtz & Stimpson. A very active little animal that uses its long proboscis to feed on the juices of sea cucumbers.

HABITAT: Ectoparasitic.

LOCALITIES: Entire.

OCCURRENCE: Fairly common.

RANGE: Florida, Texas, and West Indies.

Superfamily STROMBACEA Rafinesque, 1815

Family STROMBIDAE Rafinesque, 1815

Genus *Strombus* Linné, 1758

FIGHTING CONCH

Strombus alatus Gmelin, 1790, *Syst. Natur.*, 14th ed., p. 3513

Gk. *strombos* a spiral shell named by Aristotle; L. *alatus* winged.

SIZE: Length 3 to 4 in.

COLOR: Dark reddish brown to a lighter brown, some mottled or with zigzag markings.

SHAPE: Conical.

ORNAMENT OR SCULPTURE: Spire has 8 whorls, body whorl is four-fifths of total length of this heavy, solid shell. Wide shoulders with or without short spines. Spiral striations near base. Sutures distinct. Anomphalous.

APERTURE: Long, narrow. Lip has broad outward flare. Interior polished. Somewhat flared siphonal notch at base and a rounded notch in lip just above it.

OPERCULUM: Corneous, clawlike, does not close aperture.

PERIOSTRACUM: Thin velvetlike coating.

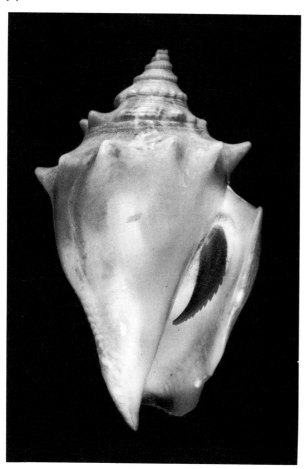

Strombus alatus

REMARKS: This grazing scavenger has well-developed eyes on long eye stalks on the head. The long, narrow foot tipped with its clawlike operculum moves the animal about in awkward leaps. With this hopping motion it can right itself and return to the water if stranded by a wave. Hurricane Carla (1961) stranded thousands on the southern half of Padre Island. Sexes are separate; the male brings his heavy shell up alongside the female until the shells touch, then fertilizes the female with a long verge that is thrust under her shell. She lays the fertilized eggs in gelatinous ribbons.

HABITAT: Intertidal to about 10 fathoms.

LOCALITIES: Entire.

OCCURRENCE: Fairly common.

RANGE: South Carolina, both sides of Florida and Texas.

GEOLOGIC RANGE: Miocene to Recent.

Superfamily CALYPTRAEACEA Blainville, 1824

Family CALYPTRAEIDAE Blainville, 1824

Genus *Crepidula* Lamarck, 1799

Subgenus *Crepidula* s.s.

COMMON ATLANTIC SLIPPER SHELL
Crepidula (Crepidula) fornicata (Linné, 1767), *Syst. Natur.*, 10th ed., p. 1257.
L. *crepidula* a small sandal, *fornicata* vaulted.

SIZE: Length up to 2 in.

COLOR: Dirty white to tan with mottlings of brown shades.

SHAPE: Limpetlike, oval, oblique. Curved to fit place of attachment.

ORNAMENT OR SCULPTURE: Smooth except for fine growth lines. Body whorl is major part of shell. Apex turned to one side.

APERTURE: Oval, oblique with thin margin. Polished interior. Deck occupies about half of aperture; white; margin is sinuous.

OPERCULUM: None.

PERIOSTRACUM: None visible.

REMARKS: Mucociliary feeding. The adult shell is sedentary and tends to pile up in stacks of up to 19 individuals that gradually diminish in size. Frequently, on dead olive shells. The stack of hermaphroditic mollusks will have the right margin of each member in contact with the right margin of the one it is on. The bottom and larger animals are female, the top are males, and those in between are in transitional

Crepidula (Crepidula) fornicata

stages. The eggs are brooded in the female mantle cavity.

HABITAT: One to 6 fathoms.
LOCALITIES: Entire.
OCCURRENCE: Common.
RANGE: Canada to Florida and Texas, introduced to California and England.
GEOLOGIC RANGE: Lower Miocene to Recent.

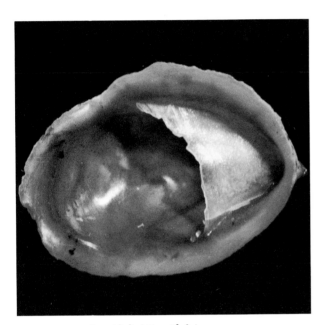

Crepidula (Crepidula) convexa

FADED SLIPPER SHELL
Crepidula (Crepidula) convexa Say, 1822, *J. Acad. Natur. Sci. Phila.* 2:226.
L. *crepidula* a small sandal, *convexus* vaulted.
SIZE: Length 12 mm.; width 8 mm.
COLOR: Translucent tan or mottled with reddish brown. Internal septum white.
SHAPE: Limpetlike, low, oval with the apex near the margin. This varies corresponding to the shape of the place of attachment.
ORNAMENT OR SCULPTURE: Smooth with centrally located apex.
APERTURE: Very large oval with thin margin. Internal deck or septum is deep seated and convex; this supports the soft parts.
OPERCULUM: None.
PERIOSTRACUM: None visible.
REMARKS: A sedentary animal that takes on the shape of its attachment site. It has the ability to form food

into pellets and store them in a special groove near the neck. Hermaphrodite. *C. convexa* looks like young *C. fornicata* and likes to make its home on pectens in bays and inlets.

HABITAT: Intertidal to moderate depths on shell, rocks, and grass.
LOCALITIES: Entire.
OCCURRENCE: Common.
RANGE: Massachusetts to Florida, Texas and the West Indies.

Subgenus *Janacus* Mörch, 1852

EASTERN WHITE SLIPPER SHELL
Crepidula (Janacus) plana Say, 1822, *J. Acad. Natur. Sci. Phila.* 2(1):226.
L. *crepidula* a small sandal, *planos* level, flat.
SIZE: Length up to 30 mm.; width to 16 mm.
COLOR: White.
SHAPE: Elongated oval, flat. Conforms to place of attachment.
ORNAMENT OR SCULPTURE: Smooth except for fine growth lines. Apex is depressed.
APERTURE: Large, oval with thin margin. Deck about half the length of shell, notched to one side. Polished.
OPERCULUM: None.
PERIOSTRACUM: Thin, yellowish.
REMARKS: Not as particular as to attachment site as are *C. fornicata* and *C. convexa*—dead shell, old bottle, piers, oysters, all will do. As a result the shape of

Crepidula (Janacus) plana

this very flat shell can be quite varied. Eggs are brooded under the shell of this hermaphrodite.

HABITAT: Intertidal to moderate depths.

LOCALITIES: Entire.

OCCURRENCE: Common.

RANGE: Canada to Florida, Gulf states, Texas, rare in West Indies.

GEOLOGIC RANGE: Lower Miocene to Recent.

Superfamily LAMELLARIACEA d'Orbigny, 1841

Family LAMELLARIIDAE d'Orbigny, 1841

Subfamily LAMELLARIINAE d'Orbigny, 1841

Genus *Lamellaria* Montagu, 1815

RANG'S LAMELLARIA

Lamellaria cf. *L. rangi* Bergh, 1853, *Monographi* 8:94.

L. dim. of *lamina* plate, leaf, layered; dedicated to P. C. A. Sander L. Rang (1784–1859), French malacologist.

SIZE: Length about 6 mm.

COLOR: Translucent, glassy, white. Nearly invisible.

SHAPE: Auriform, like *Sinum* but more globose.

ORNAMENT OR SCULPTURE: Very fragile, thin. Two and one-half to 3 whorls, last whorl very large. Surface has fine, irregular growth lines.

APERTURE: Very large, slightly oblique oval. Columella very thin.

OPERCULUM: None.

PERIOSTRACUM: None.

REMARKS: Animal envelops the shell. Looks like a small bit of milky jelly when found alive. Difficult to detect. Carnivorous on ascidians, and deposit their eggs in crevices of these colonial animals.

HABITAT: On living whip coral, other ascidians, *Hydrozoa* and *Alcyonaria.*

LOCALITIES: Port Aransas.

OCCURRENCE: Rare. This may be due to a difficulty in seeing them.

RANGE: Gulf of Mexico and Puerto Rico.

Family ERATOIDAE Gill, 1871

Subfamily TRIVIINAE Troschel, 1863?

Genus *Trivia* Gray, 1852

SUFFUSE TRIVIA

Trivia suffusa (Gray, 1832), *Cat. Shells* 134:16.

Trivia, a surname of Diana, goddess of the hunt; L. *suffuscus* brownish, dusky.

SIZE: Length 6 to 12 mm.

COLOR: Pinkish with suffused brownish mottlings and specks. Underside white.

SHAPE: Ovate.

ORNAMENT OR SCULPTURE: Dorsal groove well defined. Ribs are beaded.

APERTURE: Long, narrow, polished. Canals at both ends, touched with pink. Lip is rolled inward and sculptured with 18 to 23 fine ribs.

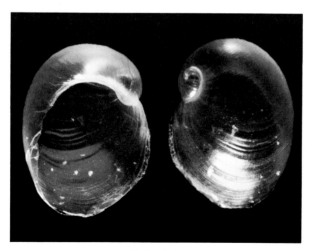

Lamellaria cf. *L. rangi*

Trivia suffusa

OPERCULUM: None.

PERIOSTRACUM: None.

REMARKS: Carnivorous on tunicates. Lays eggs in walls of tunicates. Mantle of animal envelops shell.

HABITAT: Offshore.

LOCALITIES: Extreme south.

OCCURRENCE: Rare.

RANGE: Southeastern Florida, Texas and the West Indies.

GEOLOGIC RANGE: Miocene to Recent.

Superfamily CYPRAEACEA Rafinesque, 1815

Family CYPRAEIDAE Rafinesque, 1815

Subfamily CYPRAEINAE Rafinesque, 1815

Genus *Cypraea* Linné, 1758

Subgenus *Trona* Jousseaume, 1884

ATLANTIC DEER COWRIE

Cypraea (Trona) cervus Linné, 1771, *Mant. Plant.* 2:548.

L. Cypris, Venus, *cervus* deer.

SIZE: Length 3 to 5 in.

COLOR: Polished light brown, with large, round, white spots on the dorsal side. Whitish dorsal line.

SHAPE: Elongated oval. Spire concealed.

ORNAMENT OR SCULPTURE: Smooth and polished.

APERTURE: Long, narrow. Outer lip rolled inward, brownish without spots. Edge outlined with regular, small, alternating brown and white riblets. Similar but less distinct riblets on inner lip. Purplish on interior.

OPERCULUM: None.

PERIOSTRACUM: None.

REMARKS: The mantle that envelops the shell is mottled, pinkish brown and mauve white, and entirely covered with short papillae. At rest, the shell is covered, but when touched the animal draws into the shell. The female lays eggs in a gelatinous mass of 500 to 1,500 capsules, which she broods. This carnivore also grazes on algae and colonial invertebrates. The occasional dead shell found on the beach will usually have a smooth worn place on the underside parallel to the aperture.

HABITAT: Reefs offshore.

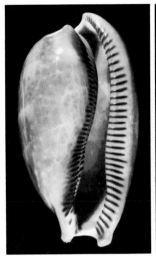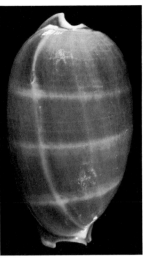

Cypraea (Trona) cervus

Mantle enveloping the shell

LOCALITIES: Port Aransas, south.

OCCURRENCE: Uncommon.

RANGE: Southeastern Florida to Yucatán and Texas.

Family OVULIDAE ?Gray, 1853

Genus *Neosimnia* Fischer, 1884

COMMON WEST INDIAN SIMNIA

Neosimnia acicularis (Lamarck, 1810), *Mus. Natur. Hist. Paris* 16(92):112.

Neosimnia acicularis

Gk. *neos* new, Simnia, according to Risso, one of the 50
    Nereids; L. dim. of *acus* needle.
SIZE: Length 12 to 18 mm.
COLOR: Glossy deep lavender or yellowish.
SHAPE: Slender, long oval.
ORNAMENT OR SCULPTURE: Polished, no dorsal ridge,
    very smooth. Spire completely covered.
APERTURE: Long, narrow, channeled at each end. Bor-
    dered by a long whitish, narrow ridge on columella
    inside aperture and the other on the edge of the body
    whorl. There is no microscopic sculpture on the pol-
    ished columella.
OPERCULUM: None.
PERIOSTRACUM: None.
REMARKS: This carnivorous animal lives on the soft
    coral, *Eugorgia virgulata*, in preference to the long

Neosimnia uniplicata

sea whip coral, *Leptogorgia setacea*, and takes on the
    color of its host.
HABITAT: On gorgonians on jetties.
LOCALITIES: Port Aransas, south.
OCCURRENCE: Uncommon.
RANGE: Southeastern United States to the West Indies
    and Texas.

SINGLE-TOOTHED SIMNIA
Neosimnia uniplicata (Sowerby, 1848), *Thes. Conch.*,
    p. 478.
Gk. *neo* new, Simnia, according to Risso, one of the 50
    Nereids; L. *uni, unus* one, *plicatus* folded.
SIZE: Length 12 to 18 mm.
COLOR: Glossy, deep reddish lavender or yellowish.
SHAPE: Slender, long oval.
ORNAMENT OR SCULPTURE: A smooth, rather thin
    shell.
APERTURE: Long and narrow, canaliculated at each
    end with a spiral plication on the posterior end.
    Only the columellar side of the aperture is bordered
    with a longitudinal ridge. Under magnification, fine
    spiral striations radiating from the upper half of
    the columellar border can be seen, especially in
    young specimens.
OPERCULUM: None.
PERIOSTRACUM: None.
REMARKS: This beautiful little shell can be obtained
    by shaking the large rolls of sea whip coral, *Lep-
    togorgia setacea*, that wash in on the outer beaches.
    The female lays her eggs on the coral.
HABITAT: Gorgonians along shore.
LOCALITIES: Entire.
OCCURRENCE: Common.
RANGE: Virginia, both sides of Florida, West Indies
    and Texas.

Genus *Cyphoma* Röding, 1798

MCGINTY'S CYPHOMA
Cyphoma mcgintyi Pilsbry, 1939, *Nautilus* 52:108.
Gk. *cyphoma* a hump, humpbacked; dedicated to
    Thomas L. McGinty (1907–).
SIZE: Length 18 to 28 mm.
COLOR: Cream with blotches of pale lavender.
SHAPE: Elongated oval.
ORNAMENT OR SCULPTURE: Distinct, rounded mid-
    dorsal ridge. Smooth and glossy. Body whorl less
    globose than *C. intermedium*.

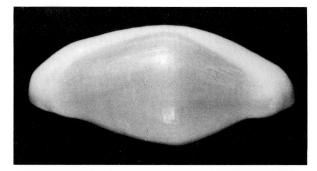

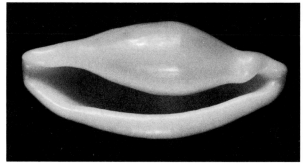

Cyphoma mcgintyi

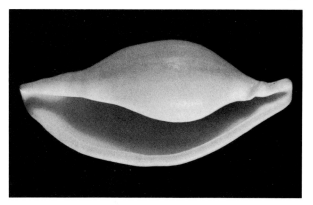

Cyphoma intermedium

APERTURE: Long, narrow. Outer lip thickened. Interior pale pink.

OPERCULUM: None.

PERIOSTRACUM: None.

REMARKS: Several specimens of this animal with its cream body spotted with blackish dots and bars lived in an aquarium for months until they were consumed by a sea anemone. Lives and feeds on gorgonians.

HABITAT: On gorgonians in warm tropical seas below low water.

LOCALITIES: Port Aransas, south.

OCCURRENCE: Uncommon.

RANGE: Lower Florida Keys, Bahamas and Texas.

WEAK-RIDGE CYPHOMA

Cyphoma intermedium (Sowerby, 1828), *Species Conch.* 1:9.

Gk. *cyphoma* a hump, humpbacked; L. *inter* between, *medius* middle.

SIZE: Length 30 mm.

COLOR: Pale reddish yellow when fresh, white when dead.

SHAPE: Elongated oval.

ORNAMENT OR SCULPTURE: Smooth. Dorsal ridge is weak or absent. Body whorl is swollen. Shell moderately thick.

APERTURE: Long, narrow, semilunar. Rather extended canals on either end. Outer lip slightly thickened. A strong plication on upper portion of twisted columella.

OPERCULUM: None.

PERIOSTRACUM: None.

REMARKS: These bisexual animals live on gorgonians with the colorful, patterned mantle wrapped around the shell. Specimens have been found in spoil banks but are rare on the Gulf beach.

HABITAT: On soft corals.

LOCALITIES: South.

OCCURRENCE: Rare.

RANGE: Greater Antilles to Brazil, Bermuda and Texas.

Superfamily NATICACEA (Swainson), Gray, 1840

Family NATICIDAE (Swainson), Gray, 1840

Subfamily POLINICINAE Gray, 1847

Genus *Polinices* Montfort, 1810

Subgenus *Neverita* Risso, 1826

SHARK'S EYE

Polinices (Neverita) duplicatus (Say, 1822), *J. Acad. Natur. Sci. Phila.* 2:247.

Gk. Polinices, son of Oedipus; L. *duplicatus* doubled.

SIZE: Diameter 1 to 2½ in.

COLOR: Porcelaneous, glossy, gray to tan often strik-

Polinices (Neverita) duplicatus

ingly marked with orange brown. Underside whitish.

SHAPE: Globose, low spire, expanded body whorl. Bay specimens are higher than Gulf ones.

ORNAMENT OR SCULPTURE: Smooth with fine growth line.

APERTURE: Large, subcircular. Outer lip thin. The funicle, a ridge of callus that begins on the outer side of the inner lip and spirals into the umbilicus, is a distinguishing feature. No siphonal canal.

OPERCULUM: Corneous, thin, and dark amber color.

PERIOSTRACUM: Thin, glossy.

REMARKS: This predator has a propodium that is thrown over the head as it plows through the sand in search of bivalves and snails. It wraps its foot about the prey and begins the slow process of drilling a neat, round hole in the shell, inserting the proboscis, and rasping out the soft parts with its radula. It will not feed unless buried and often tires before completing a hole. The female builds a col-

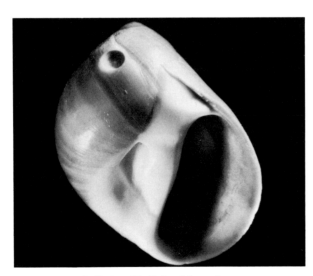

Polinices (Polinices) hepaticus

lar of mucus and sand over the margin of her aperture when she spawns. Originally described from the Texas coast.

HABITAT: Shallow waters of both bay and Gulf.

LOCALITIES: Entire.

OCCURRENCE: Common.

RANGE: Cape Cod to Florida and the Gulf states.

GEOLOGIC RANGE: Miocene to Recent.

Subgenus *Polinices* s.s.

BROWN MOON SHELL

Polinices (Polinices) hepaticus (Röding, 1798), *Mus. Boltenianum*, p. 21.

Gk. Polinices, son of Oedipus; L. *hepaticus* the liver, liver brown.

SIZE: Diameter 1 to 2 in.

COLOR: Exterior purplish brown and orange brown.

SHAPE: Globose, elongated oval.

ORNAMENT OR SCULPTURE: Polished and heavy. About 3 whorls. Body whorl dominates shell. Fine, irregular growth lines.

APERTURE: Semilunar. Columella and interior of shell white. Umbilicus white, large, and deep.

OPERCULUM: Corneous, light brown, thin, paucispiral.

PERIOSTRACUM: Thin, glossy.

REMARKS: Carnivorous and predatory. Bores holes in bivalves with radula. Buries self in sand. Deposits eggs in sand collars it builds. Common in the shallow waters of the Caribbean.

HABITAT: Shallow, sandy bottoms.

LOCALITIES: Port Aransas, south.

OCCURRENCE: Rare.

RANGE: Southeastern Florida and the West Indies and Texas.

Subfamily NATICINAE Swainson, 1840

Genus *Natica* Scopoli, 1777

Subgenus *Naticarius* Dumeril, 1805

COLORFUL ATLANTIC NATICA

Natica (Naticarius) canrena (Linné, 1758), *Syst. Natur.*, 10th ed., p. 776.

L. *natica* buttock; *canrena* from natives of Malay Archipelago, meaning unknown.

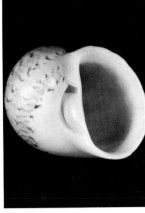

Natica (Naticarius) canrena

SIZE: Diameter 1 to 2 in.

COLOR: White or cream with variable brown markings that may be axial zigzag marks superimposed over spiral bands alternating tan and white.

SHAPE: Subglobular.

ORNAMENT OR SCULPTURE: Smooth, spire depressed, body whorl expanded. Faint waves near sutures.

APERTURE: Large, semilunar, thin outer lip. Callus large, white, entering deep umbilicus. Brownish on interior.

OPERCULUM: Hard, calcareous, 10 spiral grooves on exterior. White outside, brownish inside. Columellar margin finely serrate.

PERIOSTRACUM: Glossy, thin.

REMARKS: Lives offshore, seldom reaches the beaches. Is a predator on bivalves.

HABITAT: Sandy bottoms beyond low tide.

LOCALITIES: Entire.

OCCURRENCE: Rare.

RANGE: Southeastern United States to the West Indies and Texas.

GEOLOGIC RANGE: Miocene to Recent.

Genus *Tectonatica* Sacco, 1890

MINIATURE NATICA

Tectonatica pusilla (Say, 1822), *J. Acad. Natur. Sci. Phila.* 2:257.

Gk. *tecto* carpenter, builder; L. *natica* buttock, *pusilla* very small.

SIZE: Diameter 6 to 8 mm.

COLOR: White to fawn brown, faint reddish brown markings.

SHAPE: Subglobular, spire depressed, body whorl expanded.

ORNAMENT OR SCULPTURE: Smooth with fine growth lines, porcelaneous.

APERTURE: Large, semilunar, outer lip thin. Columella oblique. Callus is strong, practically covering umbilicus, but often has small opening next to umbilical callus.

OPERCULUM: Calcareous, smooth.

PERIOSTRACUM: Thin, glossy.

REMARKS: This tiny *Natica* can be confused with juvenile *Polinices duplicatus*, but the latter has an open umbilicus.

HABITAT: Shallow inlet areas.

LOCALITIES: Entire.

OCCURRENCE: Fairly common.

RANGE: Eastern United States, Gulf states and the West Indies.

GEOLOGIC RANGE: Pliocene to Recent.

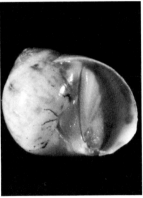

Tectonatica pusilla

Subfamily SININAE Woodring, 1928

Genus *Sinum* Röding, 1798

MACULATED BABY'S EAR

Sinum maculatum (Say, 1831), *Amer. Conch.*, p. 176.

L. *sinus* a bend or a curve, *macula* spotted.

SIZE: Diameter 1 to 2 in.

COLOR: All brown or blotched with brown.

SHAPE: Auriform, slightly elevated.

ORNAMENT OR SCULPTURE: About 3 whorls. Sutures only slightly impressed. Very weak spiral growth lines on top of whorls.

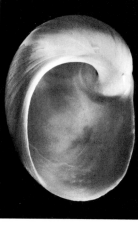

Sinum maculatum

APERTURE: Large, rounded. Outer lip sharp. Columella curved.
OPERCULUM: Corneous.
PERIOSTRACUM: Thin, yellowish brown.
REMARKS: Similar to *S. perspectivum* but shell more elevated and heavier.
HABITAT: Offshore.
LOCALITIES: Extreme south.
OCCURRENCE: Rare.
RANGE: Southeastern United States to the West Indies.

COMMON BABY'S EAR
Sinum perspectivum (Say, 1831), *Amer. Conch.*, p. 175.
L. *sinus* a curve or fold, *perspectivus* to look through, with perspective.
SIZE: Diameter 1 to 2 in.
COLOR: White.

Sinum perspectivum

SHAPE: Auriform, greatly flattened. Apex on same plane as body whorl.
MARKINGS OR SCULPTURE: About 3 whorls. Sutures slightly impressed. Many fine spiral growth lines on top of whorls.
APERTURE: Large, rounded. Outer lip sharp. Columella curved.
OPERCULUM: Corneous, minute.
PERIOSTRACUM: Thin, yellowish brown.
REMARKS: Lives on sandy bottom in shallow water. Carnivorous. Animal almost completely envelops the shell. Exudes a surprising quantity of clear mucous when touched. At low tides raccoon tracks have been found leading to and from freshly cleaned shells.
HABITAT: Along outer beaches and inlet areas in sand.
LOCALITIES: Entire.
OCCURRENCE: Common.
RANGE: Southeastern United States, Gulf states and the West Indies.
GEOLOGIC RANGE: Miocene to Recent.

Superfamily TONNACEA Suter, 1913

Family CASSIDIDAE Latreille, 1825

Genus *Phalium* Link, 1807

Subgenus *Semicassis* Mörch, 1852

SCOTCH BONNET
Phalium (Semicassis) granulatum (Born, 1780), *Test. Mus. Caes. Vind.*, p. 248.
Gk. *phalias* with white patches; L. *granulatum* granulated.
SIZE: Length 1 to 4 in.
COLOR: Background white or cream with spiral bands of regularly spaced yellowish brown squares.
SHAPE: Oval with slightly extended spire.
ORNAMENT OR SCULPTURE: Body whorl about three-fourths shell length. Some specimens have small nodules on shoulder edge. Deeply grooved spirally, the raised spiral cords are slightly convex and wider than the spaces between them. A transverse sculpture of fine lines gives an over-all reticulated pattern. Occasionally a specimen will have varices on early whorls.
APERTURE: Semilunar, length of body whorl. Outer lip thickened and reflexed with regular, small teeth

Phalium (Semicassis) granulatum

on both edges. Interior fawn colored. Parietal wall glazed, smooth with lower area pustulose. Short siphonal canal upturned to the left.

OPERCULUM: Corneous, small, thin, semilunar, light brown.

PERIOSTRACUM: Thin.

REMARKS: Will live in the passes after a storm has reopened them. The predatory animal is cream colored with close dark spots of color and feeds on sand dollars and sea urchins. Eggs are laid in a tower of horny capsules.

HABITAT: Just offshore in warm seas.

LOCALITIES: Entire, more to south.

OCCURRENCE: Fairly common.

RANGE: North Carolina to the Gulf states and West Indies.

GEOLOGIC RANGE: Pleistocene to Recent.

Genus *Cypraecassis* Stutchbury, 1837

RETICULATED COWRIE HELMET

Cypraecassis testiculus (Linné, 1758), *Syst. Natur.,* 10th ed., p. 736.

L. Cypris, Venus, *cassis* helmet, *testiculus* testicle, obovate shape.

SIZE: Length 1 to 3 in.

COLOR: Pale orangish brown with white and purplish brown blotches.

SHAPE: Narrow oval with low spire.

ORNAMENT OR SCULPTURE: Seven to 8 rather solid whorls. Body whorl almost entire length. Early

whorls spirally grooved, body whorl reticulated. Suture slightly indented.

APERTURE: Long, narrow. Outer lip thickened and reflexed, with teeth on inner margin. Parietal area heavily glazed and irregularly plicate. Parietal wall and outer lip creamy with two or three bright orange streaks.

OPERCULUM: None in adults.

PERIOSTRACUM: None visible.

REMARKS: Carnivorous predator on bivalves. Female lays eggs under rocks or broken shell in greenish brown clusters of 100 or so capsules. Never found in large colonies. Animal is light brownish orange. Usually only fragments found.

HABITAT: Reef inhabitant, offshore.

LOCALITIES: Extreme south.

OCCURRENCE: Rare.

RANGE: North Carolina to Texas and to Brazil.

Cypraecassis testiculus

Family CYMATIIDAE Iredale, 1913

Genus *Cymatium* Röding, 1798

Subgenus *Gutturnium* Mörch, 1852

KNOBBED TRITON

Cymatium (Gutturnium) muricinum (Röding, 1798), *Mus. Boltenianum,* p. 133.

Cymatium (Gutturnium) muricinum

Gk. *kymatian* dim. of *kyma* wave; L. *muricatus* pointed, murex shape.

SIZE: Length 1 to 3 in.

COLOR: Gray to brown with reddish brown colored spiral bands on the last whorl. White underside.

SHAPE: Conical with extended siphonal canal.

ORNAMENT OR SCULPTURE: Five to 7 postembryonic convex whorls. Sutures irregular and slightly indented. Spiral sculpture consists of numerous nodulose cords of unequal strength with finer threadlike striations. Transverse sculpture of 7 to 8 knobbed varices with 2 to 3 nodulose ridges in between giving the shell a rough appearance.

APERTURE: Oval with greatly thickened outer lip. Parietal area forms a large shield. Both areas glazed white. Outer lip has 7 strong teeth. Convex columella has 4 to 5 plicae near base. Siphonal canal variable in length, upturned at an angle. Interior of aperture deep reddish brown.

OPERCULUM: Corneous, unguiculate, nucleus apical, and numerous concentric growth rings.

PERIOSTRACUM: Light brown, deciduous, produced in rows of thin low blades.

REMARKS: Little is known of the feeding habits of members of this family but most are considered predatory on other mollusks and starfish. Usually found as dead shells occupied by hermit crabs.

HABITAT: Intertidal reefs in warm seas.

LOCALITIES: Extreme south.

OCCURRENCE: Rare.

RANGE: Southeastern Florida, Bermuda, West Indies and Texas.

Subgenus *Cymatriton* Clench & Turner, 1957

GOLD-MOUTHED TRITON

Cymatium (Cymatriton) nicobaricum Röding, 1798, *Mus. Boltenianum*, p. 126.

Gk. *kymatian* dim. of *kyma* wave; from Nicobar Islands in Bay of Bengal.

SIZE: Length 1½ to 2 in.

COLOR: White or gray, mottled with reddish brown. Reddish brown on spiral threads.

SHAPE: Conic ovate with siphonal canal; spire extended at 45-degree angle.

ORNAMENT OR SCULPTURE: About 7 whorls, strongly convex. Six strong spiral cords. The noduled cords are interspaced with fine threadlike cords. Five to 8 transverse varices with 3 to 5 knobs between each pair of varices. No umbilicus.

Cymatium (Cymatriton) nicobaricum

APERTURE: Obliquely oval. Bright orange with 7 single, large white teeth on the inside of outer lip. Outer lip thickened and crenulated. Teeth extend into aperture and may be single or divided. Parietal lip narrow with numerous, low lamellae. Columella curved convexly. Siphonal canal not long and usually upturned.

OPERCULUM: Broadly oval. Subcentral nucleus with concentric growth lines.

PERIOSTRACUM: Reddish brown, thin, deciduous.

REMARKS: Predatory on other mollusks and starfish. Occurs in both Indo-Pacific and western Atlantic regions. Formerly known as *C. chlorostomum* (Lamarck).

HABITAT: Offshore.

LOCALITIES: Port Aransas, south.

OCCURRENCE: Rare.

RANGE: Southeastern Florida, Bermuda, West Indies, Mexico, Texas and Indo-Pacific.

Cymatium (Septa) pileare martinianum

Subgenus *Septa* Perry, 1810

ATLANTIC HAIRY TRITON

Cymatium (Septa) pileare martinianum (d'Orbigny, 1845), *Syst. Nat.*, 10th ed., p. 749.

Gk. *kymatian* dim. of *kyma* wave; L. *pilus* hair; dedicated to Frederick W. Martini (1729–1778), German.

SIZE: Length up to 5½ in.

COLOR: Grayish brown to golden brown, banded with alternating light and dark bands.

SHAPE: Elongated conical with extended siphonal canal.

ORNAMENT OR SCULPTURE: Seven to 8 whorls, shoulder of body whorl about midway. Suture slightly indented. Spiral sculpture consists of numerous, unequal, nodulose cords. Transverse sculpture of fine lines and 3 to 5 strongly knobbed varices.

APERTURE: Elliptical, outer lip thickened into varix. Outer lip is reddish brown with 12 to 14 whitish plicae, generally paired. Parietal area is dark chocolate brown, with numerous fine, irregular, white lamellae. Columella convex. Siphonal canal short and upturned.

OPERCULUM: Corneous, unguiculate, numerous concentric growth ridges. Nucleus is marginal and brown in color.

PERIOSTRACUM: Golden brown, rough, and hairy.

REMARKS: A long veliger stage causes wide distribution. The embryonic shell is very different from the adult. Predatory and bisexual.

HABITAT: Deep water offshore on reefs.

LOCALITIES: Port Aransas, south.

OCCURRENCE: Rare.

RANGE: Florida to Tortugas, Texas, West Indies, Veracruz, Mexico, to Brazil; Indo-Pacific.

Subgenus *Linatella* Gray, 1857

POULSEN'S TRITON

Cymatium (Linatella) poulsenii (Mörch, 1877), *Mal. Blatt.*, p. 24.

Gk. *kymatian* dim. of *kyma* wave; dedicated to C. M. Poulsen, nineteenth-century Danish collector.

SIZE: Length 2 to 3 in.

COLOR: Light brown to straw yellow, occasionally banded with brown.

SHAPE: Globose conical with extended siphonal canal.

ORNAMENT OR SCULPTURE: Four postembryonic whorls, convex, body whorl slightly shouldered. Suture slightly indented. Spiral sculpture consists of

Cymatium (Linatella) poulsenii

18 to 20 flattened cords with fine thread in between. Shoulder cord might be slightly beaded. Transverse sculpture of fine growth lines; some have a thin, bladelike varix. Anomphalous.

APERTURE: Subelliptical. Outer lip crenulated and slightly expanded. Parietal area glazed. Columella arched inward and continuing as margin of siphonal canal. Canal variable, moderately long and upturned.

OPERCULUM: Corneous, thin, subcircular, with concentric growth lines around an eccentric nucleus.

PERIOSTRACUM: Thin, consists of numerous axial blades from which extend hairlike processes. Deciduous.

REMARKS: This shell is said to be rare, but at times it can be picked up by bushelfuls along the Mexican coast just south of the Rio Grande. The specimens there have thinner shells than those occasionally found on the Texas coast.

HABITAT: Offshore to 209 fathoms.

LOCALITIES: Central, south.

OCCURRENCE: Rare.

RANGE: Florida south through the West Indies to Venezuela, Mexico and Texas.

Subgenus *Monoplex* Perry, 1811

VON SALIS' TRITON

Cymatium (Monoplex) parthenopeum (von Salis, 1793), *Reisen in versch. Prov. Königreich Neapel* 1:370.

Gk. *kymation* dim. of *kyma* wave; L. Parthenop, old name of Naples.

SIZE: Length up to 5¾ in.

COLOR: Usually light brownish yellow; may have spiral bands of slightly darker brown becoming darker on the varices.

SHAPE: Conical with extended siphonal canal.

ORNAMENT OR SCULPTURE: Seven to 8 postembryonic whorls, convex and shouldered. Spire moderately extended. Suture slightly indented. Spiral sculpture consists of 5 or 6 broad, low, and often nodulose cords, with many finer threads in the interspaces and on the cords. Transverse sculpture of fine growth lines with 2 low varices in adults.

APERTURE: Subelliptical, outer lip bordered with paired teeth opposite the grooves between the external spiral cords. Parietal wall is a dark reddish brown with numerous, irregular, white plications. The anal canal is bordered with a ridge on the

Cymatium (Monoplex) parthenopeum

parietal wall. Short siphon, upturned, with columella extending into it as the parietal margin.

OPERCULUM: Corneous, unguiculate, nucleus terminal. Sculptured with concentric growth lines.

PERIOSTRACUM: Thin, brown, and produced in numerous fringed axial blades. Deciduous.

REMARKS: Syn. *C. costatus* (Born). This is sometimes found near Port Isabel but is more often found on the beach south of the Rio Grande.

HABITAT: Below low water to 35 fathoms.

LOCALITIES: Extreme south.

OCCURRENCE: Rare.

RANGE: Bermuda, Florida, Mexico, Texas, West Indies and south to Brazil.

Genus *Distorsio* Röding, 1798

Subgenus *Rhysena* Clench & Turner, 1957

ATLANTIC DISTORSIO
Distorsio (Rhysena) clathrata (Lamarck, 1816), *Tableav Encycl. Méth.* 3:4.
L. *distortus* deformed, *clathratus* barred, latticed.

SIZE: Length 1 to 3 in.

COLOR: Grayish white.

Distorsio (Rhysena) clathrata

SHAPE: Conical with siphonal canal.

ORNAMENT OR SCULPTURE: Ten irregular, convex whorls. Spire extended. Suture slightly impressed, irregular. Spiral sculpture consists of numerous low cords interspaced with fine spiral threads. Transverse sculpture of numerous cords that cross the spiral cords, producing a reticulated pattern with small knobs at the point of crossing. Seven to 9 varices.

APERTURE: Auriculate with thickened outer lip. Outer lip has 10 denticles; the third below the anal canal is the largest and is opposite the deep parietal embayment. Inner lip has numerous plicae. Two large parietal plicae border the posterior canal. The spiral cords are thickened in the parietal embayment, continuing as plicae on the columella, which is nearly straight and upturned. The wide, thinly glazed parietal shield is bordered by a thin varix ridge.

OPERCULUM: Small, corneous, unguiculate, submarginal nucleus with numerous concentric growth lines.

PERIOSTRACUM: Thin, yellowish brown, reticulate, with numerous fine hairlike processes over surface and coarse hairlike processes on the knobs.

REMARKS: The very distorted aperture of this peculiar shell is the basis for its name. Fragments are not too uncommon on Mustang and Padre islands.

HABITAT: Just below low water to 30 fathoms in warm seas.

LOCALITIES: Entire.

OCCURRENCE: Uncommon.

RANGE: Southeastern United States, Gulf states and the Caribbean.

GEOLOGIC RANGE: Pleistocene to Recent.

Family TONNIDAE Suter, 1913

Genus *Tonna* Brunnich, 1772

GIANT TUN SHELL
Tonna galea (Linné, 1758), *Syst. Natur.*, 10th ed., p. 734.
L. *tonna* a cask, *galea* helmet.

SIZE: Length up to 6¾ in.

COLOR: Creamy white to light coffee brown, generally uniform.

SHAPE: Globose, spire slightly extended. Thin.

ORNAMENT OR SCULPTURE: Seven to 7½ very con-

Tonna galea

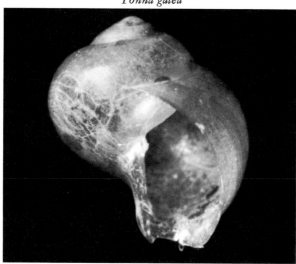

Juvenile

mella short and twisted with a ridge along its outer edge that ends at the siphonal canal.

OPERCULUM: Only in the juvenile stage.

PERIOSTRACUM: Thin, golden brown, somewhat deciduous.

REMARKS: The free-swimming, pelagic young may be found in the spring in beach drift. These embryonic shells are smooth, golden brown in color and somewhat flexible. They have 3 or 4 whorls and the aperture is closed with a tightly fitting operculum. In winter months on the upper Mexican Gulf beaches, the drift will be lined with young tuns about the size of an egg. The animal has a large foot and long proboscis, is yellowish in color, and is heavily mottled with black.

HABITAT: Sandy areas near shore.

LOCALITIES: Entire.

OCCURRENCE: Rarely taken alive in Aransas ship channel. Rare.

RANGE: North Carolina to Florida, Gulf states and West Indies. Circumtropical.

Order NEOGASTROPODA Thiele, 1925=CAENO-GASTROPODA Cox, 1959

Suborder STENOGLOSSA Troschel, 1848

Superfamily MURICACEA Rafinesque, 1815

Family MURICIDAE Rafinesque, 1815

Subfamily THAIDIDAE Suter, 1913

Genus *Thais* Röding, 1798

Subgenus *Stramonita* Schumacher, 1817

FLORIDA ROCK SHELL

Thais (*Stramonita*) *haemastoma floridana* (Conrad, 1837), *J. Acad. Natur. Sci. Phila.* 7:265.

L. Thais, wife of Ptolemaerus I of Egypt, *haema* blood, *stoma* mouth; from Florida.

SIZE: Length 2 or 3 in.

COLOR: Light gray to yellowish, mottled with a darker color in an axial pattern.

SHAPE: Conical.

ORNAMENT OR SCULPTURE: Six to 7 convex whorls. Sutures in this rather heavy shell are fine and occasionally indented. Sculpture is quite variable.

vex whorls. Body whorl dominates shell. Suture deep and channeled. Spiral sculpture consists of 19 to 21 rather broad, flattened ridges. There is usually a finer ridge between 2 of the larger ones on the upper half of the whorl. Fine axial growth lines. Umbilicate.

APERTURE: Subovate, large. Outer lip is thin and crenulate until maturity. At maturity it becomes reflexed and develops a thickened ridge well below the lip margin. The parietal area is glazed. Colu-

Thais (Stramonita) haemastoma floridana

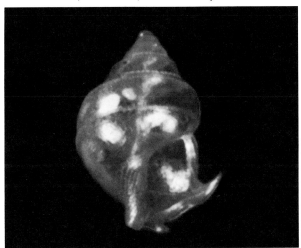

Juvenile

is short with a ridge along parietal wall. Siphonal canal short and oblique. Umbilicus closed.

OPERCULUM: Corneous, unguiculate. Underside thickened and shiny along outer edge.

PERIOSTRACUM: Deciduous.

REMARKS: Carnivorous on bivalves. Sexes separate, lays eggs in purplish capsules clustered together on rocks, cans, bottles, etc. The first hatched feed on the yolk of the unhatched eggs and emerge as larvae in the veliger stage.

HABITAT: Intertidal on rocks.

LOCALITIES: Entire.

OCCURRENCE: Common.

RANGE: North Carolina to Florida, Texas, West Indies, and Central America.

GEOLOGIC RANGE: Pleistocene to Recent.

HAY'S ROCK SHELL

Thais (Stramonita) haemastoma haysae Clench, 1927, *Nautilus* 41:6.

L. Thais, wife of Ptolemaerus I of Egypt, *haema* blood, *stoma* mouth; dedicated to Mrs. M. L. Hayes.

SIZE: Length up to 4½ in.

COLOR: Grayish with irregular mottling of a darker color in either axial or spiral pattern.

Thais (Stramonita) haemastoma haysae

Spiral sculpture may consist of incised lines and 2 rows of small nodules. Transverse sculpture of fine growth lines. Shoulders may be angled or not.

APERTURE: Subovate, interior salmon pink. Outer lip has dark brown between the denticulations that run into the interior of aperture. Parietal lip glazed, smooth, and thickened by inductura. Columella straight, may have faint plicae near base. Anal canal

SHAPE: Conical.

ORNAMENT OR SCULPTURE: Seven to 8 convex, solid whorls. Sutures usually indented; body whorl has angled shoulder. Spiral sculpture consists of numerous, coarse, incised lines with 2 rows of large nodules on the whorl shoulder. Transverse sculpture of fine growth lines.

APERTURE: Subovate. Outer lip thickened with crenulations that run into the aperture. Interior a light brownish to pinkish orange. Parietal lip glazed and thickened by inductura. At its upper edge is a ridge that runs into aperture. There may be weak plicae on the base of the straight columella. Siphonal canal short and oblique. Umbilicus generally closed.

OPERCULUM: Corneous, unguiculate. Underside thickened and shiny along outer edge.

PERIOSTRACUM: Deciduous.

REMARKS: Clench separates this from *T. h. floridana* (Conrad), but the characteristics are not always easy to define. Some authors are reluctant to separate the two. This mollusk is a serious oyster pest. Originally described from Texas.

HABITAT: Shallow water on rocks or oyster reefs.

LOCALITIES: Entire.

OCCURRENCE: Fairly common.

RANGE: Gulf of Mexico from Florida west to Texas.

GEOLOGIC RANGE: Pleistocene to Recent.

Subfamily MURICINAE Rafinesque, 1815

Genus *Murex* Linné, 1758

Subgenus *Phyllonotus* Swainson, 1833

GIANT EASTERN MUREX

Murex (Phyllonotus) fulvescens ·(Sowerby, 1834), *Conch. Illus. Cat.*, p. 7.

L. *murex* the purple shellfish, *fulvus* tawny.

SIZE: Length 5 to 7 in.

COLOR: Milky white to dirty gray with reddish brown blotches and spiral threads.

SHAPE: Conical with extended siphonal canal.

ORNAMENT OR SCULPTURE: Six to 7 convex, heavy whorls. Suture distinct and irregular. Spire short. Spiral sculpture consists of strong, brown cords that connect the corresponding spines of each varix. Between them are numerous raised threads. Transverse sculpture consists of 6 to 10 highly spinous varices. The largest spines are on shoulder of whorls; all are

Murex (Phyllonotus) fulvescens

erect, opened toward the outer lip, and irregular in height and size. There are numerous fine growth lines.

APERTURE: Oval to subcircular. Outer lip crenulated and thickened into a very spinose varix. Parietal lip glazed with a low ridge at upper part. Siphonal canal fairly short and broad. Previous canals form a series of flutings terminating in a false umbilicus. Interior porcelaneous white.

OPERCULUM: Corneous, unguiculate, thick with numerous concentric growth lines.

PERIOSTRACUM: None visible.

REMARKS: The female deposits eggs in rubbery capsules attached to some substrate. The larvae have a nonpelagic development. Aquarium observations show that the animal will consume large amounts of food before burying itself for three or four weeks. When it emerges there will be an addition to the shell with a thin edge. It feeds again, reburies itself, and builds its spiny outer lip. This occurs about three times per year. In between additions to the shell it eats little. When feeding it may grasp the victim with its foot and use muscular pressure to pull the valves open or it will bore a hole with its radula.

HABITAT: On jetties and just offshore or in inlet areas.
LOCALITIES: Entire.
OCCURRENCE: Fairly common.
RANGE: North Carolina, Florida to Texas and northern Mexico.
GEOLOGIC RANGE: Pleistocene to Recent.

APPLE MUREX
Murex (Phyllonotus) pomum (Gmelin, 1790), *Syst. Natur.*, 13th ed., p. 3527.
L. *murex* the purple shellfish, *pomum* apple.
SIZE: Length 2 to 4½ in.
COLOR: Dark brown to yellowish tan with irregular dark brown spiral bands, which are often reduced to spots.
SHAPE: Conical with slightly extended siphonal canal.
ORNAMENT OR SCULPTURE: Seven to 9 solid, convex whorls. Suture not always distinct. Spire extended. Spiral sculpture consists of a series of strong cords. These scaly cords form nodules on the ridges. Between them are several finer scaly threads. Transverse sculpture consists of 3 prominent, equidistant varices on each whorl. Each varix has a row of low, open spines and a fluted edge on the forward margin. There are several ridges between the varices.

Murex (Phyllonotus) pomum

APERTURE: Oval to subcircular, large. Interior polished and colored pink or ivory, yellow, and orange. Outer lip thin, crenulate with varix bordering outer edge. Spotted brown to correspond with spiral bands. Parietal area glazed, adheres to body whorls, except for erect edge. Siphon short and slightly recurved. There is a dark brown spot on the upper end of the parietal wall.
OPERCULUM: Corneous, unguiculate, heavy with strong growth lines.
PERIOSTRACUM: None visible.
REMARKS: This carnivore bores holes in the shells of its prey. The female lays eggs in leathery capsules that are attached to the bottom. Sometimes found in inlets with hermit crabs in dead shell.
HABITAT: Gravelly bottom, 3 to 7 fathoms.
LOCALITIES: Entire.
OCCURRENCE: Uncommon.
RANGE: North Carolina to Florida, West Indies, and Texas.
GEOLOGIC RANGE: Miocene to Recent.

Superfamily BUCCINACEA Rafinesque, 1815

Family COLUMBELLIDAE Swainson, 1840 = PYRENIDAE Suter, 1913

Genus *Anachis* H. & A. Adams, 1853

HALF-FOLDED DOVE SHELL
Anachis avara semiplicata (Stearns, 1873), *Proc. Phila. Acad. Natur. Sci.* 25:344.
L. *anachites* name given by Pliny to the diamond, *avarus* greedy, *semi* half, *plicatus* folded.
SIZE: Length 16 mm.
COLOR: Glossy, whitish, with reddish brown, irregular markings.
SHAPE: Elongate conical, fusiform.
ORNAMENT OR SCULPTURE: Eight slightly convex whorls, sutures distinct. Spire extended. Spiral sculpture consists of faint striations that become stronger on the base. Axial sculpture consists of 12 rounded ribs on upper part of body whorl with traces on penultimate whorl.
APERTURE: Narrow, length of body whorl. Outer lip thickened, thin at extreme edge, weak denticulations within. Siphonal canal short, columella convex. Parietal lip polished, white.
OPERCULUM: Corneous, oval, small, brown.

Anachis avara semiplicata

PERIOSTRACUM: Thin, brown.

REMARKS: This carnivore is not timid or nocturnal. On the jetty at Port Mansfield numerous shells were seen hanging by a fine thread of mucus attached to the rock. The animal will also spin this elastic support and hang from sea weed in water. Carnivorous. Often found in the bays clustered in algae on dead shell. Lays eggs in single, gelatinous capsule attached to seaweed. *A. transliterata* Ravenel, 1861 (not figured), is commonly found in the central and eastern range. It has been considered as a subspecies of *A. avara* Say, 1822, and reported for Texas as being *A. avara*, but Dall (1889, p. 187) considers it to be more typical of the species than *A. avara*.

HABITAT: Bay margins, grass flats, inlets.

LOCALITIES: Entire.

OCCURRENCE: Common.

RANGE: Florida, Texas.

Subgenus *Costoanachis* Sacco, 1890

FAT DOVE SHELL

Anachis (*Costoanachis*) cf. *A. obesa* (C. B. Adams, 1845), *Proc. Boston Soc. Natur. Hist.* 2:2.

L. *anachites* name given by Pliny to the diamond, *obesus* fat.

SIZE: Length 4 to 6 mm.

COLOR: Variable. Some are whitish with dark brown bands, others are solid reddish brown.

SHAPE: Ovate conical, short fusiform.

ORNAMENT OR SCULPTURE: Five convex whorls, stout and rotund. Spiral sculpture consists of strong spiral

cords that do not cross the transverse ribs, giving a reticulated pattern.

APERTURE: Oval, oblique; outer lip is thickened in adults with denticulations on the inner edge. Base of columella is denticulate.

OPERCULUM: Corneous, oval, small.

PERIOSTRACUM: Thin, light brown.

REMARKS: This shell is common in the beach drift where faded, cream-colored dead specimens may confuse the collector. Carnivorous. This may be *A. ostreicola* Melvill, 1881. A dissertation on the Columbellidae, to be published by George Radwin of the Smithsonian Institution, will aid in identifying the several *anachis* found in Texas.

HABITAT: Oyster reefs, grass flats, inlets.

LOCALITIES: Entire.

OCCURRENCE: Common.

RANGE: Virginia to Florida, the Gulf states and the West Indies.

GEOLOGIC RANGE: Pliocene to Recent.

Anachis (*Costoanachis*) cf. *A. obesa*

Genus *Mitrella* Risso, 1826

Subgenus *Astyris* H. & A. Adams, 1853

LUNAR DOVE SHELL

Mitrella (*Astyris*) *lunata* (Say, 1862), *J. Acad. Natur. Sci. Phila.* 5:213.

L. dim. of *mitra* a miter, *lunatus* crescent shaped.

SIZE: Length 5 mm.

COLOR: Glossy white to cream with numerous fine, zig-zag brown markings. Occasional specimens will have the brown markings arranged in definite spiral bands.

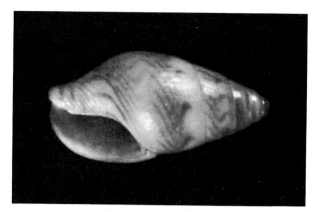

Mitrella (Astyris) lunata

SHAPE: Ovate conical.

ORNAMENT OR SCULPTURE: Smooth with about 5 flat-sided, tapering whorls. Spiral striations on base of shell.

APERTURE: Long oval, outer lip thin on edge and denticulated on interior. Columella short. Edge of siphonal canal is dark brown.

OPERCULUM: Corneous, brown.

PERIOSTRACUM: Thin, brownish.

REMARKS: At times these beautiful little shells are strikingly marked in spiral bands of white below the suture, then brown dots and a row of oblique lines. According to Olsson and Harbinson (1953) *M. lunata* should be placed in the genus *Anachis*, subgenus *Alia*. Carnivorous.

HABITAT: High-salinity shell reef just below low-tide mark, grass flats, and inlets.

LOCALITIES: Entire.

OCCURRENCE: Fairly common.

RANGE: Massachusetts to Florida, Texas and the West Indies.

GEOLOGIC RANGE: Pliocene to Recent.

Family BUCCINIDAE Rafinesque, 1815

Genus *Cantharus* Röding, 1798

Subgenus *Pollia* Sowerby, 1834

CANCELLATE CANTHARUS
Cantharus (Pollia) cancellarius (Conrad, 1846), *Proc. Acad. Natur. Sci. Phila.* 3(1):25.
Gk. *kantharos* drinking cup; L. *cancellare* a lattice.

SIZE: Length 18 to 28 mm.

COLOR: Yellowish brown.

SHAPE: Ovate conical.

ORNAMENT OR SCULPTURE: Five to 6 convex, heavy whorls. Spire conical. Spiral sculpture consists of sharp cords that form beads and cross the narrow transverse ribs, making a reticulate pattern.

APERTURE: Long oval, glossy white; outer lip thin and crenulate with fine denticulations on inner edge. Posterior canal weak or absent. Siphonal canal straight, short, and slightly upturned. One plica at base of columella.

OPERCULUM: Corneous, brown, concentric with subcentral nucleus.

PERIOSTRACUM: Yellowish brown, moderately thin, arranged in spiral rows.

REMARKS: Lives on the jetties but has not been found since Hurricane Beulah in 1967; will probably re-establish itself.

HABITAT: Shallow water in rocky places, inlet areas.

LOCALITIES: Entire.

OCCURRENCE: Common seasonally.

RANGE: Western coast of Florida to Texas and Yucatán.

GEOLOGIC RANGE: Pleistocene to Recent with related forms to Eocene.

Cantharus (Pollia) cancellarius

TINTED CANTHARUS
Cantharus (Pollia) tinctus (Conrad, 1846), *Proc. Acad. Natur. Sci. Phila.* 3(1):25.
Gk. *kantharos* drinking cup; L. *tinctus* painted.

SIZE: Length 18 to 28 mm.

Cantharus (Pollia) tinctus

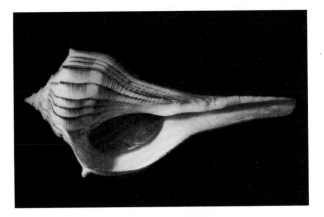

Busycon contrarium

COLOR: Variable, blue gray, yellow, chocolate, and milk white; darkest at apex.

SHAPE: Ovate conical.

ORNAMENT OR SCULPTURE: Five to 6 convex, heavy whorls. Spire conical. Spiral sculpture consists of cords with finer threads in between, crossing over the weak axial ribs. Weak nodules on whorl shoulder.

APERTURE: Oval, outer lip thickened and denticulate on the inner edge. Parietal lip glazed. A plication on the upper part borders the small posterior or abaxial canal. Siphonal canal almost straight and slightly upturned.

OPERCULUM: Corneous, pyriform.

PERIOSTRACUM: Thin, brown.

REMARKS: Can be found alive on the jetties, also dead in the rolls of sea whip coral. Less common than *C. cancellaria.*

HABITAT: Shallow water, rocks, and seaweed close to shore.

LOCALITIES: Port Aransas, south.

OCCURRENCE: Fairly common.

RANGE: North Carolina to both sides of Florida, Texas and the West Indies.

Family MELONGENIDAE Gill, 1871

Genus *Busycon* Röding, 1798

LIGHTNING WHELK
Busycon contrarium (Conrad, 1867), *Amer. J. Conch.* 3:182–185.

Gk. *busycon* a large, coarse fig; L. *contrarius* opposite, reverse.

SIZE: Length 4 to 16 in.

COLOR: Pale fawn to light yellowish gray with long axial, wavy brown streaks. Large adults usually lose color.

SHAPE: Pyriform with sinistral aperture.

ORNAMENT OR SCULPTURE: Body whorl large, spire one-fifth height of shell. Spire turreted, sutures slightly below shoulder. Fine spiral threads. Growth lines colored and corresponding with spines that circle shoulder. More spiny, more turreted, and with a higher spire than *B. sinistrum* Hollister.

APERTURE: Pyriform; outer lip thin and edged in purplish brown. Interior pale yellow to light orange. Siphonal canal is long, somewhat twisted and recurved.

OPERCULUM: Corneous, brown, concentric with subcentral nucleus.

PERIOSTRACUM: Thin, brownish.

REMARKS: This carnivorous animal can be caught in the bays with crab lines. It buries itself in the sand with the siphonal canal protruding. It feeds on mollusks, opening bivalves by chipping the valve edges with its own shell until it can insert the proboscis. The female constructs long strings of horny, disk-shaped capsules with her black colored foot, attaching them to the substrate. Hollister (1958) calls this shell *B. sinistrum* and the type found in Texas *B. pulleyi.* He suggests several new subgenera and species that await general adoption.

HABITAT: Intertidal, offshore, and in bays.

LOCALITIES: Entire.

OCCURRENCE: Common.

RANGE: South Carolina to Florida and the Gulf states.

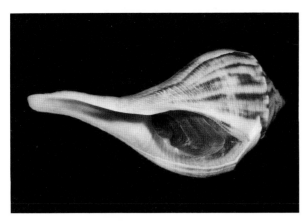

Busycon spiratum plagosus

PEAR WHELK

Busycon spiratum plagosus (Conrad, 1863), *Proc. Acad. Natur. Sci. Phila.* 14:583.

Gk. *busycon* a large coarse fig; L. *spira* coiled, *plagosus* fond of punishing.

SIZE: Length 3 to 4 in.

COLOR: Creamy with irregular brown axial lines.

SHAPE: Pyriform.

ORNAMENT OR SCULPTURE: Spire whorls are turreted, producing a step at each suture. Suture is boxlike. Sharp carina at the shoulder is finely beaded. Spiral sculpture consists of fine threads.

APERTURE: Pyriform, outer lip thin. Interior is strongly striate and rosy brown except near the lip, where it is white. Siphonal canal is long and nearly straight.

OPERCULUM: Corneous, brown, concentric with subcentral nucleus. There is an arched channel on the outer surface running from bottom to top, near the outer margin.

PERIOSTRACUM: Thin, brownish.

REMARKS: At times these whelks can be found living on sand bars in inlet areas where they will "pop up" after the tide has been out for a while. The female constructs the egg capsules in the same manner as *B. contrarium* but they are smaller and have sharply crenulated edges. Bivalves are the main source of food. This shell has been commonly referred to as *B. spiratum*, from which it differs by having a rectangular sutural channel instead of the V-shaped channel of *B. spiratum*. Hollister (1958) refers to this shell as *Busycotypus plagosus* (Conrad). The shape of the spire appears to be the characteristic used to distinguish the subgenus. Most authors lump the forms of this genus from the western Gulf of Mexico with those of the eastern Gulf, the *B. spiratus* (=*B. pyrum*) group (*spirata, spiratum* of authors).

HABITAT: Offshore in sandy bottoms to 4 fathoms.

LOCALITIES: Entire.

OCCURRENCE: Fairly common.

RANGE: Mobile Bay to Campeche Bay, Mexico.

Family NASSARIIDAE Iredale, 1916

Genus *Nassarius* Dumeril, 1806

Subgenus *Phrontis* H. & A. Adams, 1853

COMMON EASTERN NASSA

Nassarius (Phrontis) vibex (Say, 1822), *J. Acad. Natur. Sci. Phila.* 2:234.

L. *nassa* a basket for catching fish, *vibex* the mark of a blow.

SIZE: Length 12 mm.

COLOR: Gray brown to whitish with a few splotches of darker brown.

SHAPE: Ovate conical.

ORNAMENT OR SCULPTURE: Seven convex whorls. Body whorl dominates this short, heavy shell. Spiral sculpture consists of fine threads that cross about 12 transverse ribs. Sutures shallow. Apex acute.

APERTURE: Oval, small; outer lip with a thick varix, denticulate within. Columella arched, short. Parietal area well developed and glazed white. Siphonal canal short, slightly upturned.

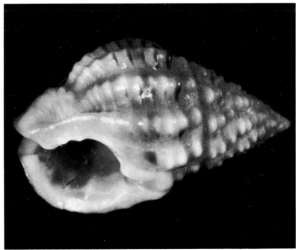

Nassarius (Phrontis) vibex

OPERCULUM: Corneous, brown, unguiculate.

PERIOSTRACUM: None visible.

REMARKS: These carnivorous scavengers have 2 long proboscises with 2 horns in front and an expandable foot with 2 tails behind. These "mud snails" are common on the flats, but move to deeper water in the winter. Eggs are laid in gelatinous capsules attached to bottom.

HABITAT: In bay and open-sound margins and inlet areas.

LOCALITIES: Entire.

OCCURRENCE: Common.

RANGE: Cape Cod to Florida, the Gulf states and West Indies.

GEOLOGIC RANGE: Pliocene to Recent.

Subgenus *Nassarius* s.s.

SHARP-KNOBBED NASSA

Nassarius (Nassarius) acutus (Say, 1822), *J. Acad. Natur. Sci. Phila.* 2:234.

L. *nassa* a basket for catching fish, *acutus* acute, sharp.

SIZE: Length 6 to 12 mm.

COLOR: Cream white to yellowish, occasionally with brown spiral thread.

SHAPE: Ovate conical.

ORNAMENT OR SCULPTURE: Seven convex whorls. Spire is pointed and longer than body whorl. Spiral sculpture consists of spiral threads that cross similar transverse ribs, giving a beaded, cancellate appearance to the shell.

APERTURE: Oval, slightly oblique. Short recurved siphonal canal.

OPERCULUM: Corneous, unguiculate, brown.

PERIOSTRACUM: None visible.

REMARKS: This scavenger is attracted by the smell of decaying flesh and by light. It feeds on debris, other mollusks, and mollusk egg capsules. The female lays eggs in gelatinous capsules that are attached to the bottom.

HABITAT: Open lagoon, inlet, and along shore.

LOCALITIES: Entire.

OCCURRENCE: Common.

RANGE: Western coast of Florida to Texas.

GEOLOGIC RANGE: Miocene to Recent.

Family FASCIOLARIIDAE Gray, 1853

Subfamily FASCIOLARIINAE Gray, 1853

Genus *Fasciolaria* Lamarck, 1799

Subgenus *Cinctura* Hollister, 1957

BANDED TULIP SHELL

Fasciolaria (Cinctura) hunteria (Perry, 1811), *Conchology* (4):50.

L. *fasciola* a band; dedicated to John Hunter, governor of New South Wales in 1811.

SIZE: Length 2 to 4 in.

COLOR: Background color cream with irregular purplish brown and orange brown mottlings. Widely spaced, rarely broken, brown spiral bands.

SHAPE: Elongate fusiform.

MARKINGS OR SCULPTURE: Seven to 9 rounded whorls. Smooth near the well-defined sutures. Spiral striations on base and fine transverse growth lines.

Nassarius (Nassarius) acutus

Fasciolaria (Cinctura) hunteria

APERTURE: Long oval, glazed white inside. Outer lip thin, brownish with numerous, raised white threads on the inner surface. Parietal area thinly glazed with white. Moderately long siphonal canal is open. The incurved columella has a strong plication toward end.

OPERCULUM: Corneous, brown, unguiculate.

PERIOSTRACUM: Thin, yellowish brown.

REMARKS: This carnivorous animal is smaller than *F. tulipa*. It can use its strong foot to jump out of the unwary collector's pocket. The eggs are placed in vase-shaped capsules attached to shell, pilings, and other structures. The male is smaller than the female. He squirts sprays of water that set up vibrations that attract the female. Hollister (1957) states that *F. hunteria* Perry ranges westward in the Gulf only to Mobile Bay and the species that occurs from there to along the Texas coast is *F. lilium* Fischer van Waldheim, 1807.

HABITAT: Inlet areas and offshore.

LOCALITIES: Entire, more to south.

OCCURRENCE: Fairly common.

RANGE: North Carolina to Florida and the Gulf states, Mexico.

GEOLOGIC RANGE: Pleistocene to Recent.

TRUE TULIP SHELL

Fasciolaria (Cinctura) tulipa (Linné, 1758), *Syst. Natur.*, 10th ed., p. 754.

L. *fasciola* a band, *tulipa* a flower.

SIZE: Length 3 to 5 in.

COLOR: Cream background with brown blotches and numerous broken spiral bands. Some specimens are reddish, orange, or mahogany brown.

SHAPE: Elongate fusiform.

ORNAMENT OR SCULPTURE: About 9 rounded whorls, suture distinct. Smooth appearing but has fine spiral striae just below suture and on base. Transverse sculpture consists of fine, irregular growth lines.

APERTURE: Long oval, flushed with orangish color. Outer lip thin, denticulate on the inner edge. Parietal area thinly glazed. Columella fairly long, curved inward with 2 oblique plicae about midway.

OPERCULUM: Corneous, brown, unguiculate.

PERIOSTRACUM: Thin, yellowish brown.

REMARKS: The carnivorous, predatory animal is a bright flame color. The female lays the fertilized eggs in leathery, vase-shaped capsules that are attached to shell, rocks, and other hard objects in clusters of 25 to 30 capsules. Rarely on the Texas coast; only old, dead shells are found.

HABITAT: Grass bottoms from littoral to 5 fathoms.

LOCALITIES: South.

OCCURRENCE: Rare.

RANGE: North Carolina to southern half of Florida, the West Indies and Texas.

Genus *Pleuroploca* P. Fischer, 1884

FLORIDA HORSE CONCH

Pleuroploca gigantea (Kiener, 1840), *Iconogr. Coq. Viv.*, p. 5.

Gk. *pleura* rib, side; L. *giganteus* gigantic.

SIZE: Length up to 2 ft.

COLOR: Dirty white to chalky salmon. Juveniles are bright orange.

SHAPE: Fusiform.

ORNAMENT OR SCULPTURE: About 8 convex whorls.

Fasciolaria (Cinctura) tulipa

Pleuroploca gigantea

Sutures distinct. Spiral sculpture consists of strong, irregularly spaced cords, with finer threads between. Transverse growth lines.

APERTURE: Oval, polished, orange colored. Outer lip thin, slightly crenulate. Columella has two plicae near base. Siphonal canal is long and upturned.

OPERCULUM: Corneous, brown, unguiculate.

PERIOSTRACUM: Heavy, dark brown. Flakes when dry.

REMARKS: The large size of this shell makes it readily identifiable. Old specimens are usually covered with calcareous bryozoa and are faded and worn. This carnivore is the largest gastropod in the Gulf of Mexico. Skin divers have taken beautiful specimens from the jetty at Port Aransas.

HABITAT: Offshore and in inlet areas.

LOCALITIES: Entire.

OCCURRENCE: Fairly common.

RANGE: North Carolina to both sides of Florida, Texas and Mexico.

GEOLOGIC RANGE: Miocene to Recent.

Superfamily VOLUTACEA Rafinesque, 1815

Family OLIVIDAE Latreille, 1825

Subfamily OLIVINAE Latreille, 1825

Genus *Oliva* Martyn, 1786

Subgenus *Ispidula* Gray, 1847

LETTERED OLIVE

Oliva (Ispidula) sayana Ravenel, 1834, *Cat. Rec. Shells*, p. 19.

L. *oliva* olive; dedicated to Thomas Say (1787–1834), collector.

SIZE: Length 2 to 2½ in.

COLOR: Polished cream-colored background with numerous brownish zigzag markings.

SHAPE: Elongated oval.

ORNAMENT OR SCULPTURE: Five to 6 whorls. Body whorl dominates shell. Spire short, acute. Sutures deep.

APERTURE: Long, narrow, purplish within. Outer lip thin. Siphonal canal is an oblique notch at the base. White columella is plicated.

OPERCULUM: None.

PERIOSTRACUM: None.

REMARKS: This carnivorous predator plows along just

Oliva (Ispidula) sayana

under the sand. Skin divers will find it in inlet areas. It is nocturnal, and the shell is covered with the propodium and lateral folds of the foot.

HABITAT: Inlets and offshore.

LOCALITIES: Entire.

OCCURRENCE: Common.

RANGE: North Carolina to Florida and the Gulf states.

GEOLOGIC RANGE: Miocene to Recent.

Genus *Olivella* Swainson, 1831

COMMON RICE OLIVE

Olivella dealbata (Reeve, 1850), *Conch. Icon.* 6:25.

L. *oliva* olive, *ella* dim. suffix, *dealbatus* whitened.

SIZE: Length 6 to 9 mm.

COLOR: Glossy white or cream. Body whorl faintly marked with brownish zigzag streaks. Color variable.

SHAPE: Elongate oval.

Olivella dealbata

ORNAMENT OR SCULPTURE: Smooth, sutures distinct and slightly canaliculated. The fasciole at the base of the shell is white and bounded with a fine raised thread.

APERTURE: Long, narrow, about three-fourths the length of the body whorl. Outer lip thin. Parietal inductura is well developed. Columella is slightly concave with 7 to 9 weak oblique plications. The siphonal notch not as pronounced as in *O. minuta*.

OPERCULUM: Corneous, brown, thin.

PERIOSTRACUM: None.

REMARKS: At times in winter months these shells can be found by the thousands in the drift on the outer beaches. They will readily float when you wash and screen the beach drift. This species is closely related to *O. floralia* Duclos and may be only a form of it. Carnivorous.

HABITAT: Inlet areas in the sand.

LOCALITIES: Entire, more to east.

OCCURRENCE: Common.

RANGE: North Carolina to both sides of Florida, West Indies and Texas.

Olivella (Niteoliva) minuta

Subgenus *Niteoliva* Olsson, 1956

MINUTE DWARF OLIVE
Olivella (Niteoliva) minuta (Link, 1807), *Mart. Conch. Cab.* 2:182.
L. *oliva* olive, *minutus* small.

SIZE: Length 6 to 12 mm.

COLOR: Variable. Polished grayish white background with purplish brown zigzag lines on the body whorl and fine brown lines along sutures.

ORNAMENT OR SCULPTURE: Very smooth. Apex acute. Sutures open and grooved but not as pro-

nounced as *O. dealbata*. A fine spiral line is above the base.

OPERCULUM: Corneous, thin, semiovate.

PERIOSTRACUM: None.

REMARKS: A beautiful little carnivore with nocturnal habits. It can be found plowing along just below the surface of inlet areas at low tides on moonlit nights. The propodium envelops the shell. This shell may be the *O. mutica* of Dall and Simpson.

HABITAT: Inlets and surf zone.

LOCALITIES: Entire, more to south.

OCCURRENCE: Common at times.

RANGE: West Indies, Caribbean, and Texas.

GEOLOGIC RANGE: Miocene to Recent.

Family CANCELLARIIDAE Gray, 1853

Genus *Cancellaria* Lamarck, 1799

COMMON NUTMEG
Cancellaria reticulata (Linné, 1767), *Syst. Natur.*, 10th ed., p. 112.
L. *cancellare* a lattice, *reticulatus* reticulate.

SIZE: Length 1 to 1¾ in.

COLOR: Cream background with mottlings and irregular bands of reddish brown.

SHAPE: Conical with globose body whorl.

ORNAMENT OR SCULPTURE: About 6 very convex whorls. Sutures distinct. Body whorl large; whorls of spire gradually decreasing in size toward acute

Cancellaria reticulata

apex. Strong, regular spiral cords and similar, oblique riblets give a very reticulate pattern.

APERTURE: Elongate, suboval; glazed white inside. Outer lip is thin with crenulations on the inner surface. Columella is somewhat twisted with 2 strong folds. The top fold is marked with several smaller ridges. Siphonal canal short and upturned.

OPERCULUM: None, uses mucus and sand to close aperture.

PERIOSTRACUM: None visible.

REMARKS: Found only after severe weather on the outer beaches. Carnivorous.

HABITAT: Offshore, shallow water to several fathoms.

LOCALITIES: Central, south.

OCCURRENCE: Rare.

RANGE: North Carolina to both sides of Florida and West Indies.

GEOLOGIC RANGE: Pleistocene to Recent.

Family MARGINELLIDAE Fleming, 1828

Genus *Prunum* Herrmannsen, 1852

Subgenus *Leptegouana* Woodring, 1928

COMMON ATLANTIC MARGINELLA

Prunum (Leptegouana) apicina Menke, 1828, *Syn. Méth. Moll.*, p. 87.

L. *prunum* plum, *apiculus* small, pointed.

SIZE: Length 12 mm.

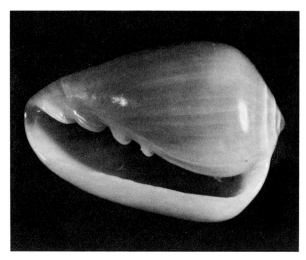

Prunum (Leptegouana) apicina

COLOR: Polished cream, yellowish, or grayish tan with several reddish brown spots on outer lip.

SHAPE: Conical, broad anteriorly.

ORNAMENT OR SCULPTURE: Smooth, spire short, and convex body whorls large.

APERTURE: Long, narrow, the length of body whorl. Outer lip thickened, notched at base. Columella with 4 strong plicae below.

OPERCULUM: None.

PERIOSTRACUM: None.

REMARKS: A carnivorous scavenger. Mantle reflected over body. Bleached dead shells are found in the southern part of the coast, only rarely reported alive. Syn. *Marginella apicina*.

HABITAT: Shallow, grassy, and inlet-influenced areas.

LOCALITIES: Port Aransas, south.

OCCURRENCE: Uncommon.

RANGE: North Carolina to Florida, the Gulf states and the West Indies.

GEOLOGIC RANGE: Pliocene to Recent.

Suborder TOXOGLOSSA Troschel, 1848

Superfamily CONACEA Rafinesque, 1815

Family TURRIDAE H. &. A. Adams, 1853. A family of small shells that baffle the experts and are discouraging to the novice. The most evident characteristics by which they are determined are the anal notch in the posterior (abaxial) margin of the outer lip and the short siphonal canal. They are variously shaped but mostly fusiform. The variation that is common to individuals within a species makes sorting a difficult operation. After the sorting takes place it becomes difficult to match the specimen with a published figure. They each have a radula but not all have an operculum.

Subfamily CLAVINAE Casey, 1904

Genus *Crassispira* Swainson, 1840

Subgenus *Crassispirella* Bartsch & Rehder, 1939

OYSTER TURRET

Crassispira (Crassispirella) tampaensis Bartsch & Rehder, 1939, *Proc. U.S. Nat. Mus.* (3070):136.

L. *crassus* thick, fat, *spira* coil, twist; from Tampa, Florida.

Crassispira (Crassispirella) tampaensis

SIZE: Length 8 to 22 mm.
COLOR: Pale yellow brown to chestnut.
SHAPE: Turriculate.
ORNAMENT OR SCULPTURE: About 10 to 12 convex, slightly shouldered whorls. Sutures distinct. Apex acute. Just below the suture is a single, strong spiral cord; about 20 beaded axial ribs begin at the base of this cord. Numerous regular spiral threads give a beaded effect to the convex ribs.
APERTURE: Narrow, oval. Outer lip thin and crenulate with posterior anal notch and a short siphonal canal at the base. A narrow callus on the parietal wall. Umbilicate.
OPERCULUM: Corneous, apical nucleus, dark.
PERIOSTRACUM: None visible.
REMARKS: This species may be synonymous with *Clathrodrillia ostrearum* (Stearns, 1872). Olsson and Harbison (1953) present evidence to support the claim that this genus is *Clathrodrillia* Dall, 1918.
HABITAT: Inlet influence to 90 fathoms.
LOCALITIES: Port Aransas.
OCCURRENCE: Rare, probably fossil.
RANGE: North Carolina to southern half of Florida, Cuba and Texas.

Subfamily MANGELLIINAE Fischer, 1883

Genus *Mangelia* (*Mangilia*) Risso, 1826

Mangelia cf. *M. atrostyla* Dall, 1889, *Bull. M.C.Z.* 18: 111.
Dedicated to naturalist G. Mangili (1767–1826); Gk. *atropous* erect; L. *stilus* slender, pointed writing instrument.

SIZE: Length 7 to 8 mm.
COLOR: Variable, from yellowish white to dark brown, wholly or in stripes and bands.
SHAPE: Turriculate.
ORNAMENT OR SCULPTURE: Six postnuclear whorls with strong shoulders. Suture deeper than *Kurtziella limonitella*. Only 8 axial ribs, the last forms a strong varix.
APERTURE: Notch shallow. Canal short.
OPERCULUM: None known.
PERIOSTRACUM: None visible.
REMARKS: Dead in beach drift.
HABITAT: Not known.
LOCALITIES: Port Aransas, south.
OCCURRENCE: Rare.
RANGE: Hatteras to Antilles and Texas.

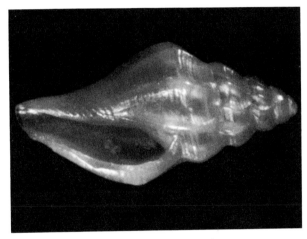

Mangelia cf. *M. atrostyla*

Mangelia cf. *M. cerina* (Kurtz & Stimpson, 1851), *Proc. Boston Soc. Natur. Hist.* 4:115.
Dedicated to naturalist G. Mangili (1767–1826); L. *cerinus* wax colored.
SIZE: Length 9 mm.
COLOR: Waxen white to ash color on upper whorls. No lines of color.
SHAPE: Turriculate.
ORNAMENT OR SCULPTURE: Spire equal to last whorl. Five postnuclear whorls, flattish, angulated on shoulder. Small larval whorls, smooth except for last, which has 4 nodulous spiral lines. Granulous spiral sculpture. Nine swollen transverse riblets. No varix. Suture appressed and undulated.
APERTURE: Narrow, oblique, about one-half length of

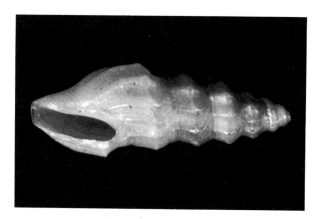

Mangelia cf. *M. cerina*

body whorl. Notch well marked and away from suture. Canal very short.

OPERCULUM: None.

PERIOSTRACUM: None visible.

REMARKS: Screen the drift in the bay areas for this little turrid and don't worry too much about the identification; it stumps the experts.

HABITAT: Mud flats between tides. Hypersaline lagoons and inlet areas.

LOCALITIES: Entire.

OCCURRENCE: Fairly common.

RANGE: North Carolina, both sides of Florida, Texas.

GEOLOGIC RANGE: Miocene to Recent.

WAX-COLORED MANGELIA

Mangelia cf. *M. cerinella* (Dall, 1889), *Proc. U.S. Nat. Mus.* 6:329.

Dedicated to naturalist G. Mangili (1767–1826); L. *cerinus* wax colored.

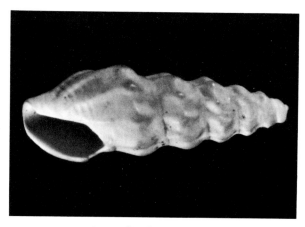

Mangelia cf. *M. cerinella*

SIZE: Length 10.5 mm.

COLOR: Whitish toward apex, ashy on intermediate, and orangish on body whorl. Never striped or spotted.

SHAPE: Turriculate. Drawn out and slender.

ORNAMENT OR SCULPTURE: Seven postnuclear whorls, angulate at periphery and sloping either way from it. Granulose spiral sculpture. Six or 7 transverse ribs. Suture less appressed and undulated than *M. cerina*.

APERTURE: Long, narrow, oblique. Hardly any indention for a notch. No canal to speak of.

OPERCULUM: None.

PERIOSTRACUM: None visible.

REMARKS: *Kurtziella cerinella* by some authors. The longest and most common Texas *Mangelia*.

HABITAT: Mud flats between tides. Hypersaline lagoons and inlets.

LOCALITIES: Entire.

OCCURRENCE: Fairly common.

RANGE: North Carolina to both sides of Florida and Texas.

GEOLOGIC RANGE: Pleistocene to Recent.

ACUTE MANGELIA

Mangelia cf. *M. oxytata* Bush, 1885, *Trans. Conn. Acad. Sci.* 6:46.

Dedicated to naturalist G. Mangili (1767–1826); L. *oxys* very sharp.

SIZE: Length 5 mm.

COLOR: Yellowish white, tinged with brown just below the suture and on anterior part of body whorl.

SHAPE: Fusiform.

Mangelia cf. *M. oxytata*

ORNAMENT OR SCULPTURE: Eight postnuclear whorls strongly angulated just below the middle, and ornamented with about 9 rather prominent, straight, transverse ribs, commencing at the periphery and extending to the suture; these, with their wide, concave interspaces, are crossed by 3 strong, rounded, equally distant threads, the third defining the suture. Where these cross the ribs nodules are formed. Nucleus smooth and glassy. Surface granulose.

APERTURE: Narrow ovate, pinched up anteriorly into a short, rather narrow, straight canal. Outer lip thickened, with a conspicuous varix, and a thick, smooth, rounded, irregularly curved, light brown edge, and a deep, narrow sinus considerably below the suture, at the angle of the shoulder. Columella slightly curved.

OPERCULUM: None.

PERIOSTRACUM: None visible.

REMARKS: This was reported by Hulings of Texas Christian University as being in the Sabine area and has recently been identified from Port Aransas.

HABITAT: Probably offshore.

LOCALITIES: East, central.

OCCURRENCE: Uncommon.

RANGE: North Carolina, the Gulf of Mexico and Texas.

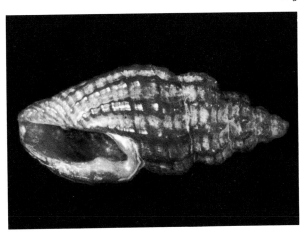

Mangelia plicosa

HABITAT: Shallow, hypersaline lagoon on grass or mud bottom.

LOCALITIES: Entire.

OCCURRENCE: Common.

RANGE: Cape Cod, Massachusetts, to western Florida and Texas.

Genus *Kurtziella* Dall, 1918

PUNCTATE MANGELIA

Kurtziella cf. *K. limonitella* (Dall, 1884), *Proc. U.S. Nat. Mus.* 6:113.

Dedicated to Lt. J. D. Kurtz, American conchologist; L. *limon* lemon colored.

SIZE: Length 9 mm.

COLOR: Whitish, lineated spirally with yellow brown. Some brown on outside of canal.

SHAPE: Turriculate.

Mangelia plicosa (C. B. Adams, 1850), *Contr. Conch.* (4):54.

Dedicated to naturalist G. Mangili (1767–1826); L. *plicare* to fold.

SIZE: Length 6 to 8 mm.

COLOR: Reddish brown. Dead shells wax colored.

SHAPE: Turriculate, spire about one-half of length.

ORNAMENT OR SCULPTURE: Six to 7 whorls. Sutures distinct. Only slightly shouldered. Spiral sculpture consists of strong, regularly spaced cords. Eleven to 12 transverse ribs made nodulose by spiral cords.

APERTURE: Semilunar. Outer lip thickened with a very pronounced posterior notch below suture. Interior dark. Parietal lip narrow. Siphonal canal short.

OPERCULUM: None.

PERIOSTRACUM: Thin, grayish.

REMARKS: Female of this little carnivore lays smooth, transparent, lens-shaped egg capsules about 0.16 mm. in diameter. The posterior notch makes this one easy to identify. This species has been placed in the genus *Pyrgocythara* by some recent workers.

ORNAMENT OR SCULPTURE: Spire trifle shorter than last whorl. Five postnuclear whorls, rounded, angulated behind periphery. Spiral sculpture granulose. Twelve narrow, transverse riblets. No varix. Ribs obsolete on fasciole. Suture hardly appressed or undulated.

APERTURE: Narrow, oblique. Notch shallow, deepest at angulation. Canal not differentiable from the aperture.

OPERCULUM: None.

PERIOSTRACUM: None visible.

REMARKS: This species will not have the little beads, or nodules, on the last nuclear whorl like *M. cerina*.

HABITAT: Offshore banks and mud flats between tides. Hypersaline lagoons and inlets.

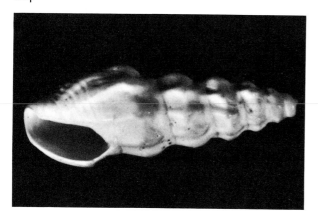

Kurtziella cf. *K. limonitella*

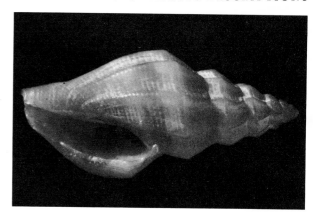

Rubellatoma diomedea

LOCALITIES: Entire.
OCCURRENCE: Fairly common.
RANGE: North Carolina to both sides of Florida and Texas.
GEOLOGIC RANGE: Miocene to Recent.

Genus *Rubellatoma* Bartsch & Rehder, 1939

REDDISH MANGELIA
Rubellatoma diomedea Bartsch & Rehder, 1939, *Proc. U.S. Nat. Mus.* (3070):131.
L. *rubellus* reddish; the shell was collected on the steamer *Albatross* (sea bird family Diomedeidae).
SIZE: Length 6 to 8 mm.
COLOR: Cream with chestnut brown bands. A narrow band below the suture and a broad band covering anterior half of body whorl.
SHAPE: Turriculate.
ORNAMENT OR SCULPTURE: About 7 whorls with angled shoulders. Fairly heavy, sutures distinct. Spiral sculpture consists of fine lines. Transverse sculpture, about 9 rounded ribs that are highest at periphery of whorl shoulder. The spaces between ribs are wider than ribs.
APERTURE: Elongate oval. Outer lip slightly thickened but edge thin, with a weak posterior notch near summit. Parietal lip narrow, polished. Siphonal canal moderately extended.
OPERCULUM: None.
PERIOSTRACUM: None visible.
REMARKS: This little carnivore is rather rare in all of its range.
HABITAT: Inlet areas to moderate depths in warm seas.

LOCALITIES: East, central.
OCCURRENCE: Rare.
RANGE: North Carolina to southeastern Florida and Texas.

Family TEREBRIDAE Mörch, 1852

Genus *Hastula* H. & A. Adams, 1853

Hastula maryleeae R. D. Burch, 1965, *Veliger* 7:242.
L. *hasta* spear; dedicated to Mary Lee Burch, the collector.
SIZE: Length 1 to 2 in.
COLOR: Variable from ivory white to dark purplish. Polished.
SHAPE: Elongate conic.
ORNAMENT OR SCULPTURE: Numerous flat-sided whorls. Sutures distinct, apex very pointed. Numerous small axial riblets near suture. The spiral sculp-

Hastula maryleeae

ture consists of microstriations. The rows of punctations typical of *Hastula* are absent.

APERTURE: Small, pear-shaped outer lip thin, with a deep, recurved siphonal notch at base.

OPERCULUM: Corneous, brown.

PERIOSTRACUM: None.

REMARKS: This shell has only been recognized as a separate species in recent years but was collected in Texas more than a hundred years ago. Originally described from Texas.

HABITAT: Sandy surf zone of warm seas.

OCCURRENCE: Common.

RANGE: From Galveston, Texas, to Veracruz, Mexico.

SALLE'S AUGER

Hastula salleana (Deshayes, 1859), *Proc. Zool. Soc. London* 27:287.

L. *hasta* spear; dedicated to A. Salle, a nineteenth-century collector.

SIZE: Length 1 to 2 in.

COLOR: Dark bluish gray or brownish. Polished.

SHAPE: Elongate conic.

ORNAMENT OR SCULPTURE: Numerous flat-sided whorls. Sutures distinct, apex very pointed. About 30 short, dark ribs below suture of each whorl. Spiral sculpture consists of microscopic rows of punctae; these are more widely spaced than in *H. cinerea* Born.

APERTURE: Small, pear shaped, dark brown within. Outer lip thin, with deep siphonal notch at base. A sharp ridge runs from this notch to midcolumella.

OPERCULUM: Corneous, brown.

PERIOSTRACUM: None.

REMARKS: These quick, burrowing carnivores live in the surf zone in mixed populations of *H. maryleeae*

Burch and the *Donax* clams. Empty shells have been found inside the small starfish, *Luida clathrata.*

HABITAT: Sandy surf zone in warm seas.

LOCALITIES: Entire, more to south.

OCCURRENCE: Common.

RANGE: Florida, west to Veracruz, Mexico.

Genus *Terebra* Bruguière, 1792

Subgenus *Strioterebaum* Sacco, 1891

COMMON ATLANTIC AUGER

Terebra (Strioterebaum) dislocata (Say, 1822), *J. Acad. Natur. Sci. Phila.* 2:235.

L. *terebra* a boring tool, auger, *dislocatus* dislocated.

SIZE: Length 1½ to 2 in.

COLOR: Grayish white to orangish white.

SHAPE: Turriculate.

ORNAMENT OR SCULPTURE: Numerous slightly convex whorls with about 25 axial ribs per whorl. Sutures distinct with a beaded spiral band just below and fine spiral striae between ribs.

APERTURE: Small, subovate. Outer lip thin with recurved siphonal notch at base. Columella short. Narrow parietal area is polished.

OPERCULUM: Corneous, thin, yellow brown.

PERIOSTRACUM: Thin, brownish.

REMARKS: This carnivore does not have a radula but contains its venom in "grooved prickles." The shell can be found under long bulges of sand at low tides.

HABITAT: Inlet area in warm seas.

LOCALITIES: Entire.

OCCURRENCE: Common.

RANGE: Virginia to Florida, Texas and the West Indies.

GEOLOGIC RANGE: Eocene? Miocene to Recent.

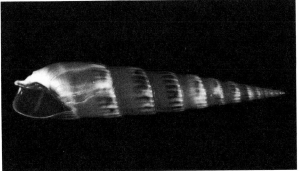

Hastula salleana

Terebra (Strioterebaum) dislocata

Terebra (Strioterebaum) protexta

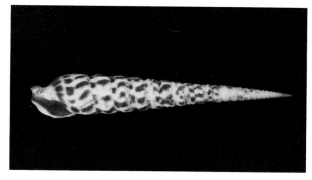

Terebra (Strioterebaum) taurinus

FINE-RIBBED AUGER

Terebra (Strioterebaum) protexta (Conrad, 1846), *Proc. Acad. Natur. Sci. Phila.* 3:26.

L. *terebra* a boring tool, auger, *pro* before, *textus* texture, fabric, structure.

SIZE: Length 18 to 25 mm.

COLOR: Brownish when living.

SHAPE: Turriculate.

ORNAMENT OR SCULPTURE: Thirteen to 15 slightly convex whorls. Sutures distinct. Spiral sculpture consists of a band below the suture similar to, but less pronounced than, that in *T. dislocata.* Convex, axial riblets are crossed by fine spiral striations.

APERTURE: Small, oblique, oval. Outer lip thin. Parietal area narrow and glossy. Columella short with upturned siphonal notch at base.

OPERCULUM: Corneous, reddish brown.

PERIOSTRACUM: Brownish.

REMARKS: Look for this in outer beach drift screenings where it may be found dead or fragmented. Alive at San Luis Pass.

HABITAT: Offshore in 1 to 50 fathoms, and inlet areas.

LOCALITIES: Entire.

OCCURRENCE: Fairly common.

RANGE: North Carolina to Florida and Texas.

GEOLOGIC RANGE: Miocene? Pliocene to Recent.

FLAME AUGER

Terebra (Strioterebaum) taurinus (Solander, 1786), *Portland Mus. Cat.,* pp. 142, 152.

L. *terebra* a boring tool, auger, *taurinus* bull.

SIZE: Length 4 to 6 in.

COLOR: Background cream with axial reddish brown flame-shaped marks in two spiral rows.

SHAPE: Slender, elongate auger, turriculate.

ORNAMENT OR SCULPTURE: About 14 flattened whorls. Sutures distinct. Two spiral incised lines between sutures. Numerous fine, wavy transverse striations.

APERTURE: Semilunar, oblique. Short. Thin outer lip. Columella convexly curved. Anterior or siphonal canal short and recurved.

OPERCULUM: Corneous, brown.

PERIOSTRACUM: Not visible.

REMARKS: Carnivorous. Animal is colored yellow. Don't expect to find a perfect specimen of this fine shell. It usually takes a hurricane to bring in the broken shells that are greatly prized. Originally described from the Texas coast.

HABITAT: Deeper water offshore.

LOCALITIES: Central, south.

OCCURRENCE: Rare.

RANGE: Southeastern Florida, the Gulf of Mexico and the West Indies.

Subclass EUTHYNEURA Spengel, 1881 = OPISTHOBRANCHIA Milne Edwards, 1848

Family PYRAMIDELLIDAE Gray, 1840. This large family of minute gastropods is very confusing to both novice and expert. A powerful microscope is needed in any attempt to separate them. They were originally classified in the Streptoneura or Prosobranchia with the Eulimidae but recent study has shown they are more closely related to the Euthyneura or Opisthobranchia. They are less removed from the prosobranchs than other opisthobranch families, with the exception of the Acteonidae, and have undergone less evolutionary change. As opposed to other opisthobranchs, they have operculums, shells in which the head-foot mass can retract, and an auricle anterior to the ventricle.

Most are suspected of being ectoparasites that feed on their hosts with an oral sucker through which they draw the blood. They are hermaphrodites, laying eggs in gelatinous masses on their hosts.

The many-whorled shells are distinguished by nuclear whorls that turn sinistrally and are at right angles to the later whorls (heterostrophic). The columella has one or more folds. There is no radula and the corneous operculum is oval and few whorled, and has an apical nucleus.

The family Pyramidellidae is represented in Texas by four genera and probably a fifth that until recently was primarily known in fossil records. The *Pyramidella* are the largest and easiest to separate. The elongate-conic shell has regularly increasing, inflated whorls, a columella with one to three plicae, and an entire (holostomate) outer lip. The *Odostomia* are short, subconic or ovate. The complexity of the sculpture and the poor figures available make the *Turbonilla* the most difficult to identify as to species. Until an expert monographs the *Turbonilla*, their correct identity is only a matter of speculation. The shell is cylindro conic, many whorled, and usually slender with a single columellar fold that varies in strength. The outer lip of this genus is entire and the shell is larger than *Odostomia* but smaller than *Pyramidella*. The *Sayella* are elongate conic ovate, with microscopic spiral striations.

The fifth and rather obscure genus is *Eulimastoma* Bartsch, 1916. The minute shell is elongate conic with a single columellar fold, a pronounced angulation at the whorl base, no sculpture, and whorls that are slightly convex to flat sided. The heterostrophic protoconch is usually immersed in the first postnuclear whorl. Living *Eulimastoma* have been described as *Odostomia*, to which they are closely related, but current study will soon correct this confusion (Corgan, In press).

When malacologists get around to studying this family more closely a great step will have been made. In the past its members have been overlooked because of their size; much of the material is lost in screening. The names assigned to the various members figured here are only "educated guesses" hesitatingly applied.

It is easy to avoid these tiny shells after a few trial identifications and one becomes tempted to agree with Dr. James Lewis, who left an amusing notation on the usually stuffy labels that accompany mollusks in the great study collections when he penciled, "What the hell, who can tell?" (*Nautilus* 55 [4]:119).

Pyramidella (Lonchaeus) crenulata

Genus *Pyramidella* Lamarck, 1799

Subgenus *Lonchaeus* Mörch, 1875

NOTCHED PYRAM
Pyramidella (Lonchaeus) crenulata (Holmes, 1859), *Post-Plio. Fos. S. Car.*, p. 88.
L. *pyramis* pyramid, *crenulatus* finely notched.
SIZE: Length 15 mm.
COLOR: Pale brown; cream to white when dead. Polished.
SHAPE: Elongate conic.
ORNAMENT OR SCULPTURE: Numerous flat-sided whorls. Sutures distinct, V-shaped channels. Spire acute. Body whorl rounded at base. Posterior margin of each whorl is delicately crenulated. A weak basal line meets outer lip at suture. Umbilicated.
APERTURE: Small, rather auriform, entire. Outer lip thin. Columella sinuous, two oblique folds.
OPERCULUM: Corneous, semicircular. Notched to fit crenulations on columella.
PERIOSTRACUM: None visible.
REMARKS: The largest of the pyrams on this coast. This ectoparasite has no gills or radula. Hermaphroditic.
HABITAT: Inlets and hypersaline lagoons.
LOCALITIES: Entire.
OCCURRENCE: Common.
RANGE: Florida, Texas.
GEOLOGIC RANGE: Upper Pliocene to Recent.

Genus *Sayella* Dall, 1885

DARK PYRAM
Sayella cf. *S. livida* Rehder, 1835, *Nautilus* 48(4):129.
Dedicated to Thomas Say; L. *lividus* black and blue.

Sayella cf. *S. livida*

SIZE: Length 3.5 to 4 mm.

COLOR: Straw yellow with a wide subsutural white band.

SHAPE: Elongate ovate.

ORNAMENT OR SCULPTURE: About 6½ moderately convex whorls. Smooth except for microscopic growth lines and spiral sculpture. Whorls closely adpressed at suture. Body whorl half the shell length with suture considerably below the periphery of the preceding whorl, which gives the shell a constricted appearance in the middle.

APERTURE: Small, obliquely ovate. Lip thin, thickening anteriorly, proceeding into base of columella, which bears a strong fold. Area surrounding base of columella is reddish brown.

OPERCULUM: Not known.

PERIOSTRACUM: None visible.

REMARKS: The holotype was collected by J. A. Singley in Corpus Christi Bay, Texas (U.S.A.M. No. 125556).

HABITAT: Bays and inlet areas.

LOCALITIES: Central, probably entire.

OCCURRENCE: Fairly common.

RANGE: Florida and Texas.

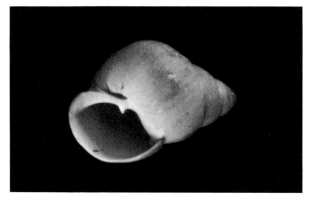

Odostomia (Odostomia) cf. *O. gibbosa*

Genus *Odostomia* Fleming, 1817

Subgenus *Odostomia* s.s.

FAT ODOSTOME

Odostomia (Odostomia) cf. *O. gibbosa* Bush, 1909, *Amer. J. Sci.* 27:475.

Gk. *odon, odontis* tooth, *stoma* a mouth; L. *gibbus* a hump.

SIZE: Length 3 mm.

COLOR: Whitish, polished.

SHAPE: Globose conic.

ORNAMENT OR SCULPTURE: About 5 very convex whorls. Shell rather fragile. Sutures distinct. Smooth except for microscopic growth striations.

APERTURE: Pear shaped, fairly large. Outer lip thin. Columella marked with a single prominent tooth near insertion.

OPERCULUM: Corneous.

PERIOSTRACUM: None visible.

REMARKS: The rounded shape of this tiny shell distinguishes it from the other *Odostomia*. Its host is not known.

HABITAT: Inlet areas.

LOCALITIES: Entire.

OCCURRENCE: Fairly common.

RANGE: Maine to southern Massachusetts and Texas.

SMOOTH ODOSTOME

Odostomia (Odostomia) laevigata (d'Orbigny, 1842), [La Sagra, *Hist. l'Ile de Cuba*], *Moll.* 1:227.

Gk. *odon, odontis* tooth, *stoma* a mouth; L. *laevis* smooth.

SIZE: Length 3 to 5 mm.

COLOR: Translucent grayish white alive, opaque white to brownish dead.

Odostomia (Odostomia) laevigata

SHAPE: Ovate conic.

ORNAMENT OR SCULPTURE: Four to 6 convex whorls. Suture fairly impressed. Spiral sculpture absent or microscopic. Nuclear whorls impressed in apex. May have a chinklike umbilicus. The form of this species is quite variable.

APERTURE: Ovate. Columellar tooth weak.

OPERCULUM: Corneous.

PERIOSTRACUM: None visible.

REMARKS: Syn. *O. acutidens* Dall.

HABITAT: Inlet-influenced areas, shallow water.

LOCALITIES: Entire.

OCCURRENCE: Probably fairly common.

RANGE: Southeastern United States to the Lesser Antilles and Texas.

Subgenus *Chrysallida* Carpenter, 1856

Ex. gr. *Odostomia (Chrysallida) dux* Dall & Bartsch, 1906, *Proc. U.S. Nat. Mus.* 30:350.

Gk. *odon, odontis* tooth, *stoma* a mouth; L. *dux* leader, guide.

SIZE: Length 1.8 mm.

COLOR: White.

SHAPE: Conical, elongate.

ORNAMENT OR SCULPTURE: Postnuclear whorls are 4. Each whorl bears 3 rows of tubercles and a smooth spiral keel just above the suture.

APERTURE: Obliquely oval or pear shaped. A plica on columella.

OPERCULUM: Corneous.

PERIOSTRACUM: None visible.

REMARKS: *O. dux* represents one very large and widely distributed species complex in *Chrysallida*. The specimen figured could be one of several that fall into this subgenus.

HABITAT: Ectoparasite, in saline bays.

LOCALITIES: East, central.

OCCURRENCE: Fairly common.

RANGE: Not defined.

HALF-SMOOTH ODOSTOME

Odostomia (Chrysallida) seminuda (C. B. Adams, 1839), *J. Boston Soc. Natur. Hist.* 2:280.

Gk. *odon, odontis* tooth, *stoma* a mouth; L. *semi* half, *nudus* naked.

SIZE: Length 4 mm.

COLOR: Whitish.

SHAPE: Conical, elongated.

ORNAMENT OR SCULPTURE: Six to 7 slightly convex whorls. Whorls shouldered, sutures distinct. Whorls sculptured between sutures by axial ribs cancelled into beads of nodules by four low, broad, equidistant ridges. Base of body whorl spirally grooved.

APERTURE: Auriform. Outer lip thick inside but edge thin, entire. Columella strong, twisted, and reflexed with an oblique plica.

OPERCULUM: Corneous.

PERIOSTRACUM: None visible.

REMARKS: This nonspecific ectoparasite is known to feed on *Crepidula* and *Pecten*. It can stand less saline waters than some of the other *Odostomia*. Holes in dead specimens indicate that gastropods feed on it in turn.

HABITAT: Ectoparasitic in bays.

LOCALITIES: Entire.

Odostomia (Chrysallida) dux   *Odostomia (Chrysallida) seminuda*

OCCURRENCE: Fairly common.

RANGE: Nova Scotia to the Gulf of Mexico.

Odostomia (Menestho) impressa

Subgenus *Menestho* Möller, 1842

IMPRESSED ODOSTOME

Odostomia (Menestho) impressa (Say, 1822), *J. Acad. Natur. Sci. Phila.* 2:244.

Gk. *odon, odontis* tooth, *stoma* a mouth; L. *impressus* pressed, imprinted.

SIZE: Length 4 mm.

COLOR: Whitish.

SHAPE: Elongate conic.

ORNAMENT OR SCULPTURE: Eight flattened whorls, shouldered above. Sutures channeled. Nuclear whorls small, partly imbedded in first succeeding turn. Spiral sculpture consists of 3 strong deeply cut grooves, the grooves cut by spiral threads. Base rounded, spirally grooved.

APERTURE: Ovate. Outer lip thin, slightly sinuous at edge, showing external sculpture within. Columella stout with strong oblique plica at insertion.

OPERCULUM: Corneous.

PERIOSTRACUM: Thin, yellowish.

REMARKS: The most common *Odostomia* on this coast. Because of its habit of feeding on *Crassostrea virginica* this little ectoparasite has been studied considerably. It will also feed on *Bittium, Crepidula,* polychaete worms, and oyster drills, into which it inserts its proboscis and sucks out the juices. Probably synonymous with *O. trifida* Totten.

HABITAT: On oyster reefs in the bays.

LOCALITIES: Entire.

OCCURRENCE: Common.

RANGE: Massachusetts Bay to the Gulf of Mexico.

DOUBLE-SUTURED (TWO-SEAMED) ODOSTOME

Odostomia (Menestho) cf. *O. bisuturalis* (Say, 1822), *J. Acad. Natur. Sci. Phila.* 2:244.

Gk. *odon, odontis* tooth, *stoma* a mouth; L. *bi* two, *sutura* a seam.

SIZE: Length 5 mm.

COLOR: White.

SHAPE: Elongate conic.

ORNAMENT OR SCULPTURE: About 8 rounded whorls. Spiral sculpture consists of a deeply marked line just below the suture. There is variation in the number of spiral lines within a population.

APERTURE: Oval, fairly large. Outer lip thin. Columella slender, strongly reflexed, with an oblique fold at insertion. This fold is not visible when aperture is viewed squarely.

OPERCULUM: Corneous.

PERIOSTRACUM: None visible.

REMARKS: This ectoparasite is not peculiar to one host and is known to feed on the algae on the outside of a shell as well as the animal inside.

HABITAT: Subtidal, prefers rocky stations.

LOCALITIES: Matagorda Peninsula, south.

OCCURRENCE: Fairly common.

RANGE: Massachusetts to Florida and Texas.

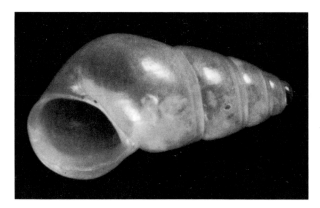

Odostomia (Menestho) cf. *O. bisuturalis*

Genus *Turbonilla* Risso, 1826, *Hist. Natur. Europe Merid.* 4:224.

Shell cylindro conic, many whorled, generally slender, columellar fold single, varying in strength, outer lip entire; shell usually smaller than in *Pyramidella* and larger than in *Odostomia.*

Type, *Turbonilla typica* Dall & Bartsch.

Key to the Subgenera of Turbonilla:

A. Spiral sculpture absent or, if present, consisting of exceedingly fine microscopic lines only.
    a. Axial ribs obsolete on the later whorls . . . . *Ptycheulimella*.
    aa. Axial ribs not obsolete on the later whorls.
        b. Axial ribs terminating at the periphery of the last whorl . . . . . *Chemnitzia*.
        bb. Axial ribs continuing over the base of the last whorl . . . . . *Turbonilla*.
AA. Spiral sculpture present, always stronger than mere microscopic lines.
    c. Spiral sculpture consisting of fine incised lines . . . . *Strioturbonilla*.
    cc. Spiral sculpture consisting of strongly incised grooves . . . . *Pyrgiscus*.

Subgenus *Pyrgiscus* Philippi, 1841

ELEGANT TURBONILLA

Turbonilla (Pyrgiscus) cf. *T. elegantula* Verrill, 1882, *Trans. Conn. Acad. Arts & Sci.* 5:538.

L. dim. of *turbo* a top, *elegans* elegant, fine.

SIZE: Length 3 to 5 mm.

COLOR: Amber, semitransparent. Spiral incisions darker.

SHAPE: Elongate conic. Slender.

ORNAMENT OR SCULPTURE: Nine rounded whorls. Sutures distinct. Twenty-two nearly perpendicular, rounded transverse ribs, which are separated by about equally wide spaces. Spaces are crossed by 5 equal, well-separated, incised spiral lines and 2 very much finer ones. Rounded base is incised with 9 unevenly spaced spiral lines.

APERTURE: Elongated oval. Columella nearly straight and slightly reflexed, with a slight fold.

OPERCULUM: Corneous.

PERIOSTRACUM: None visible.

REMARKS: The intercostal striae do not cross the ribs. Syn. *T. elegans* Verrill.

HABITAT: Probably along shore.

LOCALITIES: Port Aransas.

OCCURRENCE: Uncommon.

RANGE: West Indies and Texas.

INTERRUPTED TURBONILLA

Turbonilla (Pyrgiscus) cf. *T. interrupta* (Totten, 1835), *Amer. J. Sci.* 1(28):352.

L. dim. of *turbo* a top, *interruptus* interrupted.

SIZE: Length 6 mm.

COLOR: Brownish.

SHAPE: Slender, elongate conic.

ORNAMENT OR SCULPTURE: About 10 almost flat whorls sculptured with about 22 smooth, axial ribs separated by grooves of a little wider width, and with about 14 subequal, impressed, revolving lines arranged in pairs and entirely interrupted by the ribs. The spiral line above the periphery is heavier than the others, forming a line of deep pits. Ribs become obsolete below the middle of the body whorl. Base short and rounded with spiral lines.

APERTURE: Ovate, angular above, regularly rounded below, and about one-fifth the length of the shell. Outer lip sharp and slightly sinuous. Columella slightly curved and weakly reflexed.

OPERCULUM: Corneous.

PERIOSTRACUM: None visible.

REMARKS: The disappearance of the original shell described by Totten has contributed to the misinterpretation of this species.

HABITAT: Inlets and along shore.

LOCALITIES: Port Aransas, probably entire.

OCCURRENCE: Fairly common.

RANGE: Maine to the West Indies and Texas.

Turbonilla (Pyrgiscus) cf. *T. elegantula*

Turbonilla (Pyrgiscus) cf. *T. interrupta*

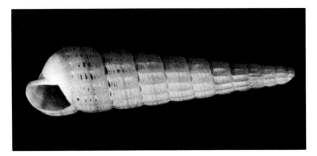

Turbonilla (Pyrgiscus) cf. *T. portoricana*

Turbonilla (Strioturbonilla) cf. *T. hemphilli*

PUERTO RICAN TURBONILLA

Turbonilla (Pyrgiscus) cf. *T. portoricana* Dall & Simpson, 1901, *Bull. U.S. Fish. Comm.* 20(1):414.

L. dim. of *turbo* a top; from Puerto Rico.

SIZE: Length 4.7 mm.

COLOR: Translucent white with a narrow yellowish brown spiral band around the whorls about one-fourth the breadth of the whorl above its suture. A pale yellow spiral band on middle of base.

SHAPE: Elongate conic.

ORNAMENT OR SCULPTURE: Ten flattened postnuclear whorls, slightly contracted at the sutures. Whorls crossed with almost vertical axial ribs, 16 on the fifth and increasing in number toward the base. Intercostal spaces are broad, wavy, and wider than ribs. Base is ribbed and crossed with 6 spiral striae.

APERTURE: Subovate. Columella oblique, outer lip well rounded, meeting columella at a right angle. Parietal callus well defined. Strong oblique fold near insertion of columella. Outer sculpture visible through outer lip.

OPERCULUM: Corneous.

PERIOSTRACUM: None visible.

HABITAT: Inlets.

LOCALITIES: Port Aransas, south.

OCCURRENCE: Uncommon.

RANGE: West Indies and Texas.

Subgenus *Strioturbonilla* Sacco, 1892

HEMPHILL'S TURBONILLA

Turbonilla (Strioturbonilla) cf. *T. hemphilli* Bush, 1900, *Proc. Acad. Natur. Sci. Phila.* 51:169.

L. dim of *turbo* a top; dedicated to H. Hemphill, nineteenth-century Florida collector.

SIZE: Length 10 to 12 mm.

COLOR: White.

SHAPE: Slender, elongate conic.

ORNAMENT OR SCULPTURE: Twelve postnuclear whorls are slightly convex and sculptured with about 20 almost perpendicular transverse ribs. The rounded ribs are separated by about equally wide, deep, concave spaces ending at the periphery of the body whorl in clean-cut ends. Base rounded and smooth. Entire surface covered with microscopic striations.

APERTURE: Squarish, somewhat expanded below; inner lip thickened and reflected.

OPERCULUM: Corneous.

PERIOSTRACUM: None visible.

HABITAT: Inlet-influenced areas.

LOCALITIES: Southern half of range, probably entire.

OCCURRENCE: Fairly common.

RANGE: Western coast of Florida, Texas.

The following *Turbonilla* are found in the drift of Nueces and Aransas bays in Texas:

Turbonilla sp. A . . . spiral sculpture, no ribs on base.
Turbonilla sp. B . . . spiral sculpture, no ribs on base.
Turbonilla sp. C . . . spiral sculpture, no ribs on base.
Turbonilla sp. D . . . spiral sculpture, no ribs on base.
Turbonilla sp. E . . . no spiral sculpture, ribs on base.
Turbonilla sp. F . . . spiral sculpture crossed with axial sculpture producing nodules.

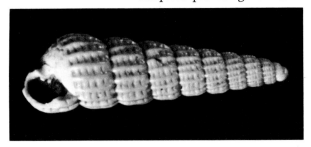

Turbonilla sp. A

Turbonilla sp. B

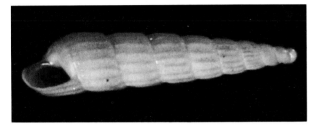

Turbonilla sp. C

Turbonilla sp. D

Turbonilla sp. E

Turbonilla sp. F

Order CEPHALASPIDEA Fischer, 1883

Superfamily ACTEONACEA d'Orbigny, 1835

Family ACTEONIDAE d'Orbigny, 1835

Genus *Acteon* Montfort, 1810

ADAMS' BABY-BUBBLE
Acteon punctostriatus (C. B. Adams, 1840), *Boston J. Natur. Hist.* 3:323.
Gk. *aktaion* a huntsman; L. *punctum* point, *striatus* striated.
SIZE: Height 3 to 6 mm.
COLOR: White, fragile.
SHAPE: Globose conical.
ORNAMENT OR SCULPTURE: Four convex whorls. Sutures deep. Spire elevated. Body whorl large with fine spiral striations over basal half.
APERTURE: Elongated pear shape. Outer lip thin. A little more than half the length of body whorl. Columella short with one strong oblique fold.
OPERCULUM: Thin, corneous.
PERIOSTRACUM: Thin.
REMARKS: The parapodia folds back over the head. The animal can withdraw into shell. This little carnivore burrows just below the surface. Hermaphroditic.
HABITAT: Inlet areas and outer beaches.
LOCALITIES: Entire.
OCCURRENCE: Fairly common.
RANGE: Cape Cod to the Gulf of Mexico and the West Indies.
GEOLOGIC RANGE: Miocene to Recent.

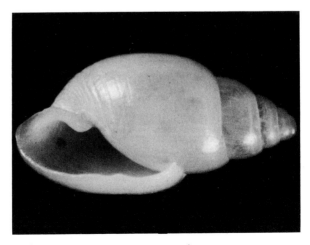

Acteon punctostriatus

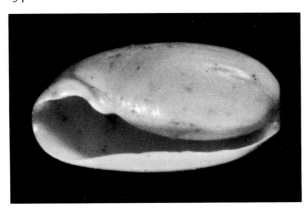

Cylichna (Cylichnella) bidentata

OPERCULUM: None.

PERIOSTRACUM: None.

REMARKS: The maculated mantle of the animal of this minute carnivore completely envelops the shell. Tiny black eyes are high on head. Crawls just under the sand leaving a trail on surface. Hermaphroditic.

HABITAT: Inlet areas near low-water mark.

LOCALITIES: Entire.

OCCURRENCE: Fairly common.

RANGE: North Carolina, Florida to Texas and West Indies.

GEOLOGIC RANGE: Pleistocene to Recent.

Genus *Cylichna* Loven, 1846

Subgenus *Cylichnella* Gabb, 1872

ORBIGNY'S BABY-BUBBLE

Cylichna (Cylichnella) bidentata (d'Orbigny, 1841), [La Sagra, *Hist. l'Ile de Cuba*], *Moll.* 1:125.

Gk. *kylix* a drinking cup; L. *bi* twice, *dens* tooth (*bidentatus* with two teeth).

SIZE: Length 2.5 to 4 mm.

COLOR: White, fragile.

SHAPE: Cylindrical.

ORNAMENT OR SCULPTURE: Smooth, spire depressed. Body whorl narrowed above and below.

APERTURE: Long, narrow. Outer lip thin and slightly flared below. Columella short with two plicae at base.

Superfamily BULLACEA Lamarck, 1801

Family BULLIDAE Lamarck, 1801

Genus *Bulla* Linné, 1758

STRIATE BUBBLE

Bulla striata Bruguière, 1792, *Encycl. Méth.*, p. 572.

L. *bulla* bubble, *striatus* striated.

SIZE: Length 18 to 25 mm.

COLOR: Whitish with small irregular mottlings of chocolate brown.

SHAPE: Oval with sunken spire.

ORNAMENT OR SCULPTURE: Delicate shell is smooth except for microscopic growth line. Spiral sculpture consists of fine grooves toward the base and within the sunken apical end.

APERTURE: Longer than body whorl, wider near base. Interior whitish. Outer lip thin. Parietal area covered with glazed white inductura.

OPERCULUM: None.

PERIOSTRACUM: None.

REMARKS: This carnivore completely envelops its shell, with the mantle looking like a lump of jelly when found alive. Burrows in the bottom. Does not like bright sun. Hermaphrodite, lays eggs in jellylike ribbons.

HABITAT: Inlet areas in shallow water.

LOCALITIES: Entire.

OCCURRENCE: Common.

RANGE: Western coast of Florida to Texas and the West Indies.

GEOLOGIC RANGE: Miocene to Recent.

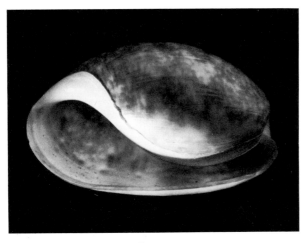

Bulla striata

Haminoea antillarum

Family ATYIDAE Thiele, 1826

Genus *Haminoea* Turton & Kingston, 1830

ELEGANT PAPER-BUBBLE
Haminoea antillarum (d'Orbigny, 1841), [La Sagra, *Hist. l'Ile de Cuba*], *Moll.* 1:124.
L. *hamus* hook; from the Antilles.
SIZE: Length 18 mm.
COLOR: Pale greenish yellow, almost translucent.
SHAPE: Rounded oval, spire insunk.
ORNAMENT OR SCULPTURE: Surface appears smooth but has fine growth striae and microscopic wavy, spiral lines.
APERTURE: Longer than body whorl, wider at bottom. Outer lip arises to the right of apical depression. Columella is extremely concave. Parietal area has narrow white inductura.
OPERCULUM: None.
PERIOSTRACUM: None.
REMARKS: This little carnivore cannot withdraw into the shell. Varies its diet with algae. Hermaphroditic.
HABITAT: Inlet-influenced areas in shallow water.
LOCALITIES: Entire.
OCCURRENCE: Fairly common.
RANGE: Gulf of Mexico to the West Indies.

CONRAD'S PAPER-BUBBLE
Haminoea succinea (Conrad, 1846), *Proc. Acad. Natur. Sci. Phila.* 3(1):26.
L. *hamus* hook, *succinum* amber.
SIZE: Length 10 mm.

COLOR: White to pale amber.
SHAPE: Cylindrical, thin; spire insunk.
ORNAMENT OR SCULPTURE: Surface covered with minute, wavy, spiral lines.
APERTURE: Longer than body whorl, wider near base. Outer lip thin, sharp. Columella concave with one weak fold above center.
OPERCULUM: None.
PERIOSTRACUM: None.
REMARKS: Sift the drift for this little shell. When alive, animal is brown. Hermaphroditic.
HABITAT: Inlet influence in shallow water.
LOCALITIES: Entire.
OCCURRENCE: Uncommon to rare.
RANGE: Florida to Texas.

Haminoea succinea

Family RETUSIDAE Thiele, 1926

Genus *Retusa* Brown, 1827

CHANNELED BARREL-BUBBLE
Retusa canaliculata (Say, 1826), *J. Acad. Natur. Sci. Phila.* 5:211.
L. *retusus* blunt, *canaliculatus* channeled.
SIZE: Length 3 to 5 mm.
COLOR: Glossy white.
SHAPE: Cylindrical with moderately elevated spire.
ORNAMENT OR SCULPTURE: Smooth except for microscopic growth lines. Suture channeled.
APERTURE: Long, narrow, wider at base. Outer lip thin. Columella is single, raised fold below parietal inductura.
OPERCULUM: None.

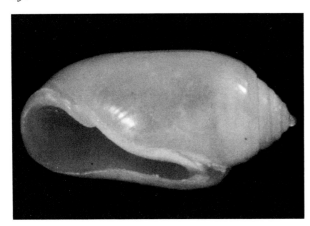

Retusa canaliculata

PERIOSTRACUM: None.

REMARKS: This may be *Acteocina candei* Orbigny, 1842 (see Wells & Wells 1962). Egg masses are gelatinous and attached by a strand to marine grasses. The veligers have a nonpelagic development that ensures a more stable population. Probably on oyster beds. Hermaphroditic. Carnivorous.

HABITAT: Enclosed moderate-salinity bays.

LOCALITIES: Entire.

OCCURRENCE: Common.

RANGE: Nova Scotia to Florida, Texas and the West Indies.

GEOLOGIC RANGE: Miocene to Recent.

Genus *Volvulella* Newton, 1891

Volvulella persimilis (Mörch, 1875), *J. Acad. Natur. Sci. Phila.* 8:189.
L. *volvere* to roll, *persimilis* very like.

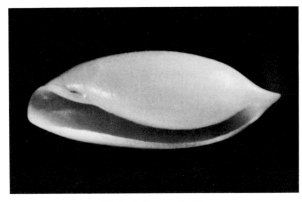

Volvulella persimilis

SIZE: Length 3 to 4 mm.

COLOR: White.

SHAPE: Cylindrical, tapering at each end.

ORNAMENT OR SCULPTURE: Body whorl dominates shell, with a sharp, spikelike apex and a tapering, rounded, anterior end, rather thin, semitransparent, somewhat lustrous, with four or five very fine, indistinct, punctate, spiral lines on each end and very indistinct, microscopic striae on the intervening surface.

APERTURE: Long, very narrow, expanded anteriorly; outer lip thin, following the curvature of the body whorl to just below the middle where it continues in a straight line and joins the inner lip in a broad curve; inner lip very thin, slightly reflected anteriorly over a slight umbilical chink.

OPERCULUM: None.

PERIOSTRACUM: Thin, pale yellow.

REMARKS: Easily overlooked if drift not screened carefully. Hermaphroditic. Syn. *V. oxytata* Bush. Genus has been called *Rhizorus*.

HABITAT: Seven to 17 fathoms, offshore.

LOCALITIES: Entire.

OCCURRENCE: Rare.

RANGE: North Carolina to southeastern Florida, Cuba, Texas.

Subgenus *Paravolvulella* Harry, 1967

Volvulella (Paravolvulella) texasiana Harry, 1967, *Veliger* 10(2):141.
L. *volvere* to roll; from Texas.

SIZE: Length 3.94 mm.

COLOR: Gray, translucent, with irregular, opaque white flecks. Ends usually stained with iron, reddish brown.

SHAPE: Cylindrical.

ORNAMENT OR SCULPTURE: Sculpture consists of a few narrow spiral bands equally spaced around the base and fainter ones on apical end. Microscopic wavy lines cover the midsection of the shell. No umbilicus. Adult shells are not tapering. A small apical spine is usually broken but has a spiral band about its base.

APERTURE: Narrow, as long as the body whorl. Outer lip and side of body whorl are flattened. Outer lip quadrate, meeting the columella at a right angle; basal end strongly arched. No columellar teeth.

OPERCULUM: None.

Volvulella (Paravolvulella) texasiana

PERIOSTRACUM: Thin, colorless.

REMARKS: A minute shell recently described by Dr. Harold Harry of Texas A&M University. Dredged offshore. Hermaphroditic.

HABITAT: Offshore, 7 fathoms.

LOCALITIES: Galveston and Port Isabel.

OCCURRENCE: Rare.

RANGE: Texas.

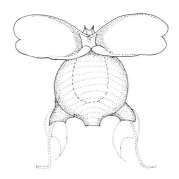

A living pteropod

Order THECOSOMATA Blainville, 1824 or PTEROPODA Cuvier, 1804

Suborder EUTHECOSOMATA Meisenheimer, 1905

Superfamily SPIRATELLACEA Thiele, 1926

Family CAVOLINIDAE d'Orbigny, 1842

Genus *Creseis* Rang, 1828

STRAIGHT-NEEDLE PTEROPOD
Creseis acicula (Rang, 1829), *Ann. Sci. Natur.* 13: 318.
Creseis, a mythological name; L. *acicula* small pin, needle.

SIZE: Length 10 mm.

COLOR: White, translucent.

SHAPE: Elongated, straight cone.

ORNAMENT OR SCULPTURE: Smooth, not coiled.

APERTURE: Small, round.

OPERCULUM: None.

PERIOSTRACUM: None.

REMARKS: These shells look like very fine tusk shells and can be easily overlooked if the drift is not carefully screened. At times they may come to shore by the thousands. The animal has a tentacular lobe on anterior fin margin. Hermaphrodites.

HABITAT: Pelagic.

LOCALITIES: Entire.

OCCURRENCE: Common, seasonally.

RANGE: Atlantic and Pacific, pelagic.

GEOLOGIC RANGE: Eocene to Recent.

Creseis acicula

Genus *Diacria* Gray, 1842

THREE-SPINED CAVOLINE
Diacria trispinosa (Lesueur, 1821), Blainville, *Dict. Sci. Natur.* 22:2.
Gk. *dia* through; L. *tri* three, *spinosa* spined.

SIZE: Length 10 mm.

COLOR: Whitish with chestnut-colored lips.

SHAPE: Lozenge shaped with extended hind part.

ORNAMENT OR SCULPTURE: Shell is somewhat compressed with three straight spines, one on either side of aperture and a longer one behind that may or may not be broken off. Longitudinally ribbed on the ventral side.

APERTURE: Compressed and thickened. Under lip curved outward.

OPERCULUM: None.

PERIOSTRACUM: None.

REMARKS: The hermaphroditic animal tends to rid itself of the long hind stalk that would hamper it in swimming. Descends during day, rises at night to feed.

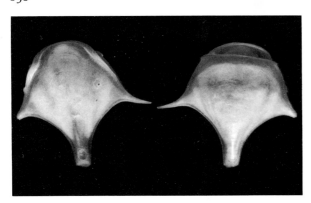

Diacria trispinosa

HABITAT: Extends into colder water than *D. quadridentata.*
LOCALITIES: Probably entire.
OCCURRENCE: Uncommon.
RANGE: Atlantic.

FOUR-TOOTHED CAVOLINE
Diacria quadridentata (Lesueur, 1821), Blainville, *Dict. Sci. Natur.* 22:81.
L. *quadri* four, *dentata* toothed.
SIZE: Length 2 to 4 mm.
COLOR: Transparent with chestnut-colored lip.
SHAPE: Globular.
ORNAMENT OR SCULPTURE: Without prominent lateral spines. Ventral side very globose. Dorsal side sculptured with rounded, radiating ribs. Lateral spines may be straight or fold back obliquely.
APERTURE: Narrow and flattened ending in slits at the lateral spines. Margins of lip thickened.
OPERCULUM: None.

PERIOSTRACUM: None.
REMARKS: This curious looking little shell is less common than the other species described. There is never an extended hind part and the shape of the shell is much more cavolinalike. Hermaphroditic.
HABITAT: Pelagic in warm waters, near surface.
LOCALITIES: Port Aransas, probably entire.
OCCURRENCE: Uncommon.
RANGE: Worldwide.

Genus *Cavolina* Abildgaard, 1791

LONG-SNOUT CAVOLINE
Cavolina longirostris (Lesueur, 1821), Blainville, *Dict. Sci. Natur.* 22:81.
Dedicated to Cavolini (1756–1810), Italian naturalist; L. *longe* long, far, *rostrum* snout.
SIZE: Length 4 to 9 mm.
COLOR: Translucent white.
SHAPE: Triangular when viewed from top.
ORNAMENT OR SCULPTURE: Ventral face of shell nearly round, sculptured with faint concentric ridges; dorsal face longitudinally ribbed, extended in front into a long, slightly folded, depressed beak; lateral spines compressed, central spine short, truncated.
APERTURE: Compressed and continued as a fissure around each side of the shell.
OPERCULUM: None.
PERIOSTRACUM: None.
REMARKS: These may be easily overlooked in the beach drift because they do not resemble the usual concept of a sea shell when living; butterflylike

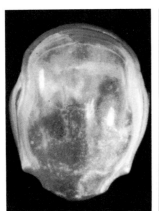 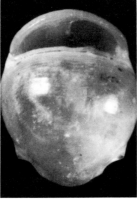

Diacria quadridentata

Cavolina longirostris

wings extend from either side of long rostrum. Hermaphrodites.

HABITAT: Pelagic.
LOCALITIES: Entire.
OCCURRENCE: Common.
RANGE: Worldwide, pelagic.

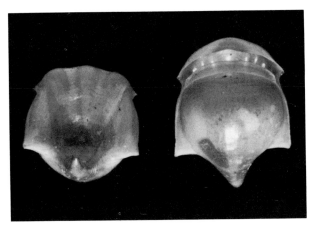

Cavolina uncinata

UNCINATE CAVOLINE
Cavolina uncinata (Rang, 1836), d'Orbigny, *Voy. Amer. Morid.*, pl. 93.
Dedicated to Cavolini (1756–1810), an Italian naturalist; L. *uncinatus* barbed, bent.

SIZE: Length 6 to 9 mm.
COLOR: Pale amber, translucent.
SHAPE: Shield shaped when viewed from top.
ORNAMENT OR SCULPTURE: Shell very inflated ventrally, surface delicately and regularly reticulated, with fine, concentric ridges in front; dorsal face with 3 low, radiating ribs, turned downward and nearly evenly rounded at the aperture; lateral spines compressed and curved backward; central spine short, stout, and upcurved.
APERTURE: Compressed and continued as a fissure around each side of shell.
OPERCULUM: None.
PERIOSTRACUM: None.
HABITAT: Pelagic.
LOCALITIES: Entire.
OCCURRENCE: Fairly common.
RANGE: Worldwide, pelagic.
GEOLOGIC RANGE: Pleistocene to Recent with related species to Miocene.

Order ANASPIDEA Fischer, 1883 (TECTIBRANCHIA of Authors)

Superfamily APLYSIACEA Lamarck, 1809 = ANASPIDEA Fischer, 1835

Family APLYSIIDAE (Lamarck, 1809)

Genus *Aplysia* Linné, 1767

SPOTTED SEA HARE
Aplysia dactylomela (Rang, 1828), *Hist. Natur. Apl.*, p. 56.
Gk. *a* without, *plasis* molding, *dactylos* finger, *mela* black.

SIZE: Length 4 to 5 in.
COLOR: Pale yellow or yellowish green with irregular violet black circles scattered over the body. Affected by food supply.
SHAPE: Elongate oval.
REMARKS: Larger than *A. willcoxi*. Live only one year. Hermaphroditic.
HABITAT: Where algae grow abundantly.

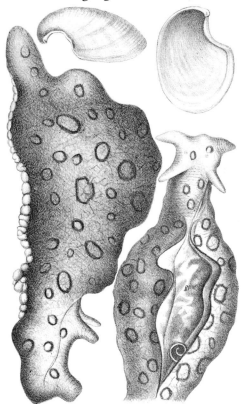

Aplysia dactylomela
(Tryon 1895, XVI, Plate 32)

LOCALITIES: Port Isabel.

OCCURRENCE: Uncommon.

RANGE: Southern half of Florida and the West Indies and Texas.

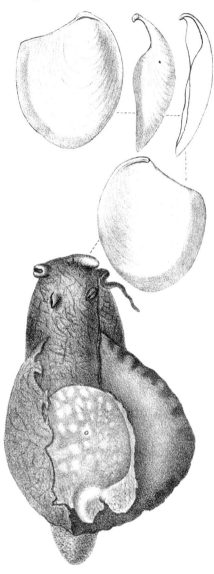

Aplysia floridensis
(Tryon 1895, XVI, Plate 37)

SOOTY SEA HARE

Aplysia floridensis (Pilsbry, 1895), *Proc. Acad. Natur. Sci. Phila.* 47:347–350.

Gk. *a* without, *plasis* molding; from Florida.

SIZE: Length 4 in.

COLOR: Uniform purple black. Inside of swimming lobes slightly lighter.

SHAPE: Elongate oval.

HABITAT: Where algae grow abundantly.

LOCALITIES: Port Isabel.

OCCURRENCE: Uncommon to rare.

RANGE: Lower Florida Keys, West Indies?, and South Texas.

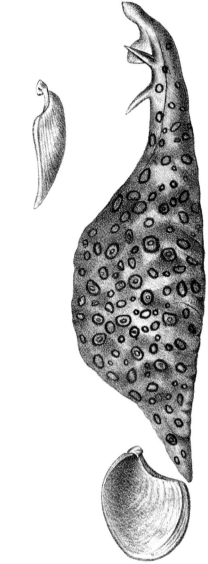

Aplysia protea
(Tryon 1895, XVI, Plate 37)

Aplysia protea (Rang, 1828), *Hist. Natur. Apl.*, p. 56.

Gk. *a* without, *plasis* molding, *proteus* variable; Proteus is a sea god who can assume any shape.

SIZE: Length 2 to 6 in.

COLOR: Variable dark colors with ring-shaped, dark

markings, often with spots within the rings. Background color yellowish green.

SHAPE: Head and neck narrower than the body. Back much swollen; swimming lobes large.

MARKINGS: Shell thin, internal. Swimming lobes on either side of body are marked on the inner side with dark patterns of color.

REMARKS: The animal is not protected by a shell. Hermaphroditic.

HABITAT: Where algae grow abundantly.

LOCALITIES: West Indies, Texas.

OCCURRENCE: Uncommon to rare.

RANGE: West Indies, Puerto Rico, and Gulf of Mexico.

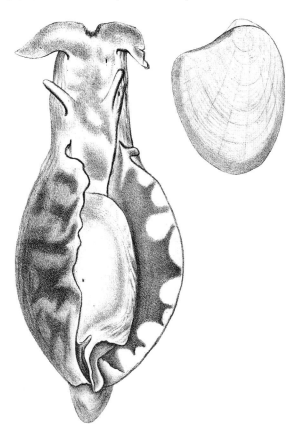

Aplysia willcoxi
(Tryon 1895, XVI, Plate 35)

WILLCOX'S SEA HARE
Aplysia willcoxi (Heilprin, 1886), *Proc. Acad. Natur. Sci. Phila.* 38:364.
Gk. *a* without, *plasis* molding; dedicated to M. Willcox, nineteenth-century malacologist.

SIZE: Length 5 to 9 in.

COLOR: Dark brown with light, irregular mottlings on the swimming lobes, head, and neck. Yellow marks on the inner border of the lobes and purplish coloration on gills and mantle.

SHAPE: Shell is a concave plate, internal. Animal is a large, slick, plastic form with neck and head narrower than the body.

MARKINGS: Shell is translucent, very fragile, pale amber colored, and very brittle when dry. It is nearly central and at the surface of the body. The head of the animal has earlike, long tentacles and a pair of shorter rhinophores behind them. There are extensions of the mantle, or "swimming lobes," used for locomotion.

APERTURE: None.

OPERCULUM: None.

PERIOSTRACUM: None.

REMARKS: This unusual creature has been seen swimming across the ship channel at Port Aransas with his head breaking the surface of the water. A skin diver will be able to observe the animals grazing on algae along the jetties. They are fun to observe in an aquarium, swimming or crawling on the bottom. When disturbed they expel a purple fluid. Eggs are laid in the grass flats of bays in tangled masses of gelatinous threads. Live only one year. Hermaphroditic.

HABITAT: Where algae grow abundantly.

LOCALITIES: Entire.

OCCURRENCE: Common.

RANGE: Cape Cod to both sides of Florida and Texas.

Genus *Bursatella* Blainville, 1817

RAGGED SEA HARE
Bursatella leachi plei (Rang, 1828), *Hist. Natur. Apl.*, p. 70.
L. *bursa* a pouch, saclike cavity; dedicated to W. E. Leach (1790–1836).

SIZE: Length 4 in.

COLOR: Olive green to gray in color, usually with white specks.

SHAPE: Elongate oval.

MARKINGS: Surface covered with numerous ragged appendages. No shell in adults.

REMARKS: First reported in Texas by Breur (1962). Eggs laid in gelatinous strands. Hermaphroditic.

HABITAT: Where algae are abundant.

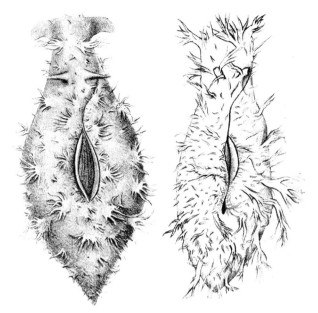

Bursatella leachi plei
(Tryon 1895, XVI, Plates 43, 44)

LOCALITIES: Port Isabel.
OCCURRENCE: Uncommon.
RANGE: Western and northwestern Florida, West Indies and Texas.

Order NUDIBRANCHIA Cuvier, 1795 or 1804

Suborder DENDRONOTOIDEA Odhner, 1936

Family SCYLLAEIDAE Fischer, 1883

Genus *Scyllaea* Linné, 1758

SARGASSUM NUDIBRANCH
Scyllaea pelagica Linné, 1758, *Syst. Natur.*, 10th ed., p. 656.
L. Scylla, sea monster on the dangerous rock opposite the whirlpool Charybdis; Gk. *pelagus* of the sea.
SIZE: Length 1 to 2 in.
COLOR: Yellowish brown to orange brown. Sargassum colored.
SHAPE: Sluglike.
ORNAMENT OR SCULPTURE: No oral tentacles. Two slender, long rhinophores. Two pairs of large foliaceous gill plumes, or cerata, on each side of body.

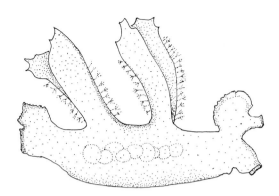

Scyllaea pelagica

APERTURE: None.
OPERCULUM: None.
PERIOSTRACUM: None.
REMARKS: This creature can be found by the hundreds clinging to sargassum, along with at least 4 other unidentified nudibranchs. It will live in an aquarium for days and is a delight to watch. Eggs are laid in yellow gelatinous strings that are more zigzag than straight. Hermaphroditic.
HABITAT: Sargassum.
LOCALITIES: Entire.
OCCURRENCE: Common.
RANGE: Southeastern United States, other warm seas.

Suborder EOLIDOIDEA Odhner, 1936

Family GLAUCIDAE Oken, 1815

Genus *Glaucus* Forster, 1777

BLUE GLAUCUS
Glaucus atlanticus Forster, 1777, *Voy. World* 1:49.
L. *glaucus* bluish gray; of the Atlantic.
SIZE: Length 2 in.
COLOR: Shades of blue.
SHAPE: Sluglike body.
ORNAMENT OR SCULPTURE: Tentacles and rhinophores small. Four clumps of bright blue frilled cerata on each side of body.
APERTURE: None.
OPERCULUM: None.
PERIOSTRACUM: None.
REMARKS: This brightly colored slug is washed ashore

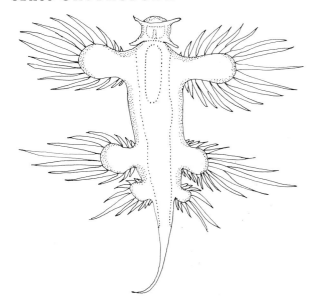

Glaucus atlanticus

by strong southeast winds in the spring when *Janthina* is coming in. Hermaphroditic. Feeds on siphonophores. Veliger has shell, foot, and an operculum.

HABITAT: Pelagic.
LOCALITIES: Entire.
OCCURRENCE: Occasionally common.
RANGE: Worldwide, pelagic in warm seas.

(Subclass PULMONATA Cuvier, 1817)

Order BASOMMATOPHORA Keferstein, 1864

Superfamily SIPHONARIACEA Gray, 1840

Family SIPHONARIIDAE Gray, 1840

Genus *Siphonaria* Sowerby, 1824

STRIPED FALSE LIMPET
Siphonaria pectinata (Linné, 1758), *Syst. Natur.*, 10th ed., p. 783.
L. *sipho* siphon, *pecten* a comb.
SIZE: Length 1 in.
COLOR: Whitish with numerous, brown bifurcating lines.
SHAPE: Conical, limpet shaped.

ORNAMENT OR SCULPTURE: Sculpture consists of numerous fine radial threads.
APERTURE: Large, circular. Interior glossy.
OPERCULUM: None.
PERIOSTRACUM: None visible.
REMARKS: This shell which is so common on the jetties of the Texas coast resembles a true limpet, but it is an air breather and very closely related to the garden snail.
HABITAT: Rocks and jetties.
LOCALITIES: Entire.
OCCURRENCE: Common.
RANGE: Eastern Florida, Texas, Mexico and St. Thomas.

Siphonaria pectinata

Superfamily ELLOBIACEA Adams, 1855

Family ELLOBIIDAE Adams, 1855

Genus *Melampus* Montfort, 1810

COFFEE MELAMPUS
Melampus bidentatus Say, 1822, *J. Acad. Natur. Sci. Phila.* 2:245.
Gk. *melos* black, *pous* foot; L. *bi* two, *dentatus* toothed.
SIZE: Length 18 mm.
COLOR: Brown with narrow cream-colored bands.
SHAPE: Ovate.
ORNAMENT OR SCULPTURE: The rather thin shell is smooth. About 5 whorls, spire low, body whorl predominates. Umbilicate.
APERTURE: Length of body whorl, expanded below.

Melampus bidentatus

 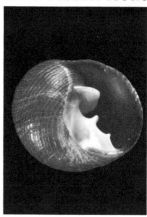

Pedipes mirabilis

Outer lip thin, crenate inside, curved into columella. Columella short with two white plicae.

OPERCULUM: None.

PERIOSTRACUM: Thin.

REMARKS: Examination of wild duck stomachs shows this little amphibious shell to be a favorite food. It must go into the water to breed. May be synonymous with *M. coffeus* Linné, 1758.

HABITAT: Under vegetation on bay and lagoon shores

LOCALITIES: Entire.

OCCURRENCE: Common.

RANGE: Florida to the West Indies and Texas.

Genus *Pedipes* Férussac, 1821

STEPPING SHELL

Pedipes mirabilis (Muhlfeld, 1818), *Mag. Ges. Natur. Freude Berlin*, p. 8.

L. *pedes* foot, relating to foot, *mirabilis* wonderful.

SIZE: Length 3 to 5 mm.

COLOR: Light to dark reddish brown.

SHAPE: Globose turbinate.

ORNAMENT OR SCULPTURE: Four to 5 strongly convex whorls. Sculpture consists of numerous incised, spiral lines. Transverse sculpture consists of irregular growth lines. Spire slightly elevated. No umbilicus. Sutures distinct.

APERTURE: Oval with the outer edge thin but thickened on the inside. Parietal area has 3 well-developed denticulations, the top one being the largest. Outer lip has one tooth directly opposite the central plica.

OPERCULUM: None.

PERIOSTRACUM: None visible.

REMARKS: This little colonial air breather can withstand some environmental changes, but the entire colony will disappear when conditions become extreme. They prefer a hard substrate.

HABITAT: On jetties.

LOCALITIES: Entire.

OCCURRENCE: Uncommon.

RANGE: Florida to the West Indies and Texas.

Pomacea sp. This conspicuous land snail is often washed up on South Texas beaches from its tropical home, probably the Veracruz area of Mexico. Due to its comparatively large size it is often thought to be a marine gastropod; for this reason it is figured with the Pulmonates so that the lucky finder of this beautiful ampullarid snail can look into it further if so desired.

Pomacea sp.

Class SCAPHOPODA Bronn., 1862

Family DENTALIIDAE Gray, 1834

Genus *Dentalium* Linné, 1758

Subgenus *Graptacme* Pilsbry & Sharp, 1897

IVORY TUSK
Dentalium (Graptacme) eboreum Conrad, 1846, *Proc. Acad. Natur. Sci. Phila.* 3:27.
L. *dens* tooth, *ebur* ivory.
SIZE: Length 1 to 2½ in.
COLOR: White.
SHAPE: Tapered cylinder. Curved.
ORNAMENT OR SCULPTURE: Smooth except for about 20 fine longitudinal lines at apical end. Apical slit deep, narrow and on the convex side.
APERTURE: Round.
OPERCULUM: None.
PERIOSTRACUM: None.
REMARKS: This very slender little shell is easily passed over because it does not fit the usual concept of a sea shell.
HABITAT: Inlet areas and near shore. Semi-infaunal.

Dentalium (Graptacme) eboreum

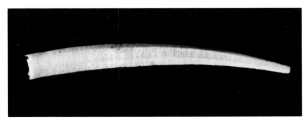

Dentalium (Dentalium) texasianum

LOCALITIES: Central, south.
OCCURRENCE: Fairly common.
RANGE: North Carolina to both sides of Florida, Texas and the West Indies.

Subgenus *Dentalium* s.s.

TEXAS TUSK
Dentalium (Dentalium) texasianum Philippi, 1848, *Z. für Malak.* 5:144.
L. *dens* tooth; from Texas.
SIZE: Length 18 to 35 mm.
COLOR: White.
SHAPE: Tapering cylinder. Curved.
ORNAMENT OR SCULPTURE: Sculptured with lengthwise ribs that have flat interspaces.
APERTURE: Hexagonal.
OPERCULUM: None.
PERIOSTRACUM: None.
HABITAT: Open-bay margins, inlets, and channel in stiff clay sediments. Semi-infaunal.
LOCALITIES: Entire.
OCCURRENCE: Common.
RANGE: North Carolina and the Gulf states.
GEOLOGIC RANGE: Pleistocene to Recent.

Family SIPHONODENTALIIDAE Simroth, 1895

Genus *Cadulus* Philippi, 1844

Subgenus *Polyschides* Pilsbry & Sharp, 1897

CAROLINA CADULUS
Cadulus (Polyschides) carolinensis Bush, 1885, *Trans. Conn. Acad. Sci.* 6(2):471.
L. *cadus* a small pail, jar; from Carolina.
SIZE: Length 10 mm.
COLOR: White.
SHAPE: Tapering cylinder. Curved.

ORNAMENT OR SCULPTURE: Smooth. Slightly swollen. Apex with 4 small slits.

APERTURE: Round.

OPERCULUM: None.

PERIOSTRACUM: None.

REMARKS: At times this small shell is abundant in the beach drift, but hard to see.

HABITAT: Near shore, 12 to 40 fathoms. Semi-infaunal.

LOCALITIES: Entire.

OCCURRENCE: Fairly common.

RANGE: North Carolina to Florida and to Texas.

GEOLOGIC RANGE: Pleistocene to Recent.

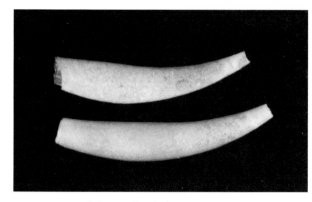

Cadulus (Polyschides) carolinensis

Class BIVALVIA Linné, 1758 = PELECYPODA Goldfuss, 1820

Subclass PALAEOTAXODONTA Korobkov, 1954

Order NUCULOIDEA Dall, 1889

Superfamily NUCULACEA Gray, 1824

Family NUCULIDAE Gray, 1824

Genus *Nucula* Lamarck, 1799

NUT CLAM
Nucula cf. *N. proxima* Say, 1822, *J. Acad. Natur. Sci. Phila.*, 1st ser. 2:270.
L. *nucula* little nut, *proximus* next of kin, nearest.

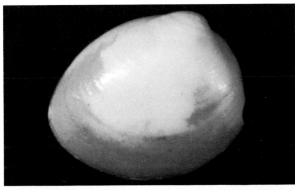

Nucula cf. *N. proxima*

SIZE: Length 6.5 mm.
COLOR: Whitish.
SHAPE: Obliquely trigonal. Equivalve, inequilateral.
ORNAMENT OR SCULPTURE: Sculpture of faint growth line and light radiating striae. Margins rounded. Ventral margin crenate.
HINGE AREA: Hinge angular with 12 comblike teeth anterior to umbones and 18 posterior.
PALLIAL LINE & INTERIOR: Inner surface pearly. Adductor muscle scars but no pallial sinus.
PERIOSTRACUM: Thin, greenish brown.
REMARKS: Parker (in litt.) reports this clam common in the Brazos River area. A deposit feeder.
HABITAT: Sandy mud bottom. Infaunal.
LOCALITIES: East, central.
OCCURRENCE: Fairly common.
RANGE: New Jersey to Florida and Texas.

Superfamily NUCULANACEA Gray, 1824

Family NUCULANIDAE H. & A. Adams, 1858

Genus *Nuculana* Link, 1807

POINTED NUT CLAM
Nuculana acuta (Conrad, 1831) *Amer. Mak. Conch.*, p. 32.
L. *nucula* little nut, *acutus* pointed.
SIZE: Length 10 mm.
COLOR: White.
SHAPE: Elongate with a pointed posterior rostrum. Equivalve. Inequilateral.
ORNAMENT OR SCULPTURE: Sculpture consists of well-defined concentric grooves that do not extend over radial ridge on rostrum.
HINGE AREA: A small triangular chondrophore and

Nuculana acuta

numerous chevron-shaped teeth on either side of umbo.

PALLIAL LINE & INTERIOR: Polished. Pallial sinus is small and rounded.

PERIOSTRACUM: Thin, yellowish brown.

REMARKS: This species is more common in southern part of range. It is less obese than *N. concentrica.* Very variable, some may lack concentric sculpture. Deposit feeder.

HABITAT: Sandy mud beyond low tide. Infaunal.

LOCALITIES: Entire.

OCCURRENCE: Fairly common.

RANGE: Cape Cod to the West Indies and Texas.

GEOLOGIC RANGE: Oligocene to Recent.

CONCENTRIC NUT CLAM

Nuculana concentrica Say, 1824, *J. Acad. Natur. Sci. Phila.* 4(1):141.

L. *nucula* little nut, *con* from *contra* opposite side, *centri-* combining form "center" from *centrum* center.

SIZE: Length 12 to 18 mm.

COLOR: Yellow white, semiglossy.

SHAPE: Rather obese and moderately rostrate. Equivalve. Inequilateral.

ORNAMENT OR SCULPTURE: Adult shells appear smooth; have very fine, concentric growth lines on the ventral half of the valves. Beaks and area just below are smooth. Radial ridge on rostrum smooth, not crossed by strong threads.

HINGE AREA: Numerous chevron-shaped teeth on either side of umbo.

PALLIAL LINE & INTERIOR: Pallial sinus is small and rounded. Polished.

PERIOSTRACUM: Thin, yellowish.

REMARKS: More common in the beach drift in the eastern part of coast than to the south. Deposit feeder.

HABITAT: Clay bottoms beyond low tide. Infaunal.

LOCALITIES: Entire.

OCCURRENCE: Fairly common.

RANGE: Northwestern Florida to Texas.

GEOLOGIC RANGE: Upper Miocene to Recent.

Nuculana concentrica

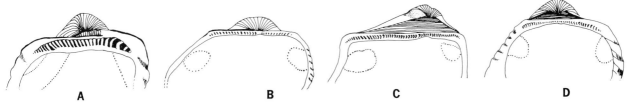

Fig. 27. Various types of ark hinges: *A, Noetia ponderosa*, beaks point posteriorly, most of ligament area anterior to beaks; *B, Barbatia domingensis*, ligament long, narrow, and posterior to beaks; less than twenty teeth in posterior half of hinge; *C, Arca imbricata*, ligament area wide, hinge long and narrow with numerous small teeth; *D, Anadara brasiliana*, beaks face each other, ligament short and transversely striate.

Subclass PTERIOMORPHIA Beurlen, 1944

Order ARCOIDA Stoliczka, 1871

Superfamily ARCACEA Lamarck, 1809

Family ARCIDAE Lamarck, 1809

Subfamily ARCINAE Lamarck, 1809

Genus *Arca* Linné, 1758

Subgenus *Arca* s.s.

MOSSY ARK
Arca (Arca) imbricata Bruguiére, 1792, *Encycl. Méth.*
   1:98.
L. *arca* box, purse, chest, *imbricare* to cover with over-
   lapping tiles.
SIZE: Height 1 in., length 2½ in.
COLOR: Whitish, concentrically marked with chestnut
   brown.

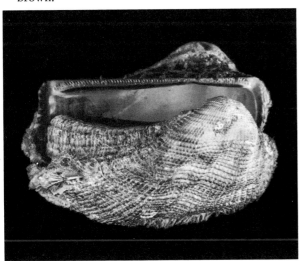

Arca (Arca) imbricata

SHAPE: Rectangular; hinge long, straight. Equivalve.
   Inequilateral.
ORNAMENT OR SCULPTURE: Numerous fine, irregular ribs crossed with growth lines giving a beaded
   appearance to sculpture. Posteriorly carinate. Large
   ventral byssal gape.
HINGE AREA: Wide, flat, pear-shaped ligamental area
   between umbones. Umbones more elevated than *A.
   zebra*. Hinge margin straight with many small trans-
   verse teeth.
PALLIAL LINE & INTERIOR: Smooth, dull purplish;
   muscle scars connected by simple pallial line. No
   siphon.
PERIOSTRACUM: Heavy, shaggy, brownish.
REMARKS: Attaches with byssal threads. Synonymous
   with *A. umbonata* Lamarck. Filter feeder.
HABITAT: On rocks or firm substrate. Byssate epifaunal
   nestler.
LOCALITIES: Entire.
OCCURRENCE: Common.
RANGE: Lake Worth, Florida, to the West Indies and
   Texas.
GEOLOGIC RANGE: Oligocene to Recent.

TURKEY WING
Arca (Arca) zebra (Swainson, 1833), *Zool. Illus.*,
   2nd ser. 3(26):118.
L. *arca* box, purse, chest, *zebra* striped.
SIZE: Length 2 to 3 in.
COLOR: Creamy white, streaked, and flecked with a
   chestnut brown in a zebra pattern.
SHAPE: Rectangular; hinge long, straight. Equivalve.
   Inequilateral.
ORNAMENT OR SCULPTURE: Ventral margin with
   wide byssal notch opposite umbones. Posterior mar-
   gin sinuate. About 26 narrow, rounded ribs with

Arca (Arca) zebra

Barbatia (Barbatia) cancellaria

finely ribbed, flat interspaces. Fine lamellar ridges across both ribs and interspaces.

HINGE AREA: Straight with wide, flat ligamental area between umbones. Numerous, small, transverse teeth.

PALLIAL LINE & INTERIOR: Smooth, white to purplish. Strong muscle and simple pallial impressions. No siphon.

PERIOSTRACUM: Heavy, brown, shaggy.

REMARKS: Occasionally found alive in the southern part of range in bamboo roots. It is more typical of the Mexican coast. Filter feeder.

HABITAT: On rocks, shell, roots in warm seas. Byssate epifaunal nestler.

LOCALITIES: South.

OCCURRENCE: Rare.

RANGE: North Carolina to Lesser Antilles and Texas.

GEOLOGIC RANGE: Upper Miocene to Recent.

Genus *Barbatia* Gray, 1847

Subgenus *Barbatia* s.s.

RED BROWN ARK

Barbatia (Barbatia) cancellaria (Lamarck, 1819), *Hist. Natur. Anim. sans Vert.* 6:41.

L. *barbatus* bearded, *cancellatus* latticelike.

SIZE: Length 1 to 1½ in.

COLOR: Red brown.

SHAPE: Oblique, rectangle. Equivalve. Inequilateral.

ORNAMENT OR SCULPTURE: Margins irregular, byssal gape on ventral edge. Numerous finely beaded ribs and irregular concentric growth lines.

HINGE AREA: Straight with narrow ligamental area between elevated umbones. Teeth obliquely inclined to center.

PALLIAL LINE & INTERIOR: Interior red brown, smooth. Simple pallial line with muscle scars. No siphon.

PERIOSTRACUM: Brown, heavy.

REMARKS: A deep-water species that is often brought in by shrimpers.

HABITAT: Offshore. Byssate epifaunal nestler.

LOCALITIES: Entire.

OCCURRENCE: Rare.

RANGE: Southern Florida, West Indies, western Gulf of Mexico.

WHITE BEARDED ARK

Barbatia (Barbatia) candida (Helbling, 1779), *Abh. Privatgessl. Bohm.* 4:129.

L. *barbatus* bearded, *candidatus* dressed in white.

SIZE: Length 1 to 2 in.

COLOR: White.

SHAPE: Oblique rectangle; hinge straight. Equivalve. Inequilateral.

ORNAMENT OR SCULPTURE: Margins irregular, slight byssal gape on ventral edge. Relatively thin for an ark. Numerous weak, slightly beaded ribs, those on the posterior dorsal area being very strongly beaded. Surface irregular.

Barbatia (Barbatia) candida

Barbatia (Acar) domingensis

HINGE AREA: Straight with narrow, lanceolate ligamental area between umbones, narrower than *Arca*. Teeth not parallel, obliquely inclined to center.

PALLIAL LINE & INTERIOR: Interior white, smooth. Simple pallial line with 2 muscle scars. No siphon.

PERIOSTRACUM: Brown, heavy, shaggy.

REMARKS: For an ark shell this one has relatively few hinge teeth. Filter feeder.

HABITAT: On rocks beyond low tide. Byssate epifaunal nestler.

LOCALITIES: Port Aransas, south.

OCCURRENCE: Rare.

RANGE: Lake Worth, Florida, to Brazil and Texas.

GEOLOGIC RANGE: Upper Miocene to Recent.

Subgenus *Acar* Gray, 1857

WHITE MINIATURE ARK

Barbatia (Acar) domingensis (Lamarck, 1819), *Hist. Natur. Anim. sans Vert.* 6:40.

L. *barbatus* bearded; from Dominican Republic.

SIZE: Length 12 to 18 mm.

COLOR: White to cream.

SHAPE: Rectangular; surface irregular. Equivalve. Inequilateral.

ORNAMENT OR SCULPTURE: Very distinctive, coarse reticulated surface. Shinglelike growth ridges.

HINGE AREA: Long, narrow ligament posterior to

beaks. Chevron-shaped teeth. Umbo curled to edge of hinge.

PALLIAL LINE & INTERIOR: Simple pallial line with 2 muscle scars. No sinus.

PERIOSTRACUM: Thin, yellowish brown.

REMARKS: Look for in holes and crevices of rocks on jetties. Filter feeder.

HABITAT: On rocks below low tide.

LOCALITIES: Southern half of coast.

OCCURRENCE: Common.

RANGE: North Carolina to Florida, Texas and Lesser Antilles.

Subgenus *Fugleria* Reinhart, 1937.

DOC BALES' ARK

Barbatia (Fugleria) tenera (C. B. Adams, 1845), *Proc. Boston Soc. Natur. Hist.* 2:9.

L. *barbatus* bearded, *tener*, thin, tender, delicate, young.

SIZE: Length 1 to 1½ in.

COLOR: White.

SHAPE: Trapezoidal, rather fat. Equivalve. Nearly equilateral.

ORNAMENT OR SCULPTURE: Thin shelled. Numerous rather evenly and finely beaded, cordlike ribs, stronger posteriorly. Small byssal gape.

HINGE AREA: Ligamental area fairly wide at the umbo end, narrowing anteriorly. Typical chevron teeth.

Barbatia (Fugleria) tenera

PALLIAL LINE & INTERIOR: Polished, white. Simple pallial line with 2 muscle scars. No siphon.

PERIOSTRACUM: Thin, brown.

REMARKS: This shell can be easily confused with *Anadara transversa*, but it is much thinner with a more "humped" and beaded surface. Filter feeder.

HABITAT: Offshore. Byssate epifaunal nestler.

LOCALITIES: Entire.

OCCURRENCE: Uncommon.

RANGE: Lake Worth, Florida, to Texas and the Caribbean and northern coast of South America.

Anadara (Cunearca) brasiliana

Subfamily ANADARINAE Reinhart, 1935

Genus *Anadara* Gray, 1847

Subgenus *Cunearca* Dall, 1898

INCONGRUOUS ARK

Anadara (Cunearca) brasiliana (Lamarck, 1819), *Hist. Natur. Anim. sans Vert.* 6:44.

Gk. *ana* without, *dara* gape; from Brazil.

SIZE: Length 1 to 2½ in.

COLOR: White.

SHAPE: Trigonal. Inequivalve. Inequilateral. Left valve overlaps right.

ORNAMENT OR SCULPTURE: Twenty-six to 28 radial ribs visible and stronger at margins. Simple pallial line with 2 muscle scars. No siphon.

HINGE AREA: Umbones well separated; ligament area wide and excavated. Oblique, comblike teeth become smaller toward center.

PALLIAL LINE & INTERIOR: Interior white, polished; ribs visible and stronger at margins. Simple pallial line with 2 muscle scars. No siphon.

PERIOSTRACUM: Thin, light brown.

REMARKS: Most common ark shell on the Texas coast, often found in pairs. Formerly known as *A. incongrua* Say. Filter feeder.

HABITAT: Offshore in shallow water. Infaunal.

LOCALITIES: Entire.

OCCURRENCE: Common.

RANGE: Southeastern United States, West Indies, and Texas.

GEOLOGIC RANGE: Miocene to Recent.

CHEMNITZ'S ARK

Anadara (Cunearca) chemnitzi (Philippi, 1851), *Z. für Malak.* 8:50.

Gk. *ana* without, *dara* gape; honoring Johan Chemnitz (1730–1800).

SIZE: Length about 1 in.

COLOR: White.

SHAPE: Trigonal. Inequivalve. Inequilateral.

ORNAMENT OR SCULPTURE: Twenty-six to 28 broad ribs, strongly recurved posteriorly, bearing barlike beads.

HINGE AREA: Umbones well separated and forward of the center of the ligamental area. Comblike teeth become smaller toward center.

PALLIAL LINE & INTERIOR: Interior white, polished;

Anadara (Cunearca) chemnitzi

ribs visible. Simple pallial line with 2 muscle scars. No siphon.

PERIOSTRACUM: Thin, light brown, scaly.

REMARKS: This shell is very similar to *A. brasiliana*, but it is heavier and smaller and more dorsally elevated. Filter feeder.

HABITAT: Offshore in shallow water. Epifaunal.

LOCALITIES: Extreme south.

OCCURRENCE: Uncommon.

RANGE: Greater Antilles to Brazil and Mexico.

GEOLOGIC RANGE: Upper Pleistocene? to Recent.

Subgenus *Lunarca* Reinhart, 1943

BLOOD ARK

Anadara (Lunarca) ovalis (Bruguière, 1789), *Encycl. Méth.* 1(1):110.

Gk. *ana* without, *dara* gape; L. *ovum* egg.

SIZE: Length 1½ to 2⅓ in.

COLOR: White.

SHAPE: Roundish to ovate. Equivalve. Inequilateral.

ORNAMENT OR SCULPTURE: Ribs have weak groove in center. Strong intercostal ribs. Left valve more heavily sculptured than right. No byssal gape.

HINGE AREA: Ligament very narrow. Umbones close

together. Comblike teeth, sloping differently at the ends than in the center.

PALLIAL LINE & INTERIOR: Simple pallial line with 2 muscle scars. No siphon. Ribs visible on polished white interior.

PERIOSTRACUM: Heavy, brown, usually worn off at umbones.

REMARKS: Frequently found alive on outer beaches. Live animal is red. This species on the Texas coast has been referred to as *A. campechiensis*. A filter feeder.

HABITAT: Just offshore. Infaunal.

LOCALITIES: Entire.

OCCURRENCE: Common.

RANGE: Cape Cod to the West Indies and the Gulf states.

GEOLOGIC RANGE: Miocene to Recent.

Anadara (Lunarca) ovalis

Subgenus *Anadara* s.s.

CUT-RIBBED ARK

Anadara (Anadara) lienosa Say, 1832, *Amer. Conch.* (4):pl. 36, fig. 1.

Gk. *ana* without, *dara* gape; L. *lien* spleen.

SIZE: Length 2½ to 5 in.

COLOR: White.

SHAPE: Obliquely rectangular. Equivalve. Inequilateral.

Anadara (Anadara) lienosa

ORNAMENT OR SCULPTURE: Very heavy shell. Thirty-five square ribs; each rib marked with a deep central groove that does not extend over umbones and is not present on more rounded posterior ribs.

HINGE AREA: Umbones incurved and flattened; hinge margin straight. Numerous, comblike teeth.

PALLIAL LINE & INTERIOR: Interior marked with delicate lines; margins crenulate. Simple pallial line and 2 muscle scars well defined. No siphon.

PERIOSTRACUM: Heavy, brown, usually worn off on upper part of shell.

REMARKS: The line down the center of the rib makes this large ark very distinctive and separates it (somewhat doubtfully) from *A. baughmani* Hertlein, a species from farther offshore, frequently brought in by shrimpers. Formerly known as *A. secticostata* Reeve and *A. lienosa floridana* Conrad. This species reviewed by Bird (1965). A filter feeder.

HABITAT: Deeper water offshore. Infaunal.

LOCALITIES: Central, south.

OCCURRENCE: Uncommon to rare.

RANGE: Southeastern United States, Texas, Greater Antilles.

GEOLOGIC RANGE: Pliocene to Recent.

TRANSVERSE ARK
Anadara (Anadara) transversa (Say, 1822), *J. Acad. Natur. Sci. Phila.* 2(1):269.

Gk. *ana* without, *dara* gape; L. *transversus* crosswise.

SIZE: Length 12 to 36 mm.

COLOR: White.

SHAPE: Transversely oblong. Left valve overlaps right. Equivalve. Inequilateral.

ORNAMENT OR SCULPTURE: Thirty to 35 rounded ribs per valve; ribs on left valve usually beaded, seldom so on right valve.

HINGE AREA: Long, narrow ligamental space separates the umbones. Numerous teeth are perpendicular to hinge line.

PALLIAL LINE & INTERIOR: Polished white interior. Simple pallial line with 2 muscle scars. No siphon.

PERIOSTRACUM: Brown, thick, usually worn off except at base of valves.

REMARKS: Unpaired valve may be hard to identify due to dissimilarity of sculpture on each valve. The long hinge distinguishes it from *A. ovalis*. A filter feeder.

HABITAT: Littoral to 6 fathoms. Inlet-influenced areas and offshore. Infaunal.

LOCALITIES: Entire.

OCCURRENCE: Common.

RANGE: South of Cape Cod to Florida and Texas to Carmen, Mexico.

GEOLOGIC RANGE: Miocene to Recent.

Anadara (Anadara) transversa

Family NOETIIDAE Stewart, 1930

Subfamily NOETIINAE Stewart, 1930

Genus *Noetia* Gray, 1840

Subgenus *Eontia* MacNeil, 1938

PONDEROUS ARK
Noetia (Eontia) ponderosa (Say, 1822), *J. Acad. Natur. Sci. Phila.* 2(1):267.
From Noe, Noah; L. *ponderosus* heavy.
SIZE: Length 2 to 2½ in.
COLOR: White.
SHAPE: Trigonal. Equivalve. Inequilateral.
ORNAMENT OR SCULPTURE: Heavy, thick shell. Thirty-two flattened square ribs with a fine line down the center. Posterior margin nearly straight, keeled. Fine concentric intercostal sculpture is absent from umbones.
HINGE AREA: Umbones well separated; sides of hinge area slope obliquely downward to a straight margin. Teeth comblike.
PALLIAL LINE & INTERIOR: White, polished. Simple pallial line and 2 strong, raised muscle scars. No siphon. No byssal threads in adults.
PERIOSTRACUM: Heavy, brown, velvetlike.
REMARKS: Often found in pairs. They do not come apart as readily as *A. brasiliana.* A filter feeder.

Side view

HABITAT: Offshore on sandy bottoms. Infaunal.
LOCALITIES: Entire.
OCCURRENCE: Common.
RANGE: Virginia to Key West, and the Gulf of Mexico.
GEOLOGIC RANGE: Miocene to Recent.

Order MYTILOIDA Férussac, 1822

Superfamily MYTILACEA Rafinesque, 1815

Family MYTILIDAE Rafinesque, 1815

Genus *Modiolus* Lamarck, 1799

TULIP MUSSEL
Modiolus americanus Leach, 1815, *Anim. Zool. Misc.* 2:32.
L. *modiolus* a small measure, or bucket on a water wheel; from America.
SIZE: Length 1 to 4 in.
COLOR: Light brown with a blush of rose red and streaks of purple.
SHAPE: Trigonal.
ORNAMENT OR SCULPTURE: Sculptured only with fine growth lines.
HINGE AREA: Anterior margin without teeth. Um-

Noetia (Eontia) ponderosa

Modiolus americanus

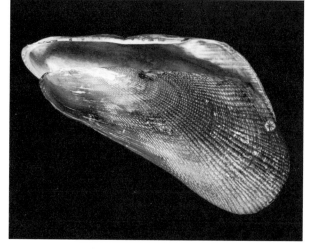

Modiolus demissus granosissimus

bones are away from the end of the shell. Ligament in groove posterior to umbones.

PALLIAL LINE & INTERIOR: Faint pallial line. Posterior muscle scar larger than anterior. No siphon.

PERIOSTRACUM: Shiny brown over anterior end; shaggy over posterior end.

REMARKS: A strikingly beautiful shell. The color is an intense red on the Texas coast and purple in Florida. This species is a synonym of *Volsella* or *Modiolus tulipa* Linné or Lamarck. A filter feeder.

HABITAT: Offshore below low-water mark. Byssate closely attached epifaunal.

LOCALITIES: Entire.

OCCURRENCE: Fairly common.

RANGE: North Carolina to the West Indies, Texas and Mexico.

RIBBED MUSSEL

Modiolus demissus granosissimus (Sowerby, 1914), *Proc. Malacol. Soc. London* 11:9.

L. *modiolus* a small measure, *demissus* low lying, humble, *granusus* full of grains.

SIZE: Length 2 to 4 in.

COLOR: Brown, shiny.

SHAPE: Elongate trigonal with rounded posterior margin.

ORNAMENT OR SCULPTURE: Numerous strong radial, beaded ribs that divide as they near posterior margin.

HINGE AREA: Long, narrow, no teeth. Ligament in a

groove posterior to umbones slightly away from the end of the shell.

PALLIAL LINE & INTERIOR: Nacreous. Rather strong pallial line with a small muscle scar at anterior end and a larger one at upper part of posterior end. No siphon.

PERIOSTRACUM: Thin, yellow brown.

REMARKS: Lives in the streams of salt marshes imbedded in soil around the roots of grasses. After storms or floods these shells will be washed onto the outer beaches in large clumps. They can live three weeks without water. Syn. *Geukensia demissus* Dillwyn. A filter feeder.

HABITAT: Salt marshes. Byssate closely attached epifaunal.

LOCALITIES: East, central.

OCCURRENCE: Common.

RANGE: Both sides of Florida, Texas and Yucatán.

Genus *Brachidontes* Swainson, 1840

Subgenus *Brachidontes* s.s.

SCORCHED MUSSEL

Brachidontes (*Brachidontes*) *exustus* (Linné, 1758), *Syst. Natur.*, 10th ed., p. 705.

Gk. *brachys* short, *dontes* teeth; L. *exustium* burn up.

SIZE: Length 18 mm. to 1½ in.

COLOR: Yellowish brown to dark brown. Interior metallic purple and white.

SHAPE: Moderately fan shaped.

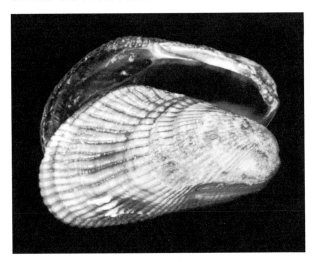

Brachidontes (Brachidontes) exustus

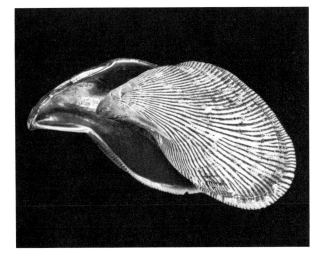

Brachidontes (Ischadium) recurvus

ORNAMENT OR SCULPTURE: Numerous rounded, radial ribs that divide as they near ventral margin and are eroded near umbones. Slight byssal gape in ventral margin.

HINGE AREA: Two small teeth at anterior end. At posterior end, beyond ligament, are 5 to 6 tiny teeth on edge of shell.

PALLIAL LINE & INTERIOR: Pallial line weak, posterior muscle scar larger than anterior. No siphon.

PERIOSTRACUM: Thin, yellowish brown.

REMARKS: This small mussel is often confused with *B. citrinus*, which is larger and has 4 anterior teeth. *B. exustus* is found attached by its byssus on the jetties and on oyster banks. It is not a brackish water inhabitant. A filter feeder.

HABITAT: On rocks or oyster reefs in saline bays and inlets. Byssate closely attached epifaunal.

LOCALITIES: Entire.

OCCURRENCE: Common.

RANGE: North Carolina to the West Indies.

GEOLOGIC RANGE: Pliocene to Recent.

Subgenus *Ischadium* Jukes-Brown, 1905

HOOKED MUSSEL

Brachidontes (Ischadium) recurvus (Rafinesque, 1820), *An. Gén. Scl. Phys. Bruxelles* 5:320.

Gk. *brachys* short, *dontes* teeth; L. *recurvus* bent, curved.

SIZE: Length 1 to 2½ in.

COLOR: Purplish gray.

SHAPE: Oval with a strong triangular hook on anterior end.

ORNAMENT OR SCULPTURE: Strong radial, rounded ribs that divide as they near ventral margin. Microscopic growth lines cross ribs.

HINGE AREA: Umbones at end of hook. Near the umbones are 3 or 4 small teeth. Ligament strong, in a large groove extending from umbones to peak of dorsal margin.

PALLIAL LINE & INTERIOR: Shiny purple and reddish brown, nacreous. Pallial line with one large posterior muscle scar. No siphon.

PERIOSTRACUM: Thin, yellow brown.

REMARKS: This species lives on low-salinity reefs of *Crassostrea virginica*. During periods of drought *B. exustus* and *Ostrea equestris* will completely replace *B. recurvus* and *C. virginica*. A filter feeder.

HABITAT: Oyster reef in low-salinity bays. Byssate closely attached epifaunal.

LOCALITIES: Entire.

OCCURRENCE: Common.

RANGE: Cape Cod to the West Indies and Texas.

GEOLOGIC RANGE: Pliocene to Recent.

Genus *Amygdalum* Megerle von Mühlfeld, 1811

PAPER MUSSEL

Amygdalum papyria (Conrad, 1846), *Proc. Acad. Natur. Sci. Phila.* 3:24.

L. *amygdalum* almond, *papyrus* paper.

Amygdalum papyria

Musculus lateralis

SIZE: Length 1 to 1½ in.

COLOR: Exterior grayish. Interior iridescent, metallic white.

SHAPE: Elongated fan shape.

ORNAMENT OR SCULPTURE: Very delicate and fragile with only fine concentric growth lines.

HINGE AREA: Umbones slightly beyond anterior end. Ligament weak and thin. No hinge teeth.

PALLIAL LINE & INTERIOR: Pallial line and muscle scar faint. No siphon.

PERIOSTRACUM: Thin, deciduous, greenish, and lacquerlike.

REMARKS: For these nestlers dig carefully in the soft mud around the roots of marine grasses. A filter feeder.

HABITAT: Saline bays and inlet-influenced area. Semi-infaunal.

LOCALITIES: Entire.

OCCURRENCE: Fairly common.

RANGE: Texas and Maryland to Florida.

Genus *Musculus* Röding, 1798

LATERAL MUSCULUS

Musculus lateralis (Say, 1822), *J. Acad. Natur. Sci. Phila.* 2:264.

L. *musculus* muscle, *later* brick, tile.

SIZE: Length 10 mm.

COLOR: Variable, light brown, pink, or greenish. Interior iridescent.

SHAPE: Oblong.

ORNAMENT OR SCULPTURE: Concentric growth lines over entire surface. Radial ribs on either end leaving

the disc section with only the finer growth lines.

HINGE AREA: Umbones not terminal. No hinge teeth.

PALLIAL LINE & INTERIOR: Smooth and nacreous. No siphon.

PERIOSTRACUM: Thin, yellowish brown to green.

REMARKS: Look in colonies of tunicates (sea pork) for this frail shell, where it is attached by its byssal threads. A filter feeder.

HABITAT: Attached to tunicates. Semi-infaunal.

LOCALITIES: Port Aransas, probably entire.

OCCURRENCE: Uncommon.

RANGE: North Carolina to Florida, West Indies and Texas.

Subgenus *Gregariella* Monterosato, 1884

ARTIST'S MUSSEL

Musculus (Gregariella) opifex (Say, 1825), *J. Acad. Natur. Sci. Phila.* 4:369.

L. *gregarius* living in flocks, *opifer* maker, or craftsman.

Musculus (Gregariella) opifex

SIZE: Length 17 mm.
COLOR: Reddish brown. Iridescent interior.
SHAPE: Oval.
ORNAMENT OR SCULPTURE: Smooth except for growth lines. Posterior dorsal area somewhat keeled.
HINGE AREA: Weak, tiny denticulations along hinge.
PALLIAL LINE & INTERIOR: Nacreous. No siphon.
PERIOSTRACUM: Long, brown, hairlike processes over keeled posterior end.
REMARKS: This beautiful shell weaves nests of byssal threads. A filter feeder.
HABITAT: Bores in rock and coral offshore. Semi-infaunal.
LOCALITIES: Entire.
OCCURRENCE: Fairly common.
RANGE: Hatteras to Brazil, Gulf of Mexico.

Lioberis castaneus

Genus *Lioberis* Dall, 1898

SAY'S CHESTNUT MUSSEL
Lioberis castaneus (Say, 1825), *J. Acad. Natur. Sci. Phila.* 2:226.
Gk. *leios* smooth; L. *berus* water snake, *castanea* chestnut colored.
SIZE: Length 25 mm.
COLOR: Bluish white.
SHAPE: Elongate oval, inflated.
ORNAMENT OR SCULPTURE: Smooth except for fine, concentric growth lines.
HINGE AREA: Umbones slightly back from anterior end. No hinge teeth.
PALLIAL LINE & INTERIOR: Smooth and nacreous. No marginal crenulation. No siphon.
PERIOSTRACUM: Shiny, brown.
REMARKS: Syn. *Botula castanea* Say, 1822. A filter feeder.
HABITAT: Open-bay margins, inlets, and along shore. Infaunal.
LOCALITIES: Central, south.
OCCURRENCE: Uncommon.
RANGE: South of Carolina to Florida Keys and the West Indies, Texas and Brazil.

Genus *Lithophaga* Röding, 1798

Subgenus *Diberus* Dall, 1898

MAHOGANY DATE MUSSEL
Lithophaga (Diberus) bisulcata (d'Orbigny, 1845), [La Sagra, *Hist. l'Ile de Cuba*], *Moll.* 2:333.
Gk. *lithos* stone, *phagen* to eat; L. *bisulcus* cloven.
SIZE: Length 1 to 1½ in.
COLOR: Mahogany brown.
SHAPE: Elongate cylindrical, coming to a point at one end.
ORNAMENT OR SCULPTURE: Weak concentric growth lines. Two oblique furrows going from dorsal margin to posterior. Covered with gray calcareous deposits.
HINGE AREA: Umbones not terminal. No teeth.
PALLIAL LINE & INTERIOR: Anterior muscle in front of umbo. Smooth, nacreous.
PERIOSTRACUM: Smooth, thin, brown, and lacquerlike.
REMARKS: Bring up rocks from jetties and carefully crush with hammer and chisel to remove these borers. A filter feeder.

Lithophaga (Diberus) bisulcata

HABITAT: In rocks in both shallow and deep water. Infaunal.

LOCALITIES: Entire.

OCCURRENCE: Common.

RANGE: St. Augustine, Florida, to Bermuda, Gulf of Mexico and West Indies to Brazil.

GEOLOGIC RANGE: Oligocene to Recent.

Subgenus *Myoforceps* P. Fischer, 1886

SCISSOR DATE MUSSEL
Lithophaga (Myoforceps) aristata (Dillwyn, 1817), *Cat. Rec. Shells* 1:303.

Gk. *lithos* stone, *phagen* to eat; L. *arista* ear of corn.

SIZE: Length 1 to 1½ in.

COLOR: Grayish white.

SHAPE: Cylindrical with one end pointed.

ORNAMENT OR SCULPTURE: Smooth. Points at posterior end that cross each other in scissor fashion are formed by a cementlike deposit over periostracum.

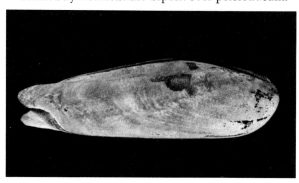

Lithophaga (Myoforceps) aristata

HINGE AREA: Umbones not terminal. No teeth.

PALLIAL LINE & INTERIOR: Interior nacreous.

PERIOSTRACUM: Shiny, thin, brown.

REMARKS: Look for in the same place as *L. bisulcata*; however, this shell is less common. A filter feeder.

HABITAT: In rocks in both shallow and deep water. Infaunal.

LOCALITIES: Central, south.

OCCURRENCE: Fairly common.

RANGE: Southern Florida and the West Indies; La Jolla, California, to Peru.

GEOLOGIC RANGE: Oligocene to Recent.

Superfamily PINNACEA Leach, 1819

Family PINNIDAE Leach, 1819

Genus *Atrina* Gray, 1840

SPINY PEN SHELL
Atrina seminuda (Lamarck, 1819), *Hist. Natur. Anim. sans Vert.* 6:131.

L. *atrium* an opening, *semi* half, *nudus* naked.

SIZE: Length up to 9 or 10 in.

COLOR: Translucent, grayish tan sometimes mottled with purple brown.

SHAPE: Wedge shaped.

ORNAMENT OR SCULPTURE: Spinose, radiating ribs with the ventral slope usually smooth. Often the ribs are smooth. Growth lines are fine.

HINGE AREA: Longest dimension of shell, straight, no teeth.

PALLIAL LINE & INTERIOR: Nacreous area is about ½ to ⅔ length of shell. Small posterior adductor muscle scar lies well within the nacreous area; it never protrudes beyond the nacreous area. Anterior adductor scar small and nearly as wide as the anterior end of shell. The scar is what separates this shell from *A. rigida*. No siphon.

PERIOSTRACUM: None visible.

REMARKS: Alive in colonies in inlet-influenced bays and offshore. It lives in the mud with only the posterior margin sticking out and attached with byssal threads. The margin of the mantle of the animal is yellow, the foot is lighter yellow. Sexes are separate. Adductor muscle is edible. Less common than *A. serrata*. A small crab is commensal with *A. seminuda*. A filter feeder.

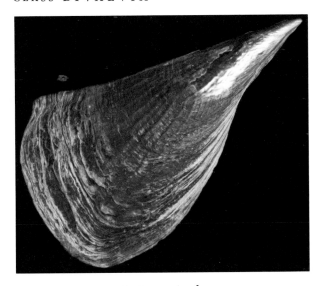

Atrina seminuda

HABITAT: Offshore and in inlet-influenced bays. Semi-infaunal.
LOCALITIES: Entire.
OCCURRENCE: Fairly common.
RANGE: Eastern United States, Brazil and Texas.
GEOLOGIC RANGE: Pleistocene to Recent.

SAWTOOTH PEN SHELL
Atrina serrata (Sowerby, 1825), *Tanh Cat. App.*, p. 5.
L. *atrium* an opening, *serra* a saw.
SIZE: Length up to 11½ in.
COLOR: Translucent light tan to medium greenish brown.

Atrina serrata

SHAPE: Wedge shaped, thin, fragile, shiny.
ORNAMENT OR SCULPTURE: Uniform sculpture of about 30 low ribs covered with fluted projections. This sculpture is much finer on ventral slope.
HINGE AREA: Straight to slightly concave without teeth.
PALLIAL LINE & INTERIOR: Nacreous layer thin and usually covering three-fourths the length of the valve. Posterior adductor muscle scar nearly circular and set well within the nacreous layer. Anterior scar small. No siphon.
PERIOSTRACUM: None visible.
REMARKS: Much thinner and spines more tubular than *A. seminuda*, and the nacreous area is larger. At times this shell is washed up on Gulf beaches by the hundreds. The adductor muscle may be prepared to eat in the same way that scallops are. These animals are harvested in Mexico for canning as scallops. A pair of colorless shrimp live commensally with *A. serrata*. Originally described from Texas coast. A filter feeder.
HABITAT: Offshore in inner shelf zone. Semi-infaunal.
LOCALITIES: Entire.
OCCURRENCE: Common.
RANGE: North Carolina to the West Indies and Texas.
GEOLOGIC RANGE: Pliocene to Recent.

Order PTERIOIDA Newell, 1965

Suborder PTERIINA Newell, 1965

Superfamily PTERIACEA Broderip, 1839

Family PTERIIDAE Broderip, 1839

Subfamily PTERIINAE Broderip, 1839

Genus *Pinctada* Röding, 1798

ATLANTIC PEARL OYSTER
Pinctada radiata (Leach, 1814), *Zool. Misc.* 1:98.
Span. *pintado* marked with spots; L. *radius* a ray, rayed.
SIZE: Length 1½ to 3 in.
COLOR: Purplish brown or black, variable.
SHAPE: Roundish with two short wings. Inequivalve. Inequilateral.
ORNAMENT OR SCULPTURE: Flat with concentrically arranged scaly projections of periostracum. Byssal gape under anterior wing of right valve.

Pinctada radiata

HINGE AREA: Straight with single lateral tooth in left valve, double laterals in right.

PALLIAL LINE & INTERIOR: Very nacreous with wide polished border around margins. No siphon.

PERIOSTRACUM: Dark brown, brittle with scalelike projections.

REMARKS: Less common than *P. colymbus*. Usually only juveniles are found. Forms lovely pearls. A filter feeder.

HABITAT: Offshore attached to rock or gorgonia, a soft coral. Byssate free-swinging epifaunal.

LOCALITIES: Southern half of coast.

OCCURRENCE: Uncommon to rare.

RANGE: Southern half of Florida and the West Indies, and Texas and Brazil.

Genus *Pteria* Scopoli, 1777

ATLANTIC WING OYSTER

Pteria colymbus (Röding, 1798), *Mus. Boltenianum,* p. 166.

Gk. *pteron* wing, *kolymbos* a diver.

SIZE: Length 1½ to 3 in.

COLOR: Brownish black with broken, radial lines of cream color. Interior highly nacreous.

SHAPE: Oval with a posterior drawn-out wing. Inequivalve. Inequilateral.

ORNAMENT OR SCULPTURE: Smooth. Byssal gape near anterior wing.

HINGE AREA: Hinge margin straight with elongated posterior ear, or wing. Two small cardinal teeth and 1 lateral tooth in each valve.

PALLIAL LINE & INTERIOR: A single, almost central muscle scar. Nacreous. No siphon.

PERIOSTRACUM: Light brown colored, shaggy with spiny projections that often project over margin.

REMARKS: On outer beaches pieces of gorgonia with attached "winged oysters" are sometimes washed up. A filter feeder.

HABITAT: Offshore and on "whistling buoy" out of Port Aransas. Byssate free-swinging epifaunal.

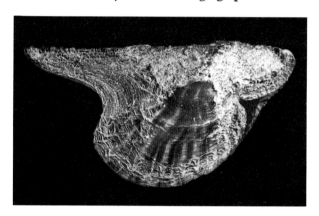

Pteria colymbus

LOCALITIES: Southern two-thirds of coast.

OCCURRENCE: Uncommon.

RANGE: Southeastern United States and the West Indies, and Texas and Brazil.

GEOLOGIC RANGE: Pliocene to Recent.

Family ISOGNOMONIDAE Woodring, 1925

Genus *Isognomon* Solander, 1786

FLAT TREE OYSTER

Isognomon alatus (Gmelin, 1791), *Syst. Natur.,* 12th ed., p. 339.

Gk. *isos gnomon* equal parallelogram; L. *alatus* winged.

SIZE: Length 2 to 3 in.

COLOR: Drab, dirty gray. Interior nacreous, purplish.

SHAPE: Flat, fan shaped.

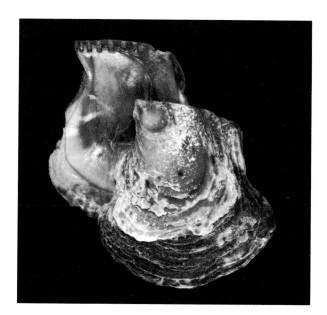

Isognomon alatus

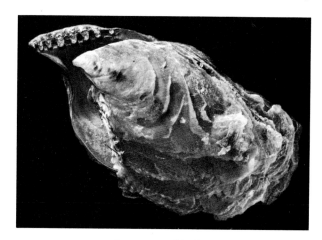

Isognomon bicolor

ORNAMENT OR SCULPTURE: Rough with flaky lamellations.

HINGE AREA: Straight hinge has 8 to 12 oblong grooves into which are set small, brown resiliums. Byssal gape on anterior margin near dorsal margin.

PALLIAL LINE & INTERIOR: Nacreous, dark. Discontinuous pallial line. No siphon.

PERIOSTRACUM: None visible.

REMARKS: On rocks in Port Aransas area. It may be missed due to its undistinguished appearance. A filter feeder.

HABITAT: On rocks in inlet-influenced areas. Byssate free-swinging epifaunal.

LOCALITIES: Port Aransas, south.

OCCURRENCE: Uncommon.

RANGE: Southern half of Florida, West Indies and Texas, southern Mexico and Central America.

TWO-TONED TREE OYSTER

Isognomon bicolor (C. B. Adams, 1845), *Proc. Boston Soc. Natur. Hist.* 2:9.

Gk. *isos gnomon* equal parallelogram; L. *bi* two, *color* hue.

SIZE: Length 12 mm. to 2 in.

COLOR: Yellowish with purple splotches. Variable.

SHAPE: Very irregular, elongated parallelogram.

ORNAMENT OR SCULPTURE: Rough with flaky lamellations.

HINGE AREA: Hinge short, straight, with 4 to 8 small, square sockets. Anterior byssal gape near dorsal margin.

PALLIAL LINE & INTERIOR: A sharply raised ridge separates the central area from the marginal area. Color of interior frequently different on opposite sides of ridge. Discontinuous pallial line. No siphon.

PERIOSTRACUM: None visible.

REMARKS: More common than *I. alatus*. A filter feeder.

HABITAT: On rocks in inlet-influenced areas. Byssate free-swinging epifaunal.

LOCALITIES: Port Aransas, south.

OCCURRENCE: Uncommon.

RANGE: Florida Keys, Bermuda, Caribbean, Texas and southern Mexico.

Superfamily PECTINACEA Rafinesque, 1815

Family PECTINIDAE Rafinesque, 1815

Subfamily PECTININAE Rafinesque, 1815

Genus *Pecten* Muller, 1776

Subgenus *Euvola* Dall, 1898

RAVENEL'S SCALLOP

Pecten (Euvola) raveneli Dall, 1898, *Trans. Wagner Free Inst. Sci.* 3(4):721.

L. *pecten* a comb; dedicated to Dr. Edmund Ravenel, nineteenth-century collector.

Pecten (Euvola) raveneli

SIZE: Length 1 to 2 in.

COLOR: Variable from pink to purple. Ribs white. Lower valve dark, upper valve white.

SHAPE: Fan shaped. Upper valve very flat. Lower valve inflated.

ORNAMENT OR SCULPTURE: About 21 rounded, radiating ribs with wider interspaces, fine concentric growth lines. Each rib has a groove down the middle.

HINGE AREA: Ears almost equal. Hinge straight. Umbones flat.

PALLIAL LINE & INTERIOR: Smooth with margin beyond pallial line showing the ribs. Simple pallial line with single muscle scar. No siphon.

PERIOSTRACUM: None visible.

REMARKS: This filter feeder is usually only washed up after storms.

HABITAT: Offshore, 12 to 35 fathoms. Free-living epifaunal.

LOCALITIES: South, mainly Port Isabel area.

OCCURRENCE: Rare.

RANGE: North Carolina to the Gulf of Mexico and the West Indies.

GEOLOGIC RANGE: Pliocene to Recent.

Subfamily CHLAMYDINAE Korobkov, 1960

Genus *Aequipecten* P. Fischer, 1886

Subgenus *Argopecten* Monterosato, 1889

CALICO SCALLOP

Aequipecten (Argopecten) gibbus (Linné, 1758), Syst. *Natur.*, 10th ed., p. 698.

L. *aequus, aequi* equal, *pecten* a comb, *gibbus* a hump.

SIZE: Length 1 to 2 in.

COLOR: Upper valve mottled with red or purple. Lower valve white with a little color.

SHAPE: Almost circular.

ORNAMENT OR SCULPTURE: About 20 squared ribs. Valves inflated.

HINGE AREA: Ears small. Hinge straight, no teeth.

PALLIAL LINE & INTERIOR: Single muscle scar. Simple pallial line. No siphon.

PERIOSTRACUM: None visible.

REMARKS: Expect only worn, faded valves. Rarely found alive at Port Isabel. *Gibbus* is rather small for the Texas beach pectens. A filter feeder.

HABITAT: Offshore on intermediate shelf. Free-living epifaunal.

LOCALITIES: Entire.

OCCURRENCE: Rare.

RANGE: Eastern United States, Gulf of Mexico and the West Indies.

Subgenus *Plagioctenium* Dall, 1898

BAY SCALLOP

Aequipecten (Plagioctenium) amplicostatus (Dall, 1898), *Trans. Wagner Free Inst. Sci.* 3(4):747.

Aequipecten (Argopecten) gibbus

Aequipecten (Plagioctenium) amplicostatus

L. *aequus, aequi* equal, *pecten* a comb, *amplius* abundant, *costa* a rib.

SIZE: Diameter 2 to 3 in.

COLOR: Upper valve mottled gray and white. Lower valve white.

SHAPE: Fan shaped, almost circular.

ORNAMENT OR SCULPTURE: Twelve to 17 high, squarish ribs. Rather inflated.

HINGE AREA: Ears not as wide as shell. Hinge straight, no teeth.

PALLIAL LINE & INTERIOR: Single muscle scar. Simple pallial line. No siphon.

PERIOSTRACUM: None visible.

REMARKS: This shell lives in the bays and is often washed up by the hundreds after a norther. The worn valves found on outer beaches have changed color to orange or black. In the aquarium the eyes of this active bivalve are visible around the mantle edge, and it will "flap" its valves and skip about the tank. A filter feeder. See Clarke (1965). *A. irradians amplicostatus* of authors.

HABITAT: Bay, open lagoons. Free-living epifaunal.

LOCALITIES: Entire.

OCCURRENCE: Common.

RANGE: Central Texas to Tuxpan, Mexico.

GEOLOGIC RANGE: Lower Miocene to Recent.

Subgenus *Aequipecten* s.s.

ROUGH SCALLOP

Aequipecten (Aequipecten) muscosus (W. Wood, 1828), *Index Testac. Suppl.*, p. 47, pl. 2, fig. 2.

L. *aequus, aequi* equal, *pecten* a comb, *muscus* moss, covered with sponge.

SIZE: Diameter 18 to 30 mm.

COLOR: Variable from red, orange, yellow, or mottled purple.

SHAPE: Fan shaped. Inequivalve. Inequilateral.

ORNAMENT OR SCULPTURE: About 20 spinose ribs with the spoon-shaped spines arranged in three rows. Small spines present in the interspaces.

HINGE AREA: Ears equal width of shell. Hinge straight, no teeth.

PALLIAL LINE & INTERIOR: One large muscle scar. Simple pallial line. No siphon.

PERIOSTRACUM: None visible.

REMARKS: Only worn and faded valves are found on the outer beaches. A filter feeder.

HABITAT: Offshore on banks. Free-living epifaunal.

LOCALITIES: South.

OCCURRENCE: Rare.

RANGE: Florida Keys, Mexico and the West Indies to Brazil.

GEOLOGIC RANGE: Recent.

Aequipecten (Aequipecten) muscosus

Lyropecten (Nodipecten) nodosus

Genus *Lyropecten* Conrad, 1862

Subgenus *Nodipecten* Dall, 1898

LION'S PAW

Lyropecten (Nodipecten) nodosus (Linné, 1758), *Syst. Natur.*, 10th ed., p. 698.
L. *lyro* lyre shaped, *pecten* a comb, *nodus* knot.
SIZE: Diameter 3 to 6 in.
COLOR: Bright red, deep orange, or maroon red.
SHAPE: Fan shaped. Nearly equivalve. Inequilateral.
ORNAMENT OR SCULPTURE: Seven to 9 large, coarse ribs that have large, hollow nodules. The shell also has numerous cordlike riblets.
HINGE AREA: Two ears and a straight toothless hinge.
PALLIAL LINE & INTERIOR: A single large muscle scar. Simple pallial line. No siphon.
PERIOSTRACUM: None visible.
REMARKS: Halves of this striking shell are found, rarely, on the southern part of the coast, but living specimens are found by divers on banks offshore or in shrimp trawls. *L. subnodosus* Sowerby is its counterpart on the Pacific coast with one more rib. A filter feeder.
HABITAT: Offshore. Epifaunal fissure dweller.
LOCALITIES: South.
OCCURRENCE: Rare.

RANGE: Florida Keys and the West Indies, Mexico and Texas.
GEOLOGIC RANGE: Oligocene to Recent.

Family PLICATULIDAE Watson, 1930

Genus *Plicatula* Lamarck, 1801

KITTEN'S PAW

Plicatula gibbosa Lamarck, 1801, *Syst. Anim.*, p. 132.
L. *plicatus* folded, *ula* fem. dim., *gibbus* a hump.
SIZE: Length 1 in.
COLOR: Whitish with red brown lines on the ribs.
SHAPE: Fan shaped. Inequivalve.
ORNAMENT OR SCULPTURE: Rather heavy with 5 to 7 raised ribs that crenulate the margin so that valves interlock. Attached by right valve.
HINGE AREA: Two hinge teeth in each valve lock into corresponding notches in the opposite valve, as in *Spondylus americanus*. Ligament internal.
PALLIAL LINE & INTERIOR: Simple pallial line. No siphon.
PERIOSTRACUM: None visible.
REMARKS: Dead shells are not uncommon on the outer beaches and in spoil banks, but the shell is probably

Plicatula gibbosa

not living near shore at the present time. Attaches by right valve. A filter feeder.

HABITAT: Offshore on banks. Cemented epifaunal.

LOCALITIES: Entire.

OCCURRENCE: Uncommon.

RANGE: North Carolina to Florida, the Gulf states and the West Indies.

GEOLOGIC RANGE: Oligocene? Pleistocene to Recent.

Family SPONDYLIDAE Gray, 1826

Subfamily SPONDYLINAE Gray, 1826

Genus *Spondylus* Linné, 1758

ATLANTIC THORNY OYSTER

Spondylus americanus Hermann, 1781, *Der Natur. Forscher.* 16:51.

Gk. *spondylus* vertebra, spines; from America.

SIZE: Diameter 3 to 4 in.

COLOR: Variable; white with orange or yellow umbones, rose or cream.

SHAPE: Nearly circular to oval. Inequivalve.

ORNAMENT OR SCULPTURE: Adults have spines up to 2 inches in length that are erect and arranged radially. Lower valve is larger and deeper than upper.

HINGE AREA: The 2 large cardinal teeth on either side of ligament in each valve align with sockets in the opposite valve.

Beachworn specimen

PALLIAL LINE & INTERIOR: Smooth and white with one large muscle scar. No siphon.

PERIOSTRACUM: None visible.

REMARKS: The worn, faded valves found on the Texas beaches bear little resemblance to the spectacular living shell that is brought up by divers offshore. The ball and socket hinge distinguishes these eroded specimens from *Chama*. A filter feeder.

HABITAT: Offshore attached to reefs. Cemented epifaunal.

LOCALITIES: Southern half of coast.

OCCURRENCE: Uncommon.

RANGE: Florida, West Indies, Texas, Central America and Brazil.

GEOLOGIC RANGE: Pleistocene to Recent.

Superfamily ANOMIACEA Rafinesque, 1815

Family ANOMIIDAE Rafinesque, 1815

Genus *Anomia* Linné, 1758

COMMON JINGLE SHELL

Anomia simplex d'Orbigny, 1845, [La Sagra, *Hist. l'Ile de Cuba*], *Moll.* 2:367.

Gk. *anomoios* unlike; L. *simplex* simple.

SIZE: Length 1 to 2 in.

COLOR: Variable; translucent yellow or dull orange. Shiny.

Spondylus americanus

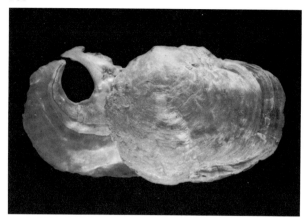

Anomia simplex

Pododesmus rudis

SHAPE: Subcircular. Upper valve more convex than flat attached valve.

ORNAMENT OR SCULPTURE: Wavy and undulating. Conforms to contours and texture of the surface on which it is attached. Lower valve has a hole near umbones from which a chitinous byssus projects to anchor the shell.

HINGE AREA: Ligament not supported with teeth or ridges.

PALLIAL LINE & INTERIOR: Nacreous interior. Pallial line not distinct. No siphon. Upper valve has large muscle scar opposite the hole and two smaller ones below.

PERIOSTRACUM: None visible.

REMARKS: This pretty little shell is often attached to an oyster or old shell. Usually only the top valve is found in the drift. Some old specimens are black from being buried in the mud. A filter feeder.

HABITAT: Hypersaline oyster or rock reef. Byssate closely attached epifaunal.

LOCALITIES: Entire.

OCCURRENCE: Common.

RANGE: Eastern United States, Gulf of Mexico and West Indies to Brazil.

GEOLOGIC RANGE: Miocene to Recent.

Genus *Pododesmus* Philippi, 1837

ROUGH JINGLE SHELL
Pododesmus rudis (Broderip, 1834), *Proc. Zool. Soc. London* 2:2.
Gk. *pous* foot, *desmus* bound; L. *rudis* rough.
SIZE: Length up to 2 in.

COLOR: Brownish, lower valve white.

SHAPE: Irregular. Upper valve convex. Lower valve flat with a hole near the umbones.

ORNAMENT OR SCULPTURE: Rough, wavy surface arranged somewhat concentrically.

HINGE AREA: Rather narrow, no hinge teeth.

PALLIAL LINE & INTERIOR: The muscle scar in the top valve is opposite the hole with a second large scar diagonally below it. No siphon.

PERIOSTRACUM: None visible.

REMARKS: C. Young only recently found this inconspicuous shell while diving off the jetties at Port Isabel. It attaches with a byssus, as does *Anomia*. A filter feeder.

HABITAT: On rock in inlet-influenced areas. Byssate closely attached epifaunal.

LOCALITIES: South.

OCCURRENCE: Rare.

RANGE: Florida, Texas and the West Indies.

GEOLOGIC RANGE: Oligocene to Recent.

Superfamily LIMACEA Rafinesque, 1815

Family LIMIDAE Rafinesque, 1815

Genus *Lima* Bruguière, 1797

Subgenus *Lima* s.s.

SPINY LIMA
Lima (Lima) lima (Linné, 1758), *Syst. Natur.*, 10th ed., p. 699.
L. *lima* file.
SIZE: Length 1 to 1½ in.

Lima (Lima) lima

Lima (Limaria) pellucida

COLOR: White.

SHAPE: Oblique oval. Elongate.

ORNAMENT OR SCULPTURE: Fragile shell. Radial ribs are spinose.

HINGE AREA: Hinge line is oblique. Anterior ear much smaller than posterior ear.

PALLIAL LINE & INTERIOR: Interior polished, pallial line simple, single muscle scar. No siphon.

PERIOSTRACUM: Thin, usually lost.

REMARKS: This is a tropical species that occasionally comes to shore attached to bamboo roots or coconuts. A filter feeder.

HABITAT: Offshore. Epifaunal byssate fissure dweller.

LOCALITIES: Port Aransas, south.

OCCURRENCE: Rare.

RANGE: Southeastern Florida and the West Indies.

Subgenus *Limaria* Link, 1807

ANTILLEAN LIMA

Lima (Limaria) pellucida C. B. Adams, 1846, *Proc. Boston Soc. Natur. Hist.* 2:103.

L. *lima* file, *pelluceo* to be transparent.

SIZE: Length 18 to 25 mm.

COLOR: Translucent white.

SHAPE: Oblique oval, elongate.

ORNAMENT OR SCULPTURE: Fragile shell gapes on either side. Radial ribs small, fine, uneven in size and distribution. Margins finely serrate.

HINGE AREA: Oblique, hinge partly external, no hinge teeth. Hinge ears almost equal in length.

PALLIAL LINE & INTERIOR: Interior polished, pallial line simple, single muscle scar. No siphon.

PERIOSTRACUM: Thin, light brown, usually lost.

REMARKS: The living mollusk has a delicate, "fringed" mantle. It uses these mantle tentacles to move about on the bottom; like the pectens, it is also able to flap its valves to swim about. It has fine byssal threads. A filter feeder.

HABITAT: Offshore. Epifaunal byssate fissure dweller.

LOCALITIES: Port Aransas, south.

OCCURRENCE: Rare.

RANGE: Southeastern United States and the West Indies, Texas.

Suborder OSTREINA Férussac, 1822

Superfamily OSTREACEA Rafinesque, 1815

Family OSTREIDAE Rafinesque, 1815

Genus *Crassostrea* Sacco, 1897

EASTERN OYSTER

Crassostrea virginica (Gmelin, 1792), *Hist. Conch.*, p. 200.

L. *crassus* thick; Gk. *ostrea* oyster; from Virginia.

SIZE: Length 2 to 6 in.

COLOR: Dull gray.

SHAPE: Most irregular and variable from oval to wierdly elongate.

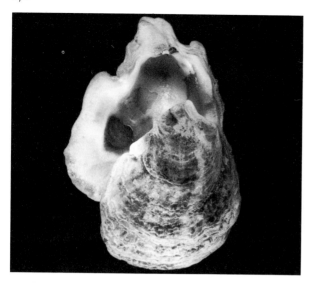

Crassostrea virginica

ORNAMENT OR SCULPTURE: Surface rough with leaf-like scales. Valve margins slightly undulating or straight. Beaks long and curved. Upper valve smaller, flatter, and smoother than lower.

HINGE AREA: Shell is attached at umbo of left valve, which is longer than that of the right valve. Both umbones have a central channel for ligamentary attachment.

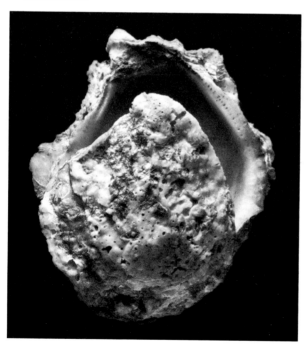

Ostrea equestris

PALLIAL LINE & INTERIOR: Muscle scar is subcentral and colored a deep purple. Interior smooth. No siphon.

PERIOSTRACUM: Eroded.

REMARKS: Bulletin No. 40 of the Texas Game and Fish Commission is recommended to those who want to know more about this valuable "shellfish." It is ovoviviparous and produces over 50 million eggs. It does not have the interior marginal teeth that *O. equestris* has. A filter feeder.

HABITAT: Brackish bays and estuaries. Cemented epifaunal.

LOCALITIES: Entire.

OCCURRENCE: Common.

RANGE: Gulf of St. Lawrence to Gulf of Mexico, West Indies.

GEOLOGIC RANGE: Miocene to Recent.

Genus *Ostrea* Linné, 1758

HORSE OYSTER OR CRESTED OYSTER
Ostrea equestris Say, 1834, *Amer. Conch.* 6:58.
Gk. *ostrea* oyster; L. *eques* horseman.

SIZE: Length 1 to 2 in.

COLOR: Dull gray to brownish. Interior gray green.

SHAPE: Rather oval, fairly constant.

ORNAMENT OR SCULPTURE: Surface rough with raised margins that are crenulated. Left valve flatter than right.

HINGE AREA: Six to 12 teeth on larger valve with corresponding cavities on smaller valve. Hinge narrow and curved.

PALLIAL LINE & INTERIOR: Muscle centrally located and scar not pigmented. No siphon.

PERIOSTRACUM: None visible.

REMARKS: Due to its economic importance this species has been thoroughly studied. Menzel's (1955) comparison with *C. virginica* should be read. This species requires a more saline environment than *C. virginica* and will replace the latter during sustained drought. It is larviparous and lays up to 1 million eggs. A filter feeder.

HABITAT: High-salinity oyster reef. Cemented epifaunal.

LOCALITIES: Entire.

OCCURRENCE: Common.

RANGE: Southeastern United States, Gulf states, and West Indies.

GEOLOGIC RANGE: Pleistocene to Recent.

Subclass HETERODONTA Neumayr, 1884

Order HIPPURITOIDA Newell, 1965

Superfamily CHAMACEA Blainville, 1825

Family CHAMIDAE Blainville, 1825

Genus *Chama* Linné, 1758

LITTLE CORRUGATED JEWEL BOX
Chama congregata Conrad, 1833, *Amer. J. Sci.*
   23:341.
Gk. *cheme* to gape; L. *congregatic* united.
SIZE: Length up to 1 in.
COLOR: Red.
SHAPE: Round. Right, or upper, valve much smaller
   than left.
ORNAMENT OR SCULPTURE: Surface covered with
   numerous low axial corrugations. Left valve is at-
   tached. Umbones twist to the right. Never has foli-
   aceous appearance of *C. macerophylla.*
HINGE AREA: Umbones turn from right to left. Left
   valve has 2 cardinal teeth, 1 heavy and rough; in
   the right are 2 widely separated small cardinal teeth.
   Both valves have 1 small posterior lateral tooth.
PALLIAL LINE & INTERIOR: Two large muscle scars
   connected by simple pallial line. No siphon.
PERIOSTRACUM: None visible.
REMARKS: Usually only badly eroded right valves are
   found in the drift. The shell is more likely to be
   attached to the valves of *Atrina serrata.* A filter
   feeder.

HABITAT: Offshore in calcareous banks. Cemented epi-
   faunal.
LOCALITIES: Entire.
OCCURRENCE: Uncommon.
RANGE: North Carolina to Florida and the West Indies
   and Texas and Central America.
GEOLOGIC RANGE: Miocene to Recent.

LEAFY JEWEL BOX
Chama macerophylla Gmelin, 1791, *Syst. Natur.*, 13th
   ed., p. 3304.
Gk. *cheme* to gape; L. *macero* soften; Gk. *phylon* race
   or tribe.
SIZE: Length 1 to 3 in.
COLOR: Variable; white, yellow, purple, or a combination of the three.
SHAPE: Roundish. Left, or lower, valve larger and
   deeper than upper.
ORNAMENT OR SCULPTURE: Surface covered with
   scalelike fronds. Inner margins of valves have tiny
   crenulations. Attached left valve larger and deeper
   than the right, which serves as a cover.
HINGE AREA: Umbones turn from right to left. Hinge
   thick with an oblique arched cardinal tooth and a
   straight furrow.
PALLIAL LINE & INTERIOR: Pallial line simple, connecting two muscle scars. No siphon.
PERIOSTRACUM: None visible.
REMARKS: Only very worn valves of this species are
   found in outer beach drift. A filter feeder.
HABITAT: Deeper calcareous banks offshore in crevices. Cemented epifaunal.

Chama congregata

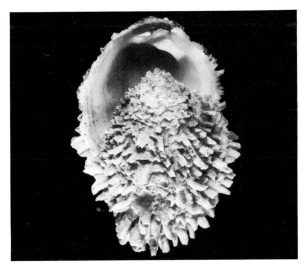

Chama macerophylla

LOCALITIES: Southern half of coast.

OCCURRENCE: Rare.

RANGE: Southeastern Florida and the West Indies and Texas to southern Mexico, Central America, and northern coast of South America.

GEOLOGIC RANGE: Pleistocene to Recent.

Genus *Pseudochama* Odhner, 1917

ATLANTIC LEFT-HANDED JEWEL BOX
Pseudochama radians Lamarck, 1819, *Hist. Natur. Anim. sans Vert.* 6:96.

Gk. *pseudes* false, *cheme* to gape; L. *radians* emit rays.

SIZE: Length 1 to 3 in.

COLOR: Cream.

SHAPE: Quadrate.

ORNAMENT OR SCULPTURE: Sculpture consists of 16 to 35 spiny radial ribs. Surface pitted. Attached right valve larger and deeper than left valve.

HINGE AREA: Umbones turn from left to right. Hinge thick.

PALLIAL LINE & INTERIOR: Pallial line simple, connecting 2 muscle scars. No siphon.

PERIOSTRACUM: None visible.

REMARKS: This is a mirror image of *C. macerophylla* Gemlin. A filter feeder.

HABITAT: Offshore on calcareous banks. Cemented epifaunal.

LOCALITIES: Entire.

OCCURRENCE: Rare.

RANGE: Southern Florida and the West Indies.

Genus *Echinochama* P. Fischer, 1887

TRUE SPINY JEWEL BOX
Echinochama cornuta (Conrad, 1866), *Amer. J. Conch.* 2(2): 105.

L. *echinus* sea urchin; Gk. *cheme* to gape; L. *cornus* horn.

SIZE: Length 1 to 1½ in.

COLOR: White.

SHAPE: Quadrate. Inflated.

ORNAMENT OR SCULPTURE: Seven to 9 radial rows of heavy, short spines. Shell heavy. Coarse granulations between ribs.

HINGE AREA: Umbones curved forward over a large, wide, heart-shaped lunule. Large cardinal tooth in left valve.

PALLIAL LINE & INTERIOR: Pallial line simple, connecting 2 muscle scars. No siphon.

PERIOSTRACUM: None visible.

REMARKS: This shell starts its life free swimming, becomes attached by the right valve for a period, and ends its life unattached. Near Port Mansfield jetty it makes up a large part of the beach shell, probably Pliocene fossils. A filter feeder.

HABITAT: Offshore on calcareous banks. Cemented epifaunal.

LOCALITIES: Entire, more to south.

OCCURRENCE: Fairly common.

RANGE: West Indies to Brazil and Texas to Carmen, Mexico.

GEOLOGIC RANGE: Miocene to Recent.

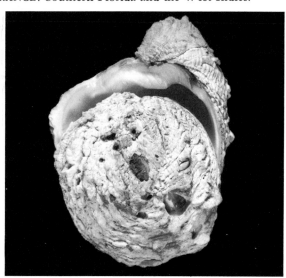

Pseudochama radians

Echinochama cornuta

Order VENEROIDA H. & A. Adams, 1858

Suborder LUCININA Dall, 1889

Superfamily LUCINACEA Fleming, 1828

Family LUCINIDAE Fleming, 1828

Subfamily LUCININAE Fleming, 1828

Genus *Anadontia* Link, 1807

BUTTERCUP LUCINA

Anadontia alba Link, 1807, *Beschr. Nat. Samml. Univ. Rostock* 3:156.

Gk. *ano* without, *dontes* teeth; L. *albus* dead white.

SIZE: Length 1½ to 2 in.

COLOR: Exterior white, interior flushed with yellowish orange.

SHAPE: Oval to circular. Inflated. Equivalve.

ORNAMENT OR SCULPTURE: Sculpture consists of weak, irregular concentric growth lines.

HINGE AREA: Hinge with weak teeth. Umbones not prominent. Hinge extended anteriorly to a faint oval lunule.

PALLIAL LINE & INTERIOR: Pallial line simple with 2 muscle scars. Anterior scar elongate and nearly parallel to pallial line. Margins smooth.

PERIOSTRACUM: Thin, brownish.

REMARKS: Unusual on northern half of coast. A filter feeder.

HABITAT: Inlet-influenced areas, bay margins, hypersaline lagoons. Infaunal.

LOCALITIES: Entire.

OCCURRENCE: Common.

RANGE: North Carolina to Florida, the Gulf states and West Indies.

CHALKY BUTTERCUP

Anadontia philippiana (Reeve, 1850), *Conch. Icon.* 5(49): pl. 5.

Gk. *ano* without, *dontes* teeth; honoring R. A. Philippi (1808–1904), German malacologist.

SIZE: Length 2 to 4 in.

COLOR: Chalky white.

SHAPE: Orbicular, inflated. Equivalve.

ORNAMENT OR SCULPTURE: Fine concentric growth lines. Interior usually pustulose.

HINGE AREA: Umbones rounded and touching. Hinge with very weak teeth.

PALLIAL LINE & INTERIOR: Pallial line is simple with 2 muscle scars. The anterior scar juts away from the line at a 30° angle instead of being parallel to it as in *A. alba*.

PERIOSTRACUM: Thin, brownish.

REMARKS: This large shell is probably no longer living on the Texas coast, but chalky pairs may be found

Anadontia alba

Anadontia philippiana

in spoil banks. According to Pulley (1953) this is *A. schrammi* (Crosse, 1876). A filter feeder.

HABITAT: Offshore at moderate depths. Infaunal.

LOCALITIES: Port Aransas south.

OCCURRENCE: Uncommon.

RANGE: North Carolina to eastern Florida, Cuba, Bermuda, and Texas.

Genus *Codakia* Scopoli, 1777

Subgenus *Codakia* s.s.

TIGER LUCINA

Codakia (Codakia) orbicularis (Linné, 1758), *Syst. Natur.*, 10th ed., p. 688.

After le Codok, a Senegal shell; Gk. *orbiculatus* circular.

SIZE: Length 2½ to 3½ in.

COLOR: White.

SHAPE: Orbicular. Compressed. Equivalve.

ORNAMENT OR SCULPTURE: Reticulate sculpture of coarse radial lines crossed by finer concentric threads.

HINGE AREA: Lunule in front of the beaks is deep, heartshaped, and nearly all on the right valve. Right valve has 2 cardinal teeth and 1 anterior lateral close to them. Left valve has 2 cardinals, a large double anterior lateral, and a small double posterior lateral.

Codakia (Codakia) orbicularis

PALLIAL LINE & INTERIOR: Pallial line simple with 2 muscle scars.

PERIOSTRACUM: Thin, brownish.

REMARKS: This species probably no longer lives on the Texas coast; the only specimens found are very chalky and worn halves, probably Pleistocene fossil. Filter feeder. Infaunal.

HABITAT: Shallow interreef flats in sand.

LOCALITIES: Port Aransas, south.

OCCURRENCE: Rare.

RANGE: Florida, Texas and the West Indies.

GEOLOGIC RANGE: Pliocene to Recent.

Lucina (Vellucina) amiantus

Genus *Lucina* Bruguière, 1797

Subgenus *Vellucina* Dall, 1901

LOVELY MINIATURE LUCINA

Lucina (Vellucina) amiantus (Dall, 1901), *Proc. U.S. Nat. Mus.* (1237): 826.

Lucina, goddess of childbirth; Gk. *amiantus* unstained.

SIZE: Length 6 to 10 mm.

COLOR: White.

SHAPE: Orbicular. Equivalve. Subequilateral.

ORNAMENT OR SCULPTURE: Rather inflated, thick-shelled. Sculptured with 8 to 9 wide, rounded ribs, which are crossed by numerous, small concentric threads.

HINGE AREA: Umbones touching, lunule small. Cardinal teeth small, not visible in adults. Laterals are well developed in right valve with sockets in left.

PALLIAL LINE & INTERIOR: Pallial line simple with 2 muscle scars. Margin crenulate.

PERIOSTRACUM: Thin, brownish.

REMARKS: Sift the drift for this little shell. It will be more abundant in the northern part of the coast. A filter feeder.

HABITAT: Inlet-influenced areas and near shore. Infaunal.

LOCALITIES: Entire.

OCCURRENCE: Common.

RANGE: North Carolina to both sides of Florida and Texas.

GEOLOGIC RANGE: Pleistocene to Recent.

Subgenus *Parvilucina* Dall, 1901

MANY-LINED LUCINA

Lucina (Parvilucina) multilineata (Tuomey & Holmes, 1857), *Post-Plio. Fos. S. Car.*, p. 61.

Lucina, goddess of childbirth; L. *multus* many, *lineas* lines.

SIZE: Length 6 to 10 mm.

COLOR: White.

SHAPE: Circular. Equivalve.

ORNAMENT OR SCULPTURE: The sculpture on this very inflated shell consists of numerous, fine, concentric threads that are stronger near the umbones.

HINGE AREA: Small cardinal and lateral teeth in right and left valves. Umbones small.

PALLIAL LINE & INTERIOR: Pallial line simple with 2 muscle scars. Interior margin finely crenulate.

PERIOSTRACUM: Thin, brownish.

REMARKS: This is another shell to screen drift for. It is more common in the southern part of the coast. A filter feeder.

HABITAT: Offshore and inlet-influenced areas. Infaunal.

LOCALITIES: Entire.

OCCURRENCE: Common.

RANGE: North Carolina to both sides of Florida and Texas.

GEOLOGIC RANGE: Pliocene to Recent.

Subgenus *Pseudomiltha* P. Fischer, 1885

FLORIDA LUCINA

Lucina (Pseudomiltha) floridana Conrad, 1833, *Amer. J. Sci.* 23:344.

Lucina, goddess of childbirth; from Florida.

SIZE: Length 1½ in.

COLOR: White.

SHAPE: Circular. Equivalve. Subequilateral.

ORNAMENT OR SCULPTURE: Compressed. Rather smooth except for fine, irregular growth lines.

HINGE AREA: Umbones low and pointing forward. Lunule oval. Hinge margin thick but teeth are weakly defined.

PALLIAL LINE & INTERIOR: Pallial line simple with 2 muscle scars, anterior scar elongate.

PERIOSTRACUM: Thin, brownish, deciduous.

Lucina (Pseudomiltha) floridana

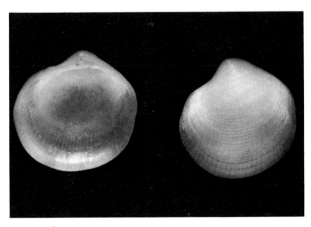

Lucina (Parvilucina) multilineata

REMARKS: Easily found dead in bay drift or by dredging. A filter feeder.

HABITAT: Open bays and inlet-influenced areas. Infaunal.

LOCALITIES: Entire, more to south.

OCCURRENCE: Common.

RANGE: Western coast of Florida to Texas.

GEOLOGIC RANGE: Upper Miocene to Recent.

Genus *Phacoides* Gray, 1847

Subgenus *Phacoides* s.s.

THICK LUCINA

Phacoides (Phacoides) pectinatus (Gmelin, 1791), *Syst. Natur.*, 13th ed., p. 3236.

Gk. *phakas* bean shaped; L. *pecten* a comb.

SIZE: Length 1 to 2½ in.

COLOR: Pale yellow to orange.

SHAPE: Oval. Lenticular.

ORNAMENT OR SCULPTURE: Posterior dorsal slope is rostrate, anterior slope less rostrate. Sculpture of unequally spaced lamellate ridges with finer lines in interspaces.

HINGE AREA: Ligament partially visible from the outside. Lunule small. Anterior and posterior lateral teeth strong, cardinals weak.

Phacoides (Phacoides) pectinatus

PALLIAL LINE & INTERIOR: Pallial line simple with 2 muscle scars, anterior scar very elongate.

PERIOSTRACUM: Thin, deciduous.

REMARKS: This species has a long slender foot used for digging. A filter feeder.

HABITAT: Open-bay margins and hypersaline lagoons. Infaunal.

LOCALITIES: Entire.

OCCURRENCE: Common.

RANGE: North Carolina to Florida, Texas and the West Indies, Central America to Brazil.

GEOLOGIC RANGE: Pliocene to Recent.

Family UNGULINIDAE H. & A. Adams, 1857

Genus *Diplodonta* Bronn, 1831

Subgenus *Phlyctiderma* Dall, 1899

PIMPLED DIPLODON

Diplodonta (Phlyctiderma) semiaspera Philippi, 1836, *Wiegm. Arch.* 2(2):225.

Gk. *diplos* twofold, *dontes* teeth; L. *semi* half, *asper* rough.

SIZE: Length up to 12 mm.

COLOR: Chalky white.

SHAPE: Orbicular, inflated. Equivalve.

ORNAMENT OR SCULPTURE: Thin shell is marked with numerous concentric rows of microscopic pimples.

HINGE AREA: There are 2 cardinal teeth in each valve, laterals are absent. The left anterior and right posterior ones are split.

PALLIAL LINE & INTERIOR: Simple pallial line with 2 elongate muscle scars.

PERIOSTRACUM: Thin, yellowish brown.

REMARKS: This shell may be found in holes in jetty rocks, where they are not visible until the rock is broken open, also in old *Crassostrea* and *Mercenaria* shells. They build nests around themselves of mud and sand held together with mucus. It is not known how the hole has been excavated, because the shells are not distorted. A filter feeder.

HABITAT: Open-bay centers, jetties, and inlet-influenced areas. Infaunal.

LOCALITIES: Entire.

OCCURRENCE: Fairly common.

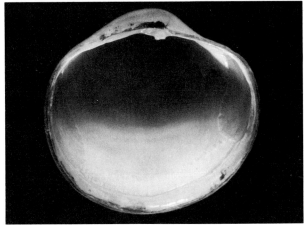

Diplodonta (Phlyctiderma) semiaspera

RANGE: North Carolina to Florida, Texas, West Indies, and Yucatán.
GEOLOGIC RANGE: Pliocene to Recent.

Diplodonta (Phlyctiderma) cf. *D. soror* C. B. Adams, 1852, *Contr. Conch.* 12:247.
Gk. *diplos* twofold, *dontes* teeth; L. *soror* sister.
SIZE: Length 8 to 18 mm.
COLOR: Translucent white.
SHAPE: Orbicular, inflated.
ORNAMENT OR SCULPTURE: Smooth with only fine concentric growth lines. Microscopic roughness on posterior slope, which is slightly compressed.
HINGE AREA: Two cardinal teeth in each valve, no laterals.
PALLIAL LINE & INTERIOR: Pallial line simple with 2 muscle scars.
PERIOSTRACUM: Thin; yellowish brown.

Diplodonta (Phlyctiderma) cf. *D. soror*

REMARKS: Possibly synonymous with *D. punctatus* (Say, 1822), *J. Acad. Natur. Sci. Phila.*, p. 308. Parker (in litt.) still holds that this is *D. punctata*, which he reported as living in the bays of Texas in 1960. A filter feeder.
HABITAT: Inlet-influenced areas. Infaunal.
LOCALITIES: Entire, more to south.
OCCURRENCE: Fairly common.
RANGE: Texas to Jamaica.

Superfamily GALEOMMATACEA Gray, 1840

Family KELLIIDAE Clark, 1851

Genus *Mysella* Angas, 1877

ATLANTIC FLAT LEPTON
Mysella planulata (Stimpson, 1851), *Shells of New England*, p. 17.
Gk. dim. of *mya* a sea mussel; L. *planus* level, flat, plain.

Mysella planulata

SIZE: Length 3.5 mm.

COLOR: White.

SHAPE: Oblong oval. Flattened. Equivalve. Inequilateral.

ORNAMENT OR SCULPTURE: Smooth with only fine concentric growth lines.

HINGE AREA: Small pointed beaks are three-fourths the distance back from the anterior end. Dorsal margin depressed in front and back of beaks.

PALLIAL LINE & INTERIOR: Two, suboval, almost equal adductor muscle scars. Pallial line simple.

PERIOSTRACUM: Thin, brown.

REMARKS: Easily overlooked and confused with *Aligena texasiana*. The figure in Parker (1959) is *Aligena* not *Mysella*.

HABITAT: Attaches to pilings, bouys, and grasses in bays and shallow water.

LOCALITIES: Entire.

OCCURRENCE: Fairly common.

RANGE: Greenland to Texas and West Indies.

Family LEPTONIDAE Gray, 1847

Genus *Lepton* Turton, 1822

Lepton cf. *L. lepidum* Say, 1826, *J. Acad. Natur. Sci. Phila.* 5:221.

Gk. *leptos* small, fine; L. *lepidum* graceful, charming.

SIZE: Length 5 mm.

COLOR: White, translucent.

SHAPE: Trigonal. Equilateral. Equivalve.

ORNAMENT OR SCULPTURE: Smooth and glassy appearing but with numerous microscopic, longitudinal striations that curve toward the anterior edge on the anterior margin and toward the posterior edge on the posterior margin.

HINGE AREA: Cardinal teeth obsolete. Lateral teeth prominent.

PALLIAL LINE & INTERIOR: Pallial line simple.

PERIOSTRACUM: None visible.

REMARKS: Commensal with other invertebrates. Attached to the host by means of a byssus. A filter feeder.

HABITAT: Along shore.

LOCALITIES: Entire.

OCCURRENCE: Fairly common.

RANGE: Charleston Harbor, South Carolina, and Texas.

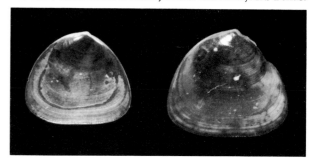

Lepton cf. *L. lepidum*

Superfamily: CYAMIACEA Philippi, 1845

Family: SPORTELLIDAE Dall, 1899

Genus: *Aligena* Lea, 1843

TEXAS LEPTON

Aligena texasiana Harry, 1969, *Veliger* 11(3):168.

L. *aliger* winged; from Texas.

Aligena texasiana

SIZE: Length 4.81 mm.

COLOR: Chalky white.

SHAPE: Subtrigonal. Inflated. Equivalve. Almost equilateral.

ORNAMENT OR SCULPTURE: Smooth, except for microscopic, concentric growth lines. Both posterior and anterior ends are rounded. Ventral margin is almost straight, but shows a slight depression about midway. Valves very thin.

HINGE AREA: Beaks are rounded and touching. No escutcheon or lunule. A single tooth in each valve, projecting beyond the midline and larger in right valve.

PALLIAL LINE & INTERIOR: Two, suboval adductor scars, about equal in size. Pallial line simple but broken into a series of subtriangular marks.

PERIOSTRACUM: Thin, tan.

HABITAT: Shallow bays. Probably commensal.

LOCALITIES: Entire.

OCCURRENCE: Fairly common.

RANGE: Louisiana and Texas.

Superfamily **CARDITACEA** Fleming, 1820

Family **CARDITIDAE** Fleming, 1820

Subfamily **CARDITINAE** Fleming, 1820

Genus *Cardita* Bruguière, 1792

Subgenus *Carditamera* Conrad, 1838

BROAD-RIBBED CARDITA

Cardita (Carditamera) floridana (Conrad, 1838), *Fos. Med. Tert.*, p. 12.

Gk. *kardia* heart; from Florida.

SIZE: Length 1 to 1½ in.

COLOR: Whitish with small bars of chestnut brown arranged concentrically on the ribs.

SHAPE: Oval. Equivalve. Inequilateral.

ORNAMENT OR SCULPTURE: Fifteen to 18 strong radiating ribs, beaded by transverse growth lines. Beaks close together.

HINGE AREA: External hinge ligament; hinge oblique, strong cardinal teeth. Right valve has 1 anterior lateral, left valve 1 posterior lateral tooth. Lunule small, deeply indented under beaks.

Cardita (Carditamera) floridana

PALLIAL LINE & INTERIOR: Interior smooth with 2 muscle scars. No siphon.

PERIOSTRACUM: Grayish, fairly heavy.

REMARKS: When alive this shell attaches itself by a byssus. On the Texas coast dead shells are common, but to date none have been found alive. Pairs are often found in spoil banks, indicating that it once lived here in the grass flats.

HABITAT: Inlet-influenced areas and hypersaline lagoons. Infaunal.

LOCALITIES: Southern half of coast.

OCCURRENCE: Common.

RANGE: Southern half of Florida and South Texas to Mexico.

GEOLOGIC RANGE: Pliocene to Recent.

Crassinella lunulata

Suborder ASTARTEDONTINA Korobkov, 1953

Superfamily CRASSATELLACEA Férussac, 1821

Family CRASSATELLIDAE Férussac, 1821

Subfamily CRASSATELLINAE Férussac, 1821

Genus *Crassinella* Guppy, 1874

LUNATE CRASSINELLA
Crassinella lunulata (Conrad, 1834), *J. Acad. Natur. Sci. Phila.* 7(1):133.
L. *crassius* thick, *ella* dim. suffix, *luna* moon.

SIZE: Length 6 to 8 mm.

COLOR: Whitish or pinkish. Interior brown.

SHAPE: Trigonal. Slightly inequivalve.

ORNAMENT OR SCULPTURE: Beaks at the middle. The valves are lapped in such a way that the posterior dorsal margin of the left valve is more visible than that of the right. The flattened valves are sculptured with well-developed concentric ribs.

HINGE AREA: Ligament is internal. There are hinge teeth on either side, 2 cardinals in each and 1 anterior lateral in the right, 1 posterior lateral in the left.

PALLIAL LINE & INTERIOR: Simple pallial line joining the 2 adductor muscle scars.

PERIOSTRACUM: None visible.

REMARKS: You must screen the drift for this distinctively shaped shell. Originally described from Texas. A filter feeder.

HABITAT: Inlet-influenced areas and channels on shelly bottom. Infaunal.

LOCALITIES: Entire.

OCCURRENCE: Fairly common.

RANGE: Southeastern United States and Texas, West Indies.

GEOLOGIC RANGE: Miocene to Recent.

Superfamily CARDIACEA Lamarck, 1809

Family CARDIIDAE Lamarck, 1809

Subfamily CARDIINAE Lamarck, 1809

Genus *Dinocardium* Dall, 1900

GIANT ATLANTIC COCKLE
Dinocardium robustum (Solander, 1786), *Portland Mus. Cat.*, p. 58.

Dinocardium robustum

Gk. *deinos* huge, *kardia* heart; L. *robustus* hard and
strong.

SIZE: Length 3 to 4 in.

COLOR: Pale tan, mottled with red brown. Posterior
slope mahogany brown. Interior salmon pink.

SHAPE: Obliquely ovate. Inflated. Equivalve.

ORNAMENT OR SCULPTURE: Thirty-two to 36 rounded, radial smoothish ribs.

HINGE AREA: Umbones rounded. Heavy external ligament. One cardinal tooth in each valve, with 2 anterior lateral and 1 posterior lateral in right valve
and complementary arrangement in left valve.

PALLIAL LINE & INTERIOR: Two muscle scars connected by simple pallial line. Margins crenulate.

PERIOSTRACUM: Thin, brownish.

REMARKS: This is the largest cockle shell on the Texas
coast. Storms will often wash great numbers in alive.
The long slender foot of the animal is blackish.
Juveniles can be found living in mud flats in the
bays around Aransas Pass and adults at San Luis
Pass. A shallow filter feeder.

HABITAT: Close to shore and in inlet-influenced areas.
Infaunal.

LOCALITIES: Entire.

OCCURRENCE: Common.

RANGE: Virginia to northern Florida, Texas and Carmen, Mexico.

GEOLOGIC RANGE: Miocene to Recent.

Genus *Laevicardium* Swainson, 1840

COMMON EGG COCKLE

Laevicardium laevigatum (Linneé, 1758), *Syst. Natur.*,
10th ed., p. 680.

L. *laevi* smooth; Gk. *kardia* heart.

SIZE: Length 1 to 2 in.

COLOR: Cream colored or pale yellow variably mottled
with brown.

SHAPE: Obliquely egg shaped.

ORNAMENT OR SCULPTURE: Smooth, polished with
obscure radiating ribs.

HINGE AREA: Umbones rounded. Cardinal and lateral
teeth present. Interior cream colored with finely
serrated margins.

PALLIAL LINE & INTERIOR: Two muscle scars connected by simple pallial line. Ventral margin crenulate.

PERIOSTRACUM: Thin, brownish, usually lost.

REMARKS: This shell is another that is common on the

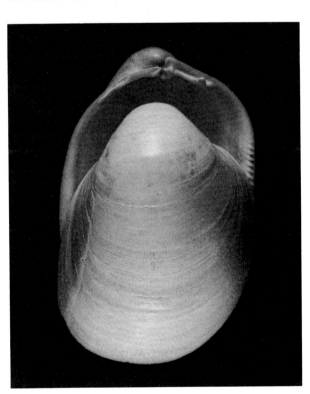

Laevicardium laevigatum

upper Mexican coast and is a straggler on the southern part of the Texas coast. A filter feeder.

HABITAT: Offshore. Infaunal.

LOCALITIES: Central, south.

OCCURRENCE: Uncommon.

RANGE: North Carolina to both sides of Florida, Mexico, West Indies, Texas, Central America to Brazil.

GEOLOGIC RANGE: Oligocene to Recent.

MORTON'S EGG COCKLE

Laevicardium mortoni (Conrad, 1830), *J. Acad. Natur. Sci. Phila.* 6:259.

L. *laevis* smooth; Gk. *kardia* heart; honoring Samuel G. Morton (1799–1851), American malacologist.

SIZE: Length 16 to 25 mm.

COLOR: Exterior cream colored, irregularly patterned with brown. Interior yellow with brown mottlings.

SHAPE: Oval, inflated. Equivalve.

ORNAMENT OR SCULPTURE: Smooth, polished with faint concentric lines that are microscopically pimpled. Shell thin.

HINGE AREA: Cardinal and lateral teeth present. Anterior laterals prominent.

PALLIAL LINE & INTERIOR: Two muscle scars connected by simple pallial line. Margins may or may not be crenulated.

PERIOSTRACUM: Thin, brownish, with blisterlike elevations.

REMARKS: This little bivalve can move about in the shallow waters with surprising speed. Ducks love it. A filter feeder.

HABITAT: Shallow water in inlet-influenced areas and hypersaline lagoons. Infaunal.

LOCALITIES: Entire, more common in south.

OCCURRENCE: Common.

RANGE: Cape Cod, Massachusetts, to Florida, West Indies, and Texas to Tecolutla, Mexico.

GEOLOGIC RANGE: Miocene to Recent.

Genus *Trachycardium* Mörch, 1853

Subgenus *Trachycardium* s.s.

PRICKLY COCKLE

Trachycardium (Trachycardium) isocardia (Linné, 1758), *Syst. Natur.*, 10th ed., p. 679.

Gk. *trachys* rough, *kardia* heart, *iso* equal.

SIZE: Height up to 3 in.

COLOR: Exterior light cream with blotches of red brown. Interior has wide band of salmon pink along margins.

Laevicardium mortoni

Trachycardium (Trachycardium) isocardia

183

SHAPE: Oval. Elongated heart shaped from the side.

ORNAMENT OR SCULPTURE: Thirty-one to 37 strong, radiating ribs that have imbricated scales.

HINGE AREA: Ligament external and posterior. The cardinal hinge teeth are arched. Umbones prominent, nearly central.

PALLIAL LINE & INTERIOR: Pallial line simple, connecting 2 muscle scars that are white. Margins crenulate.

PERIOSTRACUM: Thin, brownish.

REMARKS: This species lives offshore and is common along the Mexican Gulf beaches, but is only an occasional visitor to the Texas coast. Port Isabel is the most likely spot to find it. A filter feeder.

HABITAT: Offshore. Infaunal.

LOCALITIES: Central, south.

OCCURRENCE: Uncommon.

RANGE: West Indies to northern coast of South America and Mexico and Texas.

GEOLOGIC RANGE: Miocene to Recent.

Trachycardium (Dallocardia) muricatum

Subgenus *Dallocardia* Stewart, 1930

YELLOW COCKLE

Trachycardium (Dallocardia) muricatum (Linné, 1758), *Syst. Natur.*, 10th ed., p. 680.

Gk. *trachys* rough, *kardia* heart; L. *muricatus* from *murex* pointed.

SIZE: Height 2 in.

COLOR: Light cream with blotches of red brown or shades of yellow. Interior white.

SHAPE: Subcircular. Equivalve. Inflated.

ORNAMENT OR SCULPTURE: Thirty to 40 sharply scaled ribs. The scales on anterior slope are on anterior side of ribs, while on the central and posterior slope they are on posterior side of ribs. The separation is marked by several ribs with double rows of spines.

HINGE AREA: Umbones prominent, nearly central. Ligament external. Cardinal and lateral teeth present.

PALLIAL LINE & INTERIOR: Interior margin crenate. Two equal-sized adductor muscle scars are connected by indistinct simple pallial line.

PERIOSTRACUM: Thin, brownish, heaviest at margins.

REMARKS: After a sustained freeze these cockles will pop out of the sandy mud in the bays. The double siphons are short and the foot is well developed for digging. A filter feeder.

HABITAT: Inlet-influenced areas and bay margins. Infaunal.

LOCALITIES: Entire.

OCCURRENCE: Fairly common.

RANGE: North Carolina to Florida, Texas and the West Indies to Brazil.

GEOLOGIC RANGE: Pleistocene to Recent.

Superfamily MACTRACEA Lamarck, 1809

Family MACTRIDAE Lamarck, 1809

Subfamily MACTRINAE Lamarck, 1809

Genus *Anatina* Schumacher, 1817

SMOOTH DUCK CLAM

Anatina anatina Spengler, 1802, *Skr. Nat. Selsk. Copenhagen* 5(2):92–128.

L. *anatis* a little duck.

SIZE: Length 2 to 3 in.

COLOR: White.

SHAPE: Trigonal. Inequilateral.

ORNAMENT OR SCULPTURE: Thin, fragile, and gaping

Anatina anatina

posteriorly. Fairly smooth, except for fine growth lines and concentric ribs near the umbones. Posterior end has a distinct radial rib.

HINGE AREA: Hinge has a prominent chondrophore and 3 small cardinal teeth anterior to chondrophore. A lateral tooth is posterior to chondrophore. Umbones high and pointed backward.

PALLIAL LINE & INTERIOR: Anterior adductor scar elongate, posterior scar rounded. Pallial sinus narrow and deep.

PERIOSTRACUM: Thin and straw colored. Usually lost.

REMARKS: Seldom found except in the winter. Less common than *Raeta plicatella*. Syn. *Labiosa lineata* (Say, 1822), according to Harry (1969). A filter feeder.

HABITAT: Two to 12 fathoms offshore. Infaunal.

LOCALITIES: Entire.

OCCURRENCE: Fairly common.

RANGE: North Carolina to northern two-thirds of Florida, Texas and Mexico.

GEOLOGIC RANGE: Miocene to Recent.

Genus *Raeta* Gray, 1853

CHANNELED DUCK CLAM
Raeta plicatella (Lamarck, 1818), *Hist. Natur. Anim. sans Vert.* 5:470.
Derivation of *Raeta* unknown; L. *plicatella* little folds.

SIZE: Length 2 to 3 in.

COLOR: White.

SHAPE: Trigonal. Inequilateral.

ORNAMENT OR SCULPTURE: Thin, fragile, and gaping posteriorly. Sculpture consists of evenly spaced, rounded concentric ribs with fine striations in the intercostal spaces. Fine radial threads.

HINGE AREA: Hinge has a prominent chondrophore and small, irregular cardinal teeth. A single lateral tooth posterior to chondrophore. Umbones high and pointed backward.

PALLIAL LINE & INTERIOR: Anterior adductor scar elongate, posterior scar rounded. Pallial sinus narrow and extending from posterior scar to about midshell almost parallel to pallial line, somewhat pointed.

PERIOSTRACUM: Thin, yellowish, usually wrinkled.

REMARKS: This species is more common on the shores in the winter. Synonymous with *A. canaliculata* (Say, 1822). A filter feeder.

HABITAT: Probably does not burrow, but lives on its side on the sandy bottom of the outer surf zone.

LOCALITIES: Entire.

OCCURRENCE: Fairly common.

RANGE: North Carolina to Florida, Texas and the West Indies and Mexico.

GEOLOGIC RANGE: Miocene to Recent.

Raeta plicatella

Mactra fragilis

Genus *Mactra* Linné, 1767

FRAGILE ATLANTIC MACTRA

Mactra fragilis Gmelin, 1792, *Syst. Natur.*, 13th ed., p. 3261.

Gk. *maktra* a kneading trough; L. *fragilis* brittle, fragile.

SIZE: Length 1½ to 2½ in.

COLOR: White.

SHAPE: Oval. Subequilateral. Equivalve.

ORNAMENT OR SCULPTURE: Margins rounded. Sculpture consists of fine, irregular growth lines. Posterior slope with 2 radial ridges. Fairly large posterior gape.

HINGE AREA: Umbones rounded, almost central. Two ligaments, 1 external, and an inner cartilaginous ligament housed in a spoon-shaped chondrophore posterior to the bifid cardinal tooth.

PALLIAL LINE & INTERIOR: Polished white interior with lightly impressed scars of about equal size and shape. Pallial sinus short and broadly rounded.

PERIOSTRACUM: Light brown, thin; heaviest on posterior slope.

REMARKS: This species has not been as common since the killing freeze in January 1962. Syn. *M. brasiliana* Lamarck, 1818. A filter feeder.

HABITAT: Open-bay margins. Infaunal.

LOCALITIES: Entire, more in south.

OCCURRENCE: Fairly common.

RANGE: North Carolina to Florida, Texas and the West Indies.

GEOLOGIC RANGE: Pliocene to Recent.

Genus *Mulinia* Gray, 1837.

DWARF SURF CLAM

Mulinia lateralis (Say, 1822), *J. Acad. Natur. Sci. Phila.* 2:309.

A meaningless name invented by Gray; L. *lateralis* side.

SIZE: Length 8 to 12 mm.

COLOR: Whitish to cream.

SHAPE: Trigonal. Inflated. Inequilateral.

ORNAMENT OR SCULPTURE: Smooth except for fine growth lines. The posterior slope is marked with a distinct radial ridge.

HINGE AREA: Umbones high and almost central. Chondrophore, bifid cardinals, and lateral teeth make up the hinge complex.

PALLIAL LINE & INTERIOR: Anterior adductor scar more elongate than posterior scar. Pallial sinus short, rounded, and oblique.

PERIOSTRACUM: Thin, light brown.

REMARKS: This clam is the most abundant and ubiq-

Mulinia lateralis

uitous bivalve on the Texas coast due to its ability to withstand a wide range of salinities. The juveniles are thin and opalescent, coming to shore in vast numbers in the winter months. A staple in the diet of the black drum. A filter feeder.

HABITAT: In clayey sediments in every type of assemblage. Infaunal.

LOCALITIES: Entire.

OCCURRENCE: Very common.

RANGE: Maine to northern Florida and to Texas, Mexico.

GEOLOGIC RANGE: Miocene to Recent.

Genus *Rangia* Des Moulins, 1832

COMMON RANGIA

Rangia cuneata (Gray, 1831), Sowerby, *Gen. Shells*, no. 36, fig.1–4.

Dedicated to French malacologist Sander Rang; L. *cuneus* wedge.

SIZE: Length 1 to 1½ in.

COLOR: Whitish.

SHAPE: Obliquely ovate. Equivalve. Inequilateral.

ORNAMENT OR SCULPTURE: Heavy, thick shell is sculptured with fine concentric growth lines.

HINGE AREA: Deeply excavated chondrophore, cardi-nals, and 2 lateral teeth. Posterior lateral is very long, reaching almost to ventral margin, easily separating it from *R. flexuosa*. Umbones prominent and nearer anterior end.

PALLIAL LINE & INTERIOR: Anterior adductor scar is smaller than posterior scar. Pallial sinus is small but distinct, directed forward and upward.

PERIOSTRACUM: Heavy, smooth, brown.

REMARKS: This shell may be found a few miles up the Brazos River, but it is a marine shell. A filter feeder.

HABITAT: River-influenced areas. Infaunal.

LOCALITIES: East and central; more prevalent centrally.

OCCURRENCE: Common.

RANGE: Northwestern Florida to Texas and to Alvarado, Mexico.

GEOLOGIC RANGE: Pliocene to Recent.

Subgenus *Rangianella* Conrad, 1868

BROWN RANGIA

Rangia (Rangianella) flexuosa (Conrad, 1839), *Amer. J. Sci.* 38:92.

Dedicated to French malacologist Sander Rang; L. *flexuosus* bent.

Rangia cuneata

Rangia (Rangianella) flexuosa

SIZE: Length 1 to 1½ in.

COLOR: Whitish.

SHAPE: Obliquely ovate. Fairly wedge shaped. Inequilateral.

ORNAMENT OR SCULPTURE: Thick, heavy shell with sculpture consisting of fine growth lines. Long posterior slope is keeled.

HINGE AREA: Umbones prominent. Chondrophore present with cardinal teeth, but laterals are much shorter than *R. cuneata*.

PALLIAL LINE & INTERIOR: Two rounded adductor muscle scars. Pallial sinus almost obsolete.

PERIOSTRACUM: Heavy, smooth, brown.

REMARKS: This shell can withstand very low salinity, as does *R. cuneata*, but it is more marine than the latter. Juvenile specimens can easily be confused with *Mulinia lateralis*. A filter feeder.

HABITAT. River-influenced areas. Infaunal.

LOCALITIES: Entire, more to east.

OCCURRENCE: Uncommon.

RANGE: Louisiana to Texas and Veracruz, Mexico.

GEOLOGIC RANGE: Pleistocene to Recent.

Genus *Spissula* Gray, 1837

ATLANTIC SURF CLAM

Spissula solidissima similis (Say, 1822), *J. Acad. Natur. Sci. Phila.* 2:309.

L. *spissus* thick, solid, *solidus* solid, *similis* like.

Spissula solidissima similis

SIZE: Length 4 to 5 in.

COLOR: Yellowish white.

SHAPE: Oval.

ORNAMENT OR SCULPTURE: Smooth except for fine, concentric growth lines.

HINGE AREA: Large, shallow, triangular umbones acute and more anteriorly located. Chondrophore. Two cardinal teeth, those in left valve fused at upper ends. The lateral teeth have an opposite deep socket.

PALLIAL LINE & INTERIOR: Adductor muscle scars are rounded and above middle of valve. Pallial sinus short, rounded, and almost parallel to pallial line.

PERIOSTRACUM: Thin, yellowish brown. Heavier on the slopes.

REMARKS: An edible bivalve but does not occur in commercial quantities. Syn. *S. s. raveneli* (Conrad, 1831). A filter feeder.

HABITAT: Near shore, 2 to 12 fathoms. Infaunal.

LOCALITIES: Entire.

OCCURRENCE: Fairly common.

RANGE: Cape Cod, both sides of Florida and Texas.

GEOLOGIC RANGE: Upper Miocene to Recent.

Family MESODESMATIDAE Deshayes, 1839

Genus *Ervilia* Turton, 1822

CONCENTRIC ERVILIA

Ervilia cf. *E. concentrica* Gould, 1862, *Otia Conch.*, p. 329.

L. *ervilia* a small lentil, *concentrica* with concentric lines.

SIZE: Length 5 to 6 mm.

COLOR: White.

SHAPE: Elliptical. Equilateral.

ORNAMENT OR SCULPTURE: Sculpture consists of fine, numerous concentric lines.

HINGE AREA: Umbones central. Resilium small and internal. Cardinal tooth bifid, laterals small.

PALLIAL LINE & INTERIOR: Muscle scars faintly impressed. Pallial sinus rounded, broad, and short.

PERIOSTRACUM: None visible.

REMARKS: Due to its similarity to juvenile *Mulinias* this tiny shell may have been long overlooked on the Texas coast. A filter feeder.

HABITAT: Near shore and open bays. Infaunal.

LOCALITIES: Entire.

Ervilia cf. *E. concentrica*

OCCURRENCE: Fairly common.
RANGE: North Carolina to both sides of Florida, the West Indies and Texas.
GEOLOGIC RANGE: Pliocene to Recent.

Superfamily SOLENACEA Lamarck, 1809

Family SOLENIDAE Lamarck, 1809

Genus *Ensis* Schumacher, 1817

JACKKNIFE CLAM
Ensis minor Dall, 1900, *Trans. Wagner Free Inst. Sci.* 3(5):955.
L. *ensis* sword, *minor* small.
SIZE: Length up to 3 in.
COLOR: Shell white. Purplish interior.
SHAPE: Cylindrical. Equivalve.
ORNAMENT OR SCULPTURE: This long narrow shell is smooth and fragile.
HINGE AREA: The left valve has 2 vertical cardinal

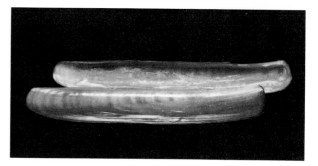

Ensis minor

teeth, and each valve has a long, low posterior tooth. Teeth in the less-pointed end.
PALLIAL LINE & INTERIOR: Two adductor muscle scars; shallow pallial sinus.
PERIOSTRACUM: Lacquerlike brownish green periostracum.
REMARKS: A favorite food of wading birds. This genus is a filter feeder.
HABITAT: Enclosed lagoon and bay margins. Infaunal.
LOCALITIES: Entire.
OCCURRENCE: Common.
RANGE: Both sides of Florida to Texas.
GEOLOGIC RANGE: Miocene to Recent.

Genus *Solen* Linné, 1758

GREEN JACKKNIFE CLAM
Solen viridis Say, 1822, *J. Acad. Natur. Sci. Phila.* 2:316.
Gk. *solen* a channel; L. *viridis* green.
SIZE: Length up to 2 in.
COLOR: White.

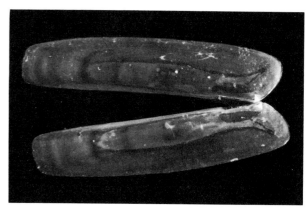

Solen viridis

SHAPE: Long, narrow, flattened cylindrical. Equivalve.

ORNAMENT OR SCULPTURE: Smooth. Dorsal edge straight, ventral edge curved. Fragile.

HINGE AREA: Hinge with a single projecting tooth at the very end of the valve.

PALLIAL LINE & INTERIOR: Two muscle scars and a pallial sinus.

PERIOSTRACUM: Thin, lacquerlike, greenish brown.

REMARKS: A western Louisiana species that is easily confused with *Ensis minor* but is much shorter than *E. minor*. Figure is that of a juvenile. A filter feeder.

HABITAT: Inlets, near shore. Infaunal.

LOCALITIES: East.

OCCURRENCE: Uncommon.

RANGE: Rhode Island to northern Florida and the Gulf states.

GEOLOGIC RANGE: Pleistocene to Recent.

Superfamily DONACACEA Fleming, 1828

Family DONACIDAE Fleming, 1828

Genus *Donax* Linné, 1758

FAT LITTLE DONAX

Donax tumidus Philippi, 1849, *Z. für Malak.* 5:147.

Gk. *donax* a dart; L. *tumidus* swollen.

SIZE: Length 8 to 12 mm.

COLOR: Whitish with pale blue, pink, or yellow blushes. Seldom rayed as is *D. variabilis*.

Donax tumidus

SHAPE: Unequally triagonal. Inflated.

ORNAMENT OR SCULPTURE: Glossy with fine concentric growth lines. Radial threads on blunt posterior end are heavily beaded.

HINGE AREA: Left valve overlaps right on ventral teeth. Ligament external. Two cardinal teeth.

PALLIAL LINE & INTERIOR: Interior smooth. Pallial sinus large, rounded.

PERIOSTRACUM: None visible.

REMARKS: The reduced winter populations of *D. tumidus* and *D. variabilis roemeri* mix, but in the summer they separate and *D. tumidus* stays in deeper water. Originally described from Texas coast. A filter feeder.

HABITAT: Surf zone in the sand. Infaunal.

LOCALITIES: Entire sandy portion.

OCCURRENCE: Fairly common.

RANGE: Northern shores of the Gulf of Mexico, Mexico.

GEOLOGIC RANGE: Pleistocene to Recent.

COQUINA SHELL

Donax variabilis texasiana Philippi, 1847, *Z. für Malak.* 4:77; Roemer, *Texas*, p. 452.

Gk. *donax* a dart; L. *variabilis* variable.

SIZE: Length 12 to 18 mm.

COLOR: Variable, often rayed, pink, purple, yellow, white, bluish, or mauve.

SHAPE: Unequally trigonal or wedge shaped. Equivalve.

ORNAMENT OR SCULPTURE: Glossy with fine concentric growth lines and radial striae that become stronger on the blunt posterior end.

Donax variabilis texasiana

HINGE AREA: Umbones low. External ligament behind umbones. Cardinal and lateral teeth present.

PALLIAL LINE & INTERIOR: Two small muscle scars. Large, rounded pallial sinus adjoining posterior scar and extending to middle of shell. Margins finely crenate.

PERIOSTRACUM: None visible.

REMARKS: Syn. *D. variabilis roemeri* Philippi, 1848. A delicately flavored chowder is made from this small shell (see Chapter 4). Loesch (1957) gives an interesting study of *Donax*. These animals spend the summer at the water's edge and die off in the fall. They feed on minute organisms on the grains of sand. Originally described from Texas coast. A filter feeder.

HABITAT: Surf-zone sand. Infaunal.

LOCALITIES: Entire sandy portion.

OCCURRENCE: Common.

RANGE: Texas to Alvarado, Mexico.

Superfamily TELLINACEA, Blainville, 1824

Family TELLINIDAE Blainville, 1824

Genus *Macoma* Leach, 1819

CONSTRICTED MACOMA
Macoma constricta (Bruguière, 1792), *Actes Soc. Hist. Natur.* 1(3):126.

Macoma constricta

Macoma, a euphonic name invented by Leach; L. *constrictus* drawn tight.

SIZE: Length 1 to 2½ in.

COLOR: White.

SHAPE: Subquadrate. Inequivalve. Inequilateral.

ORNAMENT OR SCULPTURE: Sculptured with irregular, fine, concentric growth lines. A low radial ridge marks the posterior slope, which is flexed to the right.

HINGE AREA: Umbones rounded. Ligament long, narrow. Cardinal teeth weak. No laterals.

PALLIAL LINE & INTERIOR: Muscle scars and pallial line weak. Two rather high muscle scars. Pallial sinus extended to near anterior scar, convex above and correspondingly curved below.

PERIOSTRACUM: Thin, grayish brown. Heavier toward margins.

REMARKS: This species is more tolerant of extremes in salinities and temperatures than the other macomas on the coast of Texas. A deposit feeder.

HABITAT: Open-bay margins and centers. Infaunal.

LOCALITIES: Entire.

OCCURRENCE: Common.

RANGE: Florida to Texas and the West Indies.

GEOLOGIC RANGE: Pleistocene to Recent.

SHORT MACOMA
Macoma brevifrons (Say, 1834), *Amer. Conch.*, part 7, pl. 64.

Macoma, a euphonic name invented by Leach; L. *brevifrons* short frond.

SIZE: Length 1 in.

COLOR: White to pale peach color.

SHAPE: Oval. Inequilateral.

ORNAMENT OR SCULPTURE: Shell smooth except for fine growth lines. Very weak radial ridge posteriorly.

HINGE AREA: Umbones small, pointed, anterior to middle. Ligament, brown and small. External. Two cardinal teeth in each valve. Posterior tooth in left much smaller than the others. No lateral teeth.

PALLIAL LINE & INTERIOR: Scars and pallial line hardly visible. Elongate anterior scar and small round posterior scar. Pallial sinus large, rounded, almost confluent with pallial line.

PERIOSTRACUM: Thin, yellowish brown, heavier on posterior slope.

REMARKS: This macoma is much the same color and size as *Tellina tampaensis*, but shape and hinge area differ. Until a monograph on the macomas is pub-

Macoma brevifrons

Macoma mitchelli

lished, *M. brevifrons* and *M. aurora* Hanly, 1844, will be considered synonymous on the basis of published descriptions. A deposit feeder.

HABITAT: Open-bay margins. Inlet-influenced areas. Infaunal.

LOCALITIES: Central.

OCCURRENCE: Fairly common.

RANGE: North Carolina south to Brazil and Texas.

MITCHELL'S MACOMA

Macoma mitchelli Dall, 1895, *Nautilus* 9:33.

Macoma, a euphonic name invented by Leach; dedicated to J. D. Mitchell, early Texas collector.

SIZE: Length 22 mm.

COLOR: White.

SHAPE: Elongate, subquadrate. Inequilateral.

ORNAMENT OR SCULPTURE: Smooth except for faint concentric growth lines. A weak radial ridge marks the posterior slope.

HINGE AREA: Umbones low, pointed, more posterior of center. Cardinal teeth very weak. No laterals.

PALLIAL LINE & INTERIOR: Posterior muscle scar larger and more rounded than anterior. Pallial sinus large, dorsally convex, gently sloping to pallial line before reaching anterior scar.

PERIOSTRACUM: Thin, brownish.

REMARKS: You need to look near rivers in brackish water for this small deposit feeder. Originally described from Texas coast.

HABITAT: River-influenced areas, estuaries. Infaunal.

LOCALITIES: Central and east.

OCCURRENCE: Common.

RANGE: Mississippi to Central Texas.

GEOLOGIC RANGE: Pleistocene to Recent.

TAGELUS-LIKE MACOMA

Macoma tageliformis Dall, 1900, *Proc. U.S. Nat. Mus.* 23(1210):300.

Macoma, a euphonic name invented by Leach; in the form of the bivalve *Tagelus*.

SIZE: Length up to 2½ in.

COLOR: Dull white.

SHAPE: Oblong. Inequilateral.

ORNAMENT OR SCULPTURE: Sculpture consists of fine, irregular growth lines. Posterior slightly flexed to the right. Heavier than the other macomas.

HINGE AREA: Umbones toward posterior, pointed. Ligament external, dark brown. Cardinal teeth fairly strong, 2 in each valve. Posterior tooth in left valve thin, often obsolete. No laterals.

PALLIAL LINE & INTERIOR: Anterior adductor scar elongate, posterior scar rounded. Large, convex, rounded pallial sinus one-fourth confluent with pallial line.

PERIOSTRACUM: Thin, brownish, heavier at the margins.

REMARKS: Longer and larger than most macomas. Originally described from Corpus Christi Bay. A deposit feeder.

Macoma tageliformis

Macoma tenta

HABITAT: Near shore, 2 to 11 fathoms in silty clay. Infaunal.

LOCALITIES: Entire.

OCCURRENCE: Uncommon.

RANGE: Louisiana, Texas to Tuxpan, Mexico, and the West Indies.

GEOLOGIC RANGE: Pleistocene to Recent.

TENTA MACOMA

Macoma tenta (Say, 1834), *Amer. Conch.*, part 7, pl. 65.

Macoma, a euphonic name invented by Leach; L. *tentus* stretched out.

SIZE: Length 12 to 25 mm.

COLOR: White, slightly iridescent.

SHAPE: Elongate. Oblong. Inequilateral.

ORNAMENT OR SCULPTURE: Smooth except for microscopic growth lines. Thin. Posterior margin truncated and flexed to the right; marked with a radial ridge.

HINGE AREA: Umbones small, sharp. Ligament small, brown. Two cardinal teeth in left and 1 in right valve. One posterior lateral tooth.

PALLIAL LINE & INTERIOR: Anterior adductor scar elongate, posterior rounded. Pallial sinus nearly half confluent and almost reaching anterior scar.

PERIOSTRACUM: Thin, brownish.

REMARKS: This little deposit feeder lives in mud. A selective deposit feeder.

HABITAT: Open-bay margins. Inlet influence. Shallow hypersaline lagoons. Infaunal.

LOCALITIES: Entire.

OCCURRENCE: Common.

RANGE: Cape Cod to Florida, Texas and the West Indies.

GEOLOGIC RANGE: Miocene? Pliocene to Recent.

Genus *Strigilla* Turton, 1822

Subgenus *Pisostrigilla* Olson, 1961

WHITE STRIGILLA

Strigilla (*Pisostrigilla*) *mirabilis* (Philippi, 1841), *Arch. für Natur.* 7(1):260.

L. *strigil* a scraping tool, *mirabilis* wonderful.

SIZE: Length 8 mm.

COLOR: White, translucent, shiny.

SHAPE: Oval, inflated. Inequivalve. Equilateral.

ORNAMENT OR SCULPTURE: Sculpture consists of fine growth lines crossed by oblique lines that meet the ventral margin at about 45 degrees. Posterior slope is patterned with 4 or more zig-zag rows of lines.

HINGE AREA: Umbones rounded, almost central. Two

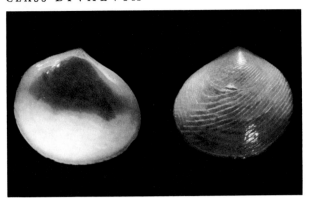

Strigilla (Pisostrigilla) mirabilis

Tellina (Angulus) tampaensis

cardinals and 2 lateral teeth present. Left posterior cardinal is very thin and fragile.

PALLIAL LINE & INTERIOR: The large pallial sinus runs forward but does not touch the anterior muscle scar. Weak cruciform muscle scars near ventral margin.

PERIOSTRACUM: None visible.

REMARKS: The typical oblique sculpture of this tiny *Strigilla* makes it stand out from the other minute bivalves. A deposit feeder.

HABITAT: Near shore and in inlet areas. Infaunal.

LOCALITIES: Entire.

OCCURRENCE: Fairly common.

RANGE: Southeastern United States, Texas and the West Indies.

GEOLOGIC RANGE: Miocene to Recent.

Genus *Tellina* Linné, 1758

Subgenus *Angulus* Megerle von Mühlfeld, 1811

TAMPA TELLIN

Tellina (Angulus) tampaensis Conrad, 1866, *Amer. J. Conch.* 2:281.

Gk. *telline* a kind of shellfish; from Tampa, Florida.

SIZE: Length 1 in.

COLOR: Smooth white, rarely suffused with pale peach coloration. Interior white, shining.

SHAPE: Ovate subtrigonal. Inequivalve. Inequilateral.

ORNAMENT OR SCULPTURE: Anterior margin broadly rounded; posterior dorsal margin steeply sloping. Sculpture consists of concentric lines separated by narrow, well-defined sulci.

HINGE AREA: Ligament brown, external. Cardinal teeth are present, but no true lateral teeth are produced.

PALLIAL LINE & INTERIOR: Adductor muscle scars well impressed. Anterior scar is longer, narrower, and higher than the posterior scar. Pallial sinus descends to the pallial line in a short, straight drop some distance from the anterior scar.

PERIOSTRACUM: None visible.

REMARKS: A delicate little deposit-feeding clam.

HABITAT: Hypersaline lagoons from 6-in. depth to shore. Infaunal.

LOCALITIES: Entire.

OCCURRENCE: Fairly common.

RANGE: Southern half of Florida to Texas, Bahamas, and Cuba.

SAY'S TELLIN

Tellina (Angulus) texana Dall, 1900, *Proc. U.S. Nat. Mus.* 23:295.

Gk. *telline* a kind of shellfish; from Texas.

SIZE: Length 16.5 mm.

COLOR: White with opalescent interior.

SHAPE: Subelliptical to subtrigonal. Inequivalve. Inequilateral.

ORNAMENT OR SCULPTURE: Anterior margin rounded; posterior dorsal margin elongate and steeply inclined. Sculpture consists of weak, finely incised, closely spaced concentric sulci.

HINGE AREA: Ligament yellowish brown, strong, ex-

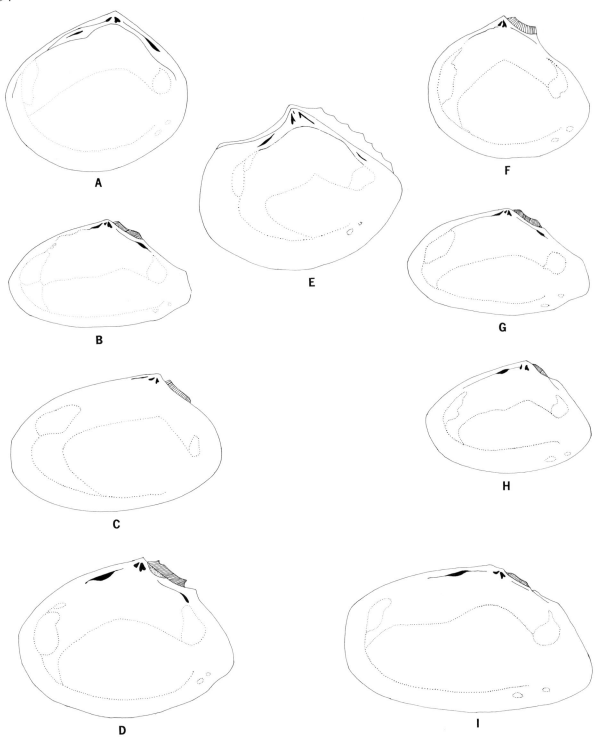

Fig. 28. Diagrammatic illustrations of the internal surface of the right valve of members of the family Tellinidae found in Texas, showing the dental configuration and muscle scars. The tellin hinge always has two cardinal teeth beneath the umbones, one of which is bifurcate. There is slight variation in these teeth, rendering them unimportant in identification. The presence of lateral teeth distinguishes the tellins from the macomas, which have no laterals. Two cruciform muscle scars are found on each valve posterior to and closely aligned with the end of the pallial line; they are found only in the superfamily Tellinacea. *A, Tellina aequistriata* j.a.; *B, Tellina alternata*; *C, Tellina iris* j.a.; *D, Tellina lineata*; *E, Tellidora cristata* j.a.; *F, Tellina tampaensis*; *G, Tellina tayloriana*; *H, Tellina texana*; *I, Tellina versicolor* (after Boss 1968).

Tellina (Angulus) texana

Tellina (Angulus) versicolor

ternal. Cardinal teeth but no true lateral teeth. Umbones posterior to middle and blunt.

PALLIAL LINE & INTERIOR: Adductor muscle scars fairly well impressed. Anterior scar elongate and rounded below. Posterior scar rounded. Pallial sinus convex above, gently inclined and slightly concave anteriorly, falling in an arch to the pallial line near but not touching the anterior scar.

PERIOSTRACUM: None visible.

REMARKS: Originally described from Texas. Syn. *T. sayi* Dall. A deposit feeder.

HABITAT: Bay centers. Infaunal.

LOCALITIES: East, central.

OCCURRENCE: Fairly common.

RANGE: New Jersey to southern half of Florida, Texas and Cuba.

GEOLOGIC RANGE: Lower Pliocene to Recent.

DeKAY's DWARF TELLIN

Tellina (Angulus) versicolor DeKay, 1843, *Natur. Hist. N.Y.* 1:209.

Gk. *telline* a kind of shellfish; L. *versicolor* changing colors.

SIZE: Length 17.5 mm.

COLOR: Translucent with red, white, or pink rays. Shiny.

SHAPE: Elongate. Subelliptical.

ORNAMENT OR SCULPTURE: Anterior dorsal margin elongate and slightly inclined; posterior dorsal margin steeply inclined, short, and slightly concave. The posterior is marked with a rounded keel. Sculpture consists of widely spaced, strongly incised concentric sulci; no radial sculpture.

HINGE AREA: Umbones just posterior to the middle, depressed and pointed. Ligament yellowish brown, external. The right valve has cardinal teeth and weak laterals. The left valve has cardinal teeth but no laterals.

PALLIAL LINE & INTERIOR: Anterior adductor scar elongate and rounded below; posterior scar rounded. Pallial sinus rises gently posteriorly, convex above and arches down to the pallial line very near the anterior scar, at times touching it.

PERIOSTRACUM: None visible.

REMARKS: Without magnification this shell can be confused with *T. iris*, but it lacks the distinctive oblique sculpture of *T. iris*. A deposit feeder.

HABITAT: Offshore in sand. Infaunal.

LOCALITIES: Entire.

OCCURRENCE: Fairly common.

RANGE: New York to southern half of Florida, Texas and the West Indies.

GEOLOGIC RANGE: Pliocene to Recent.

Tellina (Eurytellina) alternata

Subgenus *Eurytellina* Fischer, 1887

ALTERNATE TELLIN

Tellina (Eurytellina) alternata Say, 1822, *J. Acad. Natur. Sci. Phila.* 2:275.

Gk. *telline* a kind of shellfish; L. *alternus* alternating.

SIZE: Length 2¾ in.

COLOR: Glossy white, often with slight blushes of pink or yellow.

SHAPE: Elongate subtrigonal. Inequilateral. Inequivalve.

ORNAMENT OR SCULPTURE: The outline is narrow posteriorly with a slight truncation. Sculpture consists of incised concentric lines separated by broad bands. The left valve has broader bands and fewer lines than the right. A posterior ridge occurs in the right valve. Every alternate striation disappears at angle of keel.

HINGE AREA: Umbones slightly posterior to center, small, scarcely elevated. Ligament is strong, brown, and exterior. Three cardinal teeth in right valve. Both valves have an internal rib extending from the umbo to the anterior muscle scar.

PALLIAL LINE & INTERIOR: Adductor muscle scars well impressed. Pallial sinus curves upward toward umbones and extends anteriorly almost to the muscle scar.

PERIOSTRACUM: Thin, yellowish brown.

REMARKS: Until recently this species was considered a synonym of *T. tayloriana*, the beautiful pink tellin, and they are closely related. The valves of *T. alternata* are somewhat more inflated than those of *T. tayloriana*. The range of this deposit feeder is wider than that of the pink tellin.

HABITAT: In sand near shore, bay margins, and inlets. Infaunal.

LOCALITIES: Entire, more to east.

OCCURRENCE: Fairly common.

RANGE: North Carolina, Florida and the Gulf states.

GEOLOGIC RANGE: Miocene to Recent.

ROSE PETAL TELLIN

Tellina (Eurytellina) lineata Turton, 1819, *Conch. Dict.*, p. 168.

Gk. *telline* a kind of shellfish; L. *linea* line.

SIZE: Length 1½ in.

COLOR: Pink to white.

SHAPE: Elongate subtrigonal. Inequilateral. Equivalve.

ORNAMENT OR SCULPTURE: Anterior margin well rounded. Posterior dorsal margin steeply sloping. Sculpture consists of close, weak, concentric sulci separated by low, narrow bands. Posterior ridge is present but not well developed, stronger on left valve. There is a twist to the right at the posterior end.

HINGE AREA: Ligament dark brown, short, wide, and sunken. Umbones slightly raised, pointed, inflated, and located just posterior to middle. Both cardinal and lateral teeth are present. Posterior lateral well developed. A variable anterior rib between umbones and anterior scar.

PALLIAL LINE & INTERIOR: Adductor muscle scars well impressed. Pallial sinus convex above, not rising above the adductor muscle scars and extending closer to the anterior scar than does that of *T. tayloriana*.

Tellina (Eurytellina) lineata

PERIOSTRACUM: Thin, brownish.

REMARKS: The shape and color of this shell are variable, but the strong twist of the valves to the right posteriorly and the umbones that point to the back are consistent. The specimens found on the Texas coast appear to be very old, and no living shells have been reported. A deposit feeder.

HABITAT: Offshore or dead in spoil banks. Infaunal.

LOCALITIES: Central.

OCCURRENCE: Uncommon.

RANGE: All of Florida, Texas, and the West Indies.

Tellina (*Eurytellina*) *tayloriana* Sowerby, 1867, Reeve, *Conch. Icon.* 17:pl. 30, fig. 168.

Gk. *telline* a kind of shellfish; dedicated to Thomas Lambe Taylor (1802–1874), English collector.

SIZE: Length 2½ in.

COLOR: Glossy pink.

SHAPE: Elongate subtrigonal. Inequilateral. Inequivalve.

ORNAMENT OR SCULPTURE: The outline narrows posteriorly and is truncated at the end. Sculpture consists of incised concentric lines, separated by broad bands. The sculpture is stronger and closer on the right valve.

HINGE AREA: Ligament is brown and external. Three cardinal teeth and 2 laterals in the right valve. A heavy rib extends from the umbones to the anterior adductor scar in both valves.

PALLIAL LINE & INTERIOR: Pallial sinus variable, but usually about equal in each valve; flattened across top and extending almost to anterior scar, where it drops rather abruptly to the pallial line.

PERIOSTRACUM: Thin, yellowish brown.

REMARKS: The beautiful pink color is the main distinction between this and *T. alternata.* Parker (1960) believes that it lives offshore at greater depths than *T. alternata.* The right valve of *T. tayloriana* is much flatter than that of *T. alternata.* Dr. Harry (in litt.) finds no difference in the two species. A deposit feeder.

HABITAT: Near shore, 2 to 12 fathoms. Infaunal.

LOCALITIES: Entire.

OCCURRENCE: Common.

RANGE: Gulf coast of Texas and Mexico.

Subgenus *Merisca* Dall, 1900

Tellina (*Merisca*) *aequistriata* (Say, 1824), *J. Acad. Natur. Sci. Phila.* 4:145.

Gk. *telline* a kind of shellfish; L *aequi* equal, *striata* striped.

SIZE: Length 18 to 25 mm.

COLOR: White.

Tellina (Eurytellina) tayloriana

Tellina (Merisca) aequistriata

SHAPE: Moderately oval. Inequilateral.

ORNAMENT OR SCULPTURE: Sculpture consists of numerous, sharp, concentric ridges. Left valve has 1 posterior radial ridge, right valve has 2. Posterior margin is narrow and flexed.

HINGE AREA: Umbones small, sharp. Ligament small. Weak hinge area has 2 long laterals in the left valve.

PALLIAL LINE & INTERIOR: Muscle scars small. Dorsal line of pallial sinus meets the pallial line near anterior scar.

PERIOSTRACUM: Very thin, yellowish brown.

REMARKS: Syn. *Quadrans lintea* Conrad. A very delicate little shell distinguished by the pronounced posterior twist. A deposit feeder.

HABITAT: Offshore in sand, 12 to 35 fathoms. Infaunal.

LOCALITIES: Entire.

OCCURRENCE: Uncommon.

RANGE: North Carolina to both sides of Florida, Texas, and the West Indies.

GEOLOGIC RANGE: Miocene to Recent.

Subgenus *Scissula* Dall, 1900

IRIS TELLIN

Tellina (Scissula) iris Say, 1822, *J. Acad. Natur. Sci. Phila.* 2:302.

Gk. *telline* a kind of shellfish; L. *iris* the rainbow.

SIZE: Length 15.3 mm.

COLOR: Transparent to translucent suffused with pink. Two white rays often occur in posterior quarter.

SHAPE: Elongate, elliptical. Inequilateral. Equivalve.

ORNAMENT OR SCULPTURE: Anterior dorsal margin long and gently sloping to rounded anterior margin. Posterior margin is obliquely truncated and posterior slope is slightly keeled. Sculpture consists of faint growth lines that are more developed posteriorly. These are crossed by well-developed, widely spaced oblique lines.

HINGE AREA: Umbones posterior to the middle, small, and slightly pointed. Ligament light yellow brown, weak, external. Left valve has cardinal teeth but no true lateral teeth. Right valve has cardinal teeth and an anterior lateral tooth but no posterior lateral.

PALLIAL LINE & INTERIOR: Adductor scars weak. Anterior scar irregularly quadrate; posterior scar rounded. Pallial sinus rises abruptly posteriorly, descends gently, and arches to the pallial line. Well separated from anterior scar.

Tellina (Scissula) iris

PERIOSTRACUM: None visible.

REMARKS: This delicate little shell is one of the most common in the drift on the Texas coast. A deposit feeder.

HABITAT: Near shore and inlet areas. Infaunal.

LOCALITIES: Entire.

OCCURRENCE: Common.

RANGE: North Carolina to Florida, Gulf of Mexico and Bermuda.

Genus *Tellidora* H. & A. Adams, 1856

WHITE-CRESTED TELLIN

Tellidora cristata (Recluz, 1842), *Rev. Zool. Soc. Cuvier* 5:270.

Gk. *telline* a kind of shellfish; L. *cristatus* crested.

SIZE: Length up to 1½ in.

COLOR: White.

SHAPE: Subtrigonal, very compressed. Inequilateral. Inequivalve.

ORNAMENT OR SCULPTURE: Anterior and posterior dorsal margins have triangular spines. Sculpture

Tellidora cristata

Sanguinolaria cruenta

consists of strong, narrow, concentric ridges. Spines form a deep lunule and escutcheon.

HINGE AREA: Umbones central, acute, and elevated. Ligament brown, short, and partially internal. Two cardinal teeth in each valve and a strong, triangular, anterior lateral tooth in the right valve.

PALLIAL LINE & INTERIOR: Adductor scars well impressed. Anterior scar more elongate than posterior. Pallial sinus short, widely separated from anterior scar; arches down to the pallial line near the posterior end.

PERIOSTRACUM: None visible.

REMARKS: The dorsal spines make this little clam unique. Deposit feeder.

HABITAT: Inlet and channel. Bay margins in sandy bottoms. Infaunal.

LOCALITIES: Entire.

OCCURRENCE: Uncommon.

RANGE: North Carolina to western Florida and Texas.

GEOLOGIC RANGE: Pliocene to Recent.

Family PSAMMOBIIDAE Fleming, 1828

Genus *Sanguinolaria* Lamarck, 1799

ATLANTIC SANGUIN
Sanguinolaria cruenta (Solander, 1786), *Portland Mus. Cat.*, p. 53.
L. *sanguis* blood, *cruentare* stain with blood.
SIZE: Length 1½ to 2 in.

COLOR: White. Umbones and area just below a bright orangish red fading into white ventrally.

SHAPE: Subovate. Inequilateral. Inequivalve.

ORNAMENT OR SCULPTURE: A thin, gaping shell sculptured with microscopic growth lines. The left valve slightly more compressed than the right.

HINGE AREA: Hinge teeth are near the center of the dorsal margin. Two small cardinals in each valve. Ligament external.

PALLIAL LINE & INTERIOR: Large pallial sinus with a U-shaped hump at the top.

PERIOSTRACUM: Light brown, usually lost.

REMARKS: The color fades rather quickly after this burrowing clam is washed onto the shore. Pulley (1952) states this species is adventitious and does not normally live in Texas.

HABITAT: Offshore. Infaunal.

LOCALITIES: Central, south.

OCCURRENCE: Uncommon.

RANGE: Southern Florida, the Gulf states, West Indies.

Family SOLECURTIDAE d'Orbigny, 1846

Genus *Solecurtus* Blainville, 1824

CORRUGATED RAZOR CLAM
Solecurtus cumingianus (Dunker, 1861), *Proc. Zool. Soc. London* 29:425.
Gk. *solen* a pipe; L. *curtus* short; dedicated to Hugh Cuming (1791–1865), English collector.

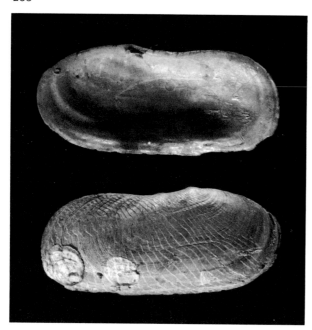

Solecurtus cumingianus

SIZE: Length 1 to 2 in.

COLOR: White, dull.

SHAPE: Elongate cylindrical. Flattened. Equivalve.

ORNAMENT OR SCULPTURE: Surface of shell sculptured with irregular growth lines crossed by fine, oblique lines. Both rounded ends gape.

HINGE AREA: External ligament is posterior to umbones. Right valve has 2 strong, cardinal teeth. Left valve with 1 cardinal tooth.

PALLIAL LINE & INTERIOR: The pallial sinus extends forward to a point below the cardinal teeth.

PERIOSTRACUM: Yellowish gray, thin.

REMARKS: Dead halves on the beach of mid–Padre Island. A deposit feeder.

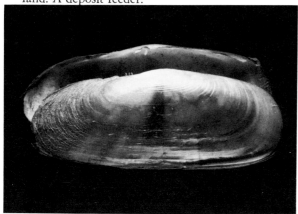

Tagelus (Mesopleura) divisus

HABITAT: Offshore. Infaunal.

LOCALITIES: South.

OCCURRENCE: Uncommon.

RANGE: Southeast United States, West Indies, Texas, Mexico, Colombia, and Brazil.

GEOLOGIC RANGE: Pliocene to Recent.

Genus *Tagelus* Gray, 1847

Subgenus *Mesopleura*, Conrad, 1867

PURPLISH TAGELUS

Tagelus (Mesopleura) divisus (Spengler, 1794), *Skr. Nat. Selsk. Copenhagen* 3(2):96.

From le Tagal, a name arbitrarily given to a Senegalese shell by Adamson in 1757; L. *dividere* to divide.

SIZE: Length 1 to 1½ in.

COLOR: Whitish purple with a strong purple radial streak about midshell.

SHAPE: Flattened cylindrical, elongate. Equivalve.

ORNAMENT OR SCULPTURE: Thin, fragile shell is unsculptured. Gaping at both ends.

HINGE AREA: Umbones posterior of center, suppressed. Cardinal but no lateral teeth in each valve. The purple ray marks the position of the weak, internal radial rib just anterior to the teeth.

PALLIAL LINE & INTERIOR: Two muscle scars. Deep pallial sinus but does not extend to cardinal teeth.

PERIOSTRACUM: Shiny, thin, chestnut brown.

REMARKS: At low tide one can see herons feeding on these fragile shells in the bays. A deposit feeder.

HABITAT: Open sound. Open-lagoon margin. Infaunal.

LOCALITIES: Entire.

OCCURRENCE: Common.

RANGE: Cape Cod to southern Florida, Gulf states and the Caribbean.

GEOLOGIC RANGE: Miocene to Recent.

STOUT TAGELUS

Tagelus (Mesopleura) plebeius (Solander, 1786), *Portland Mus. Cat.*, pp. 42, 101, 156.

From le Tagal, a name arbitrarily given to a Senegalese shell by Adamson in 1757; L. *plebeius* common.

SIZE: Length 2 to 3½ in.

COLOR: White.

SHAPE: Elongate, rectangular. Equivalve.

ORNAMENT OR SCULPTURE: The strong shell is sculp-

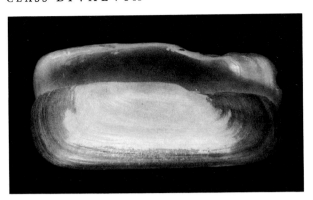

Tagelus (Mesopleura) plebeius

tured with faint, irregular growth lines. Gapes at each end. Weak radial ridge on posterior slope.

HINGE AREA: Umbones posterior of center, suppressed. Cardinal, but no lateral teeth in each valve. Lacks the internal ribs of *T. divisus*.

PALLIAL LINE & INTERIOR: Two muscle scars; deep pallial sinus.

PERIOSTRACUM: Shiny, moderately thin, greenish brown.

REMARKS: This deposit feeder uses its strong foot to bury itself upright in the mud. It can take lower salinities than *T. divisus*.

Abra aequalis

HABITAT: Enclosed lagoon. Bay margins. Infaunal.

LOCALITIES: Entire.

OCCURRENCE: Common.

RANGE: Cape Cod to southern Florida and the Gulf states.

GEOLOGIC RANGE: Miocene to Recent.

Family SCROBICULARIIDAE H. & A. Adams, 1856

Genus *Abra* Lamarck, 1818

COMMON ATLANTIC ABRA
Abra aequalis (Say, 1822), *J. Acad. Natur. Sci. Phila.* 2:307.

Gk. *habros* graceful, delicate; L. *aequalis* even.

SIZE: Diameter 6 mm.

COLOR: White.

SHAPE: Orbicular.

ORNAMENT OR SCULPTURE: Smooth, polished. Anterior margin of right valve grooved.

HINGE AREA: Umbones small, pointed. Two small cardinal teeth in right valve, 1 weak. Lateral teeth absent in left valve, 1 anterior lateral in right valve. Elongate chondrophore extending posteriorly from the cardinal teeth.

PALLIAL LINE & INTERIOR: Pallial sinus large; directed forward and upward.

PERIOSTRACUM: Thin, clear, yellowish. Usually lost.

REMARKS: At times this little shell is on the beach in great quantities. A deposit feeder.

HABITAT: Open sound. Lagoon centers. Near shore in clayey sediments. Infaunal.

LOCALITIES: Entire.

OCCURRENCE: Common.

RANGE: North Carolina to Texas and on to Tampico, Mexico.

GEOLOGIC RANGE: Miocene to Recent.

Family SEMELIDAE Stoliczka, 1870

Genus *Cumingia* G.B. Sowerby I, 1833

TELLIN-LIKE CUMINGIA
Cumingia tellinoides (Conrad, 1831), *J. Acad. Natur. Sci. Phila.* 6:258.

Cumingia tellinoides

Dedicated to naturalist Hugh Cuming (1791–1865);
L. *tellinoides* tellin shaped.
SIZE: Length 12 to 18 mm.
COLOR: White.
SHAPE: Trigonal.
ORNAMENT OR SCULPTURE: Thin shell is sculptured
with fine, slightly raised growth lines. The posterior
end is slightly pointed and flexed. There is a radial
ridge on the posterior slope.
HINGE AREA: Umbones pointed and just posterior to
center. Both valves have 1 small, bladelike cardinal
tooth, a central spoonshaped chondrophore, and
elongated anterior and posterior laterals.
PALLIAL LINE & INTERIOR: Anterior adductor scar
elongated, posterior scar rounded. Latter is largely
confluent with pallial line below. Pallial sinus deep
and rounded.
PERIOSTRACUM: Thin, yellowish. Usually lost.
REMARKS: Dead specimens have been found in holes
in rock and oyster shells but are more apt to be found
in mud. A filter feeder.
HABITAT: Bay margins, high-salinity bays and inlet
areas. Infaunal.
LOCALITIES: Entire.
OCCURRENCE: Fairly common.
RANGE: Nova Scotia to St. Augustine, Florida, Texas
and Cuba.
GEOLOGIC RANGE: Pleistocene to Recent.

Genus *Semele* Schumacher, 1817

CANCELLATE SEMELE
Semele bellastriata (Conrad, 1837), *J. Acad. Natur.
Sci. Phila.* 7:239.
Semele, mother of Bacchus; L. *bellus* pretty, *striata*
striped.
SIZE: Length 12 to 18 mm.
COLOR: Yellowish white with reddish flecks or all
purplish gray. Interior white suffused with mauve
or purple.
SHAPE: Oval.
ORNAMENT OR SCULPTURE: Surface sculptured with
concentric ridges and radial riblets that are stronger
on the anterior and posterior slopes, giving a can-
cellate appearance at these extremities.
HINGE AREA: Umbones slightly pointed, just behind
center. A horizontal chondrophore and 2 cardinal
teeth in each valve. Right valve has 2 lateral teeth.
PALLIAL LINE & INTERIOR: Two rounded muscle
scars. Pallial sinus deep and rounded.
PERIOSTRACUM: None visible.
REMARKS: These have been found after severe northers
near the jetties at Port Aransas. A deposit feeder.

Semele bellastriata

HABITAT: Offshore. Infaunal.

LOCALITIES: Entire.

OCCURRENCE: Uncommon.

RANGE: North Carolina to southern half of Florida, West Indies and Texas.

GEOLOGIC RANGE: Pliocene to Recent.

WHITE ATLANTIC SEMELE

Semele proficua (Pulteney, 1799), Hutchins, *Dorsetshire*, p. 29.

Semele, mother of Bacchus; L. *proficuus* useful.

SIZE: Length 12 to 35 mm.

COLOR: Whitish to yellowish white. Interior yellowish, glossy, sometimes flecked with mauve.

SHAPE: Orbicular. Equivalve.

ORNAMENT OR SCULPTURE: Sculpture consists of fine, irregular growth lines and microscopic radial lines.

HINGE AREA: Umbones almost central and pointed. Hinge area has long chondrophore to house resilium, 2 small, fragile cardinal teeth; right valve has 2 lateral teeth.

PALLIAL LINE & INTERIOR: Muscle scars are rounded. Pallial sinus deep, rounded, and oblique.

PERIOSTRACUM: None visible.

REMARKS: The most common *Semele* on the Texas coast. A deposit feeder.

HABITAT: Open-bay centers. Inlet areas. Infaunal.

LOCALITIES: Entire.

OCCURRENCE: Fairly common.

RANGE: North Carolina to southern half of Florida, Texas and the West Indies, Central America to Brazil.

GEOLOGIC RANGE: Pliocene to Recent.

PURPLISH SEMELE

Semele purpuracens (Gmelin, 1791), *Syst. Natur.*, 13th ed., p. 3288.

Semele, mother of Bacchus; L. *purpura* purple.

SIZE: Length 1 to 1½ in.

COLOR: Variable. Cream with purplish or orangish flecks.

SHAPE: Oval. Equivalve.

ORNAMENT OR SCULPTURE: Thin shell sculptured with concentric striae that become weaker toward posterior margin. There are microscopic lines between striae but no radial ribs. Lines tend to converge.

HINGE AREA: Umbones posterior of center, pointed. Hinge with horizontal chondrophore, 2 cardinal teeth. Right valve has 2 lateral teeth.

PALLIAL LINE & INTERIOR: Muscle scars irregularly shaped. Pallial sinus deep and rounded.

PERIOSTRACUM: None visible.

REMARKS: Only worn shells found on beach. A deposit feeder.

Semele proficua

Semele purpuracens

HABITAT: Offshore on sand bottoms and banks. Infaunal.

LOCALITIES: Entire, more to south.

OCCURRENCE: Fairly common.

RANGE: North Carolina to southern half of Florida and the West Indies and Texas, southern Mexico, Central America to Brazil.

GEOLOGIC RANGE: Pliocene to Recent.

Suborder VENERINA Vokes, 1967

Superfamily DREISSENACEA Gray, 1840

Family DREISSENIDAE Gray, 1840

Genus *Congeria* Partsch, 1835

Subgenus *Mytilopsis* Conrad, 1857

CONRAD'S FALSE MUSSEL
Congeria (Mytilopsis) leucophaeta (Conrad, 1831), *J. Acad. Natur. Sci. Phila.* 6:263.
L. *congeria* a heap, *leucophaeta* dark colored.
SIZE: Length 18 mm.

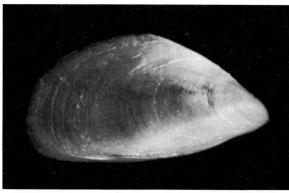

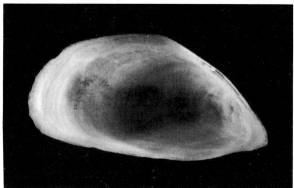

Congeria (Mytilopsis) leucophaeta

COLOR: Bluish brown to tan.

SHAPE: Mussel like.

ORNAMENT OR SCULPTURE: Exterior very rough. Anterior end much depressed. Byssal opening small.

HINGE AREA: Shelf, or septum, at beak end. There is a long thin bar under the ligament.

PALLIAL LINE & INTERIOR: Pallial line simple with 2 adductor muscle scars.

PERIOSTRACUM: Shiny, light brown.

REMARKS: The external appearance of this little shell makes it easy to confuse with a mussel, but the internal septum across the end will distinguish it. Attaches in clusters by byssal thread. A filter feeder.

HABITAT: Brackish water. Epifaunal.

LOCALITIES: Central and east.

OCCURRENCE: Fairly common.

RANGE: New York to Florida to Texas and Mexico.

GEOLOGIC RANGE: Pleistocene to Recent.

Superfamily CORBICULACEA Gray, 1847

Family CORBICULIDAE Gray, 1847

Genus *Polymesoda* Rafinesque, 1820

CAROLINA MARSH CLAM
Polymesoda caroliniana (Bosc, 1830), *Hist. Natur. Coq.* 3:37.
Gk. combining form *poly* many, *meso* middle; from Carolina.
SIZE: Length 1 to 1½ in.
COLOR: White.
SHAPE: Trigonal, inflated. Equivalve. Inequilateral.
ORNAMENT OR SCULPTURE: Rather smooth with weak concentric growth lines. Erosion of umbones is typical.
HINGE AREA: Umbones elevated. Each hinge with 3 small, almost vertical, equally sized teeth below the umbones and each hinge with 1 anterior and posterior lateral. Dark brown, long, narrow ligament is external.
PALLIAL LINE & INTERIOR: Pallial sinus narrow, ascending, fairly deep with 2 equal adductor muscle scars.
PERIOSTRACUM: Heavy, black brown, velvetlike.
REMARKS: This shell can stand very brackish to fresh water. It is common in the middle part of the coast, where it can be found alive in Lavaca Bay. It is rare on other parts of coast.

Polymesoda caroliniana

Pseudocyrena floridana

HABITAT: Estuaries. Infaunal.
LOCALITIES: All but southern tip of coast, more in Matagorda Bay and east.
OCCURRENCE: Fairly common.
RANGE: Virginia to northern half of Florida and Texas.
GEOLOGIC RANGE: Pleistocene to Recent.

Genus *Pseudocyrena* Bourguignat, 1854

FLORIDA MARSH CLAM
Pseudocyrena floridana (Conrad, 1846), *Proc. Acad. Natur. Sci. Phila.* 3:23.
Gk. *pseudes* false, *cyrenaic* relating to Cyrene, a nymph; from Florida.
SIZE: Length 1 in.
COLOR: Whitish flushed with purple pink. Interior white with wide purple margin or all purple.
SHAPE: Trigonal. Equivalve. Inequilateral.
ORNAMENT OR SCULPTURE: Smooth with weak, irregular growth lines. Not heavy. Ventral margin slightly sinuate posteriorly.
HINGE AREA: Each hinge with 3 small almost vertical teeth below the umbones and each hinge with 1 anterior and posterior lateral.

PALLIAL LINE & INTERIOR: Pallial sinus narrow, ascending, fairly deep with 2 adductor muscle scars.
PERIOSTRACUM: None visible.
REMARKS: A colorful little clam that is seldom found in the Galveston area. A filter feeder.
HABITAT: In sand in open hypersaline bays and inlets. Infaunal.
LOCALITIES: Central, south.
OCCURRENCE: Common.
RANGE: Key West to northern Florida and to Texas.
GEOLOGIC RANGE: Pleistocene to Recent.

Superfamily VENERACEA Rafinesque, 1815

Family VENERIDAE Rafinesque, 1815

Subfamily PITARINAE Stewart, 1930

Genus *Callocardia* A. Adams, 1864
Subgenus *Agriopoma* Dall, 1902

TEXAS VENUS
Callocardia (*Agriopoma*) *texasiana* (Dall, 1892), *Nautilus* 5(12):134.
L. *callosus* hardened; Gk. *kardia* heart; from Texas.
SIZE: Length 1½ to 3 in.

Callocardia (Agriopoma) texasiana

COLOR: Creamy white to dirty gray. Interior chalky white.

SHAPE: Oval elongate, inflated. Equivalve. Inequilateral.

ORNAMENT OR SCULPTURE: Smooth with only very fine concentric growth lines.

HINGE AREA: Umbones prominent, rolled in under themselves. Weak, tear-shaped lunule. Three cardinal teeth; the posterior cardinal is S-shaped in the right valve. Left anterior lateral small and fitting into a socket in right valve.

PALLIAL LINE & INTERIOR: Two small muscle scars; the anterior scar very close to margin. Pallial line is strong with a deep, triangular sinus touching the posterior scar. Margin smooth.

PERIOSTRACUM: Weak and inconspicuous.

REMARKS: Lives just below the sand. Live specimens are very rare on the Texas shores but worn halves are fairly common in the Galveston area and less so on the rest of the coast. Parker (1969: personal communication) reports this alive in the clay and mud of Matagorda and Aransas Bays. Originally described from Texas. Syn. *Pitar texasiana*.

HABITAT: Clay lagoon centers and along shore in clay. Infaunal.

LOCALITIES: Entire.

OCCURRENCE: Fairly common.

RANGE: Northwestern Florida to Texas and Mexico.

GEOLOGIC RANGE: Pleistocene to Recent.

Genus *Macrocallista* Meek, 1876

CALICO CLAM

Macrocallista maculata (Linné, 1758), *Syst. Natur.*, 10th ed., p. 686.

Gk. *macros* long in extent, *callista* a nymph; L. *maculatus* spotted.

SIZE: Length 1½ to 2½ in.

COLOR: Cream with irregular, almost checkered, brown marks. Interior white.

SHAPE: Oval. Equivalve. Inequilateral.

ORNAMENT OR SCULPTURE: Highly polished.

HINGE AREA: Umbones small. Lunule small, impressed. Two unequal cardinals. Short, blunt anterior lateral. Thin posterior lateral.

PALLIAL LINE & INTERIOR: Two muscle scars, posterior larger. Pallial sinus wider at base, angled at end. Margins smooth.

PERIOSTRACUM: Thin, glossy, light brown.

REMARKS: A south-of-the-border species, rarely found north of the Port Isabel area.

HABITAT: Offshore. Infaunal.

LOCALITIES: South.

OCCURRENCE: Uncommon.

Macrocallista maculata

RANGE: Southeastern United States, West Indies, South Texas, and Mexico.

GEOLOGIC RANGE: Miocene to Recent.

SUNRAY VENUS

Macrocallista nimbosa (Solander, 1786), *Portland Mus. Cat.*, p. 175.

Gk. *macros* long in extent, *callista* a nymph; L. *nimbus* rain cloud.

SIZE: Length 4 to 5 in.

COLOR: Pale salmon with broken, brownish radial lines.

SHAPE: Elongated oval. Equivalve. Inequilateral.

ORNAMENT OR SCULPTURE: Polished. Sculpture of inconspicuous radial and concentric lines.

HINGE AREA: Umbones depressed. Lunule impressed, oval, purplish. Long external ligament. Three cardinal teeth.

PALLIAL LINE & INTERIOR: Two muscle scars. Pallial sinus reflected, wider at base, angled at end. Margins smooth.

PERIOSTRACUM: Thin, lacquerlike, light brown.

REMARKS: This beautiful large shell must have been more abundant in the past, because tools and scrapers made from it are found in Indian middens along the bays. It lives buried in the sand when living, but Texas specimens are probably fossil.

HABITAT: May live offshore. Infaunal.

LOCALITIES: Entire, more to south.

OCCURRENCE: Uncommon.

RANGE: North Carolina to Florida and the Gulf states.

GEOLOGIC RANGE: Pliocene to Recent.

Macrocallista nimbosa

Dosinia (Dosinidia) discus

Subfamily DOSINIINAE H. & A. Adams, 1857

Genus *Dosinia* Scopoli, 1777

Subgenus *Dosinidia* Dall, 1902

DISK DOSINIA

Dosinia (Dosinidia) discus (Reeve, 1850), *Conch. Iconogr.* 6:pl. 2.

From Senegalese Dosin, a shell described by Adamson, 1757; L. *discus* quoit, flat, round object to be thrown.

SIZE: Length 2 to 3 in.

COLOR: White.

SHAPE: Lenticular, flattened. Equivalve. Inequilateral.

ORNAMENT OR SCULPTURE: Sculpture consists of numerous fine, concentric ridges.

HINGE AREA: Lunule heart shaped. Ligament strong, placed in a groove. Three cardinal teeth in each valve.

PALLIAL LINE & INTERIOR: Interior smooth and glossy. Two small muscle scars connected by pallial line with large, angular sinus extending to center of shell.

PERIOSTRACUM: Thin, yellowish brown.

REMARKS: The strong ligament is a benefit to beach-

Dosinia (Dosinidia) elegans

combers, as the pairs that are washed onto the beach in great numbers in the winter are firmly held together. The valves of this shell are often neatly punctured with a hole drilled by a predatory gastropod. A filter feeder.

HABITAT: Near shore from 2 to 12 fathoms. Infaunal.
LOCALITIES: Entire.
OCCURRENCE: Common.
RANGE: Virginia to Florida, Gulf states, Mexico and Bahamas.
GEOLOGIC RANGE: Pliocene to Recent.

ELEGANT DOSINIA
Dosinia (Dosinidia) elegans Conrad, 1846, *Amer. J. Sci.* 2(2):393.
From Senegalese Dosin, a shell described by Adamson, 1757; L. *elegans* elegant, fine.
SIZE: Length 2 to 3 in.
COLOR: Ivory.
SHAPE: Lenticular, flattened. Equivalve. Inequilateral.
ORNAMENT OR SCULPTURE: Sculpture consists of regular concentric ribs, which are fewer in number and heavier than *D. discus*.
HINGE AREA: Umbones prominent. Lunule small. Lunule partly submarginal. Three cardinal teeth in each valve with lateral teeth present. Left middle cardinal and right posterior cardinal bifid.

PALLIAL LINE & INTERIOR: Two muscle scars are connected by pallial line with a long, angular sinus touching posterior scar. Margins smooth.
PERIOSTRACUM: Thin, yellowish brown.
REMARKS: This is an inhabitant of the Texas-Mexican zone and is seldom found north of Big Shell on Padre Island. A filter feeder.
HABITAT: Offshore. Infaunal.
LOCALITIES: Southern half of coast, more near the south.
OCCURRENCE: Uncommon.
RANGE: Eastern Florida, South Texas and south to Isla Mujeres, Mexico.
GEOLOGIC RANGE: Pliocene to Recent.

Subfamily CYCLININAE Frizzell, 1936

Genus *Cyclinella* Dall, 1902

ATLANTIC CYCLINELLA
Cyclinella tenuis (Recluz, 1852), *J. Conch.* [Paris] 3:250.
Gk. *kyklos* a circle; L. *ella* dim. suffix, *tenuis* thin.
SIZE: Length 1 to 2 in.
COLOR: Whitish.
SHAPE: Circular. Flattened. Equivalve. Inequilateral.
ORNAMENT OR SCULPTURE: Resembles a *Dosinia* but is smaller and more thin shelled. Surface sculptured with very fine, irregular growth lines.

Cyclinella tenuis

HINGE AREA: Submarginal hinge ligament. Three cardinal teeth but no laterals as the *Dosinia* have. Right posterior tooth bifid.

PALLIAL LINE & INTERIOR: Two muscle scars. The anterior scar is much nearer the ventral margin than that of *Dosinia*. Pallial sinus is ascending, long, and narrow. Margins smooth.

PERIOSTRACUM: None visible.

REMARKS: Easily mistaken for *Dosinia* but much smaller. A filter feeder.

HABITAT: Inlet-influenced areas and bay margins. Infaunal.

LOCALITIES: Entire, more south.

OCCURRENCE: Fairly common.

RANGE: Eastern United States, Texas, and the West Indies to Brazil.

Subfamily GEMMINAE Dall, 1902

Genus *Gemma* Deshayes, 1853

AMETHYST GEM CLAM
Gemma cf. *G. purpurea* Lea, 1842, *Amer. J. Sci.* 3:106.

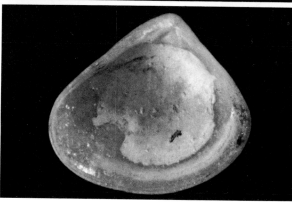

Gemma cf. *G. purpurea*

L. *gemma* precious stone, *purpurea* purplish color.

SIZE: Length 3 mm.

COLOR: Whitish with purple on the umbones and posterior areas.

SHAPE: Rounded trigonal.

ORNAMENT OR SCULPTURE: Glossy with numerous, fine, concentric ribs.

HINGE AREA: Lunule large. Two large teeth in left valve with socket between; 3 teeth in right valve.

PALLIAL LINE & INTERIOR: Small muscle scars. Pallial sinus points upward, triangular. Inner margin faintly crenulate.

PERIOSTRACUM: None visible.

REMARKS: One of the smallest bivalves. Easily overlooked unless drift is screened. A filter feeder.

HABITAT: In shallow water, sandy bottom. Infaunal.

LOCALITIES: Port Aransas area, probably other locations.

OCCURRENCE: Uncommon.

RANGE: Nova Scotia to Florida, Texas and the Bahamas. Puget Sound, Washington (introduced).

GEOLOGIC RANGE: Pleistocene to Recent.

Subfamily CHIONINAE Frizzel, 1936

Genus *Anomalocardia* Schumacher, 1817

POINTED VENUS
Anomalocardia cuneimeris (Conrad, 1846), *Proc. Acad. Natur. Sci. Phila.* 3:24.

L. *anomalus* uneven; Gk. *kardia* heart; L. *cuneus* a wedge.

SIZE: Length 12 to 18 mm.

COLOR: Variable; from white to tan with brown rays. Interior white with purplish brown at posterior margin.

SHAPE: Wedge shaped. Equivalve. Inequilateral.

ORNAMENT OR SCULPTURE: Glossy surface is sculptured with rounded concentric ridges and very faint radial lines. Posterior slope slightly rostrate.

HINGE AREA: Umbones small, lunule distinct, escutcheon depressed. Three cardinal teeth in each valve. Right anterior cardinal is small and in a horizontal position.

PALLIAL LINE & INTERIOR: Two muscle scars connected by pallial line and small, angular sinus. Margins crenulated.

Anomalocardia cuneimeris

Chione cancellata

PERIOSTRACUM: Thin, glossy, lacquerlike.

REMARKS: This thin little shell can withstand high salinity and is one of the few mollusks that can live in the central Laguna Madre and Baffin Bay. A filter feeder.

HABITAT: Both enclosed and open hypersaline lagoons. On sand. Infaunal.

LOCALITIES: Entire, more common to south.

OCCURRENCE: Common.

RANGE: Southern half of Florida, Mexico, Texas, West Indies and Central America.

GEOLOGIC RANGE: Pleistocene.

Genus *Chione* Megerle von Mühlfeld, 1811

CROSS-BARRED VENUS

Chione cancellata (Linné, 1767), *Syst. Natur.*,12th ed., p. 1130.

L. Chione, mythological personage, daughter of Boreas, *cancellatus* latticed.

SIZE: Length 1 to 1¾ in.

COLOR: White to gray, often rayed with brown. Interior glossy white with blue purple.

SHAPE: Ovate to subtrigonal. Equivalve. Inequilateral.

ORNAMENT OR SCULPTURE: Surface sculptured with numerous strong, bladelike concentric ridges and

many radial ribs. When the concentric ridges become beachworn the pattern appears very cancellate. Heavy, porcelaneous.

HINGE AREA: Escutcheon long, smooth and V-shaped. Lunule heart shaped. Three cardinal teeth in each valve; no anterior laterals.

PALLIAL LINE & INTERIOR: Two muscle scars are connected by a pallial line with a very small, triangular pallial sinus. Margins crenulate.

PERIOSTRACUM: Weak.

REMARKS: The number of dead shells found in the bay areas indicates that this species was more abundant in the past than it is today. A filter feeder.

HABITAT: Open bays, bay margins, and inlet-influenced areas. Infaunal.

LOCALITIES: Entire, more to south.

OCCURRENCE: Common.

RANGE: North Carolina to Florida, Texas and the West Indies.

GEOLOGIC RANGE: Miocene to Recent with related forms in Jurassic.

GRAY PYGMY VENUS

Chione grus (Holmes, 1858), *Post-Plio. Fos. S. Car.*, p. 37.

L. Chione, mythological personage, daughter of Boreas ; Gk. *grus, griers* gritty, granulose.

SIZE: Length 6 to 9 mm.

COLOR: Grayish white, often with a pink cast. Interior with a broad ray of purple at posterior end.

SHAPE: Oblong. Inequivalve. Inequilateral.

ORNAMENT OR SCULPTURE: Thirty to 40 fine, radial ribs that are crossed by finer, concentric threads. More heavily sculptured on posterior and anterior margins.

HINGE AREA: Lunule narrow, heart shaped, colored brown. Escutcheon narrow and sunken. Three cardinal teeth in each valve.

PALLIAL LINE & INTERIOR: Two muscle scars connected by weak pallial line with a small, oblique sinus. Margins crenulate.

PERIOSTRACUM: Thin, brownish, shaggy at posterior end.

REMARKS: This tiny shell is not easily found. A filter feeder.

Chione clenchi

Chione grus

HABITAT: Offshore in intermediate shelf assemblage. Sand and shell bottom. Infaunal.

LOCALITIES: Central, south.

OCCURRENCE: Uncommon.

RANGE: North Carolina to Key West to Louisiana and Texas to Cabo Catoche, Mexico.

GEOLOGIC RANGE: Miocene to Recent.

CLENCH'S CHIONE

Chione clenchi Pulley, 1952, *Texas J. Sci.* 4:61.

L. Chione, mythological personage, daughter of Boreas; dedicated to Wm. J. Clench of Harvard.

SIZE: Length 1 to 1½ in.

COLOR: Cream with irregular brown splotches. Some are rayed with brown.

SHAPE: Subtrigonal. Equivalve. Inequilateral.

ORNAMENT OR SCULPTURE: Sculpture consists of 12 to 15 rounded concentric ribs. Ribs not reflected dorsally or flattened on posterior slope. Ribs are sharply flexed along posterior slope, producing a knobby ridge.

HINGE AREA: Umbones recurved forward. Ligament sunken. Escutcheon of moderate size, narrow, with faint growth lines. Three cardinal teeth in each valve.

PALLIAL LINE & INTERIOR: White, sometimes with purple or brown blotch under posterior slope. Pallial line weak with small pallial sinus and 2 muscle scars. Margin finely crenulate.

PERIOSTRACUM: None visible.

REMARKS: *C. latilirata* may be the same species. The ribbing is the main difference. A filter feeder.

HABITAT: Offshore. Infaunal.

LOCALITIES: Entire, more to south.

OCCURRENCE: Fairly common.

RANGE: Texas and south along Mexico to Campeche.

Chione intapurpurea

LADY-IN-WAITING VENUS
Chione intapurpurea (Conrad, 1849), *J. Acad. Natur. Sci. Phila.* 1:209.
L. Chione, mythological personage, daughter of Boreas, *inter* within, *purpura* a purple dye.
SIZE: Length 1 to 1½ in.
COLOR: Glossy white to cream, often with irregular brown marks. Interior white, or with a purple splotch in posterior third.
SHAPE: Ovate to subtrigonal. Equivalve. Inequilateral. Inflated.
ORNAMENT OR SCULPTURE: Sculpture consists of numerous low, rounded concentric ribs. The ribs are marked with tiny serrations at the lower edge that give the interspaces a beaded appearance. Rib on posterior slope lamellate.
HINGE AREA: Lunule heart shaped with raised lamellations. Escutcheon with fine transverse lines. Ligament exterior. Three cardinal teeth in each valve, no laterals.
PALLIAL LINE & INTERIOR: Two shiny muscle scars connected by a pallial line with a short, narrow, oblique sinus.
PERIOSTRACUM: None visible.
REMARKS: This species is more readily found on the Mexican Gulf beaches. A filter feeder.

HABITAT: Near shore, 2 to 12 fathoms. Infaunal.
LOCALITIES: Central, south.
OCCURRENCE: Uncommon to rare.
RANGE: North Carolina, the Gulf states to Carmen, Mexico, and the West Indies.
GEOLOGIC RANGE: Pleistocene to Recent.

Genus *Mercenaria* Schumacher, 1817

SOUTHERN QUAHOG
Mercenaria campechiensis (Gmelin, 1790), *Syst. Natur.*, 13th ed., p. 3287.
L. *mercenarius* reward, money [Indians made wampum from it]; from Campeche, Mexico.
SIZE: Length 3 to 6 in.
COLOR: Dirty gray to whitish. Interior white, rarely with purple blotches. Porcelaneous.
SHAPE: Ovate trigonal. Equivalve. Inequilateral. Inflated.
ORNAMENT OR SCULPTURE: Sculpture of numerous, concentric growth lines; near the beaks they are farther apart. Shell very heavy.
HINGE AREA: Lunule is as wide as it is long. Three cardinals in each valve. Left middle cardinal split.

Mercenaria campechiensis

PALLIAL LINE & INTERIOR: Two muscle scars are connected by a pallial line with small, angular sinus.

PERIOSTRACUM: None visible.

REMARKS: This shell has not been used commercially as has its northern cousin (*M. mercenaria*). Hurricane Carla in 1961 brought thousands of these offshore burrowers up on the beaches. A filter feeder.

HABITAT: Offshore. Infaunal.

LOCALITIES: Entire.

OCCURRENCE: Common.

RANGE: Chesapeake Bay to Florida, Texas and Cuba.

GEOLOGIC RANGE: Miocene to Recent.

TEXAS QUAHOG

Mercenaria campechiensis texana (Dall, 1902), *Proc. U.S. Nat. Mus.* 26:378.

L. *mercenarius* reward, money; from Campeche, Mexico; from Texas.

SIZE: Length 3 to 5 in.

COLOR: Dirty white, often with brown zigzag marks. Interior white, occasionally marked with purple.

SHAPE: Ovate trigonal.

ORNAMENT OR SCULPTURE: Surface sculpture with irregular, large concentric growth lines. The central area of each valve is glossy and smooth in older specimens. Very heavy and porcelaneous.

Mercenaria campechiensis texana

HINGE AREA: Lunule is three-fourths as wide as it is long. Three cardinal teeth in each valve. Left middle cardinal split.

PALLIAL LINE & INTERIOR: Two muscle scars are connected by a pallial line with small, angular sinus. Margin faintly crenulate.

PERIOSTRACUM: None visible.

REMARKS: This species lives buried in the bays. It can be used for a chowder if relocated in clear Gulf water so that it can filter enough water to remove the flavor of the muddy bay water (see Chapter 4). A filter feeder.

HABITAT: Open bays and inlet-influenced areas. Infaunal.

LOCALITIES: Entire.

OCCURRENCE: Fairly common.

RANGE: Northern Gulf of Mexico to Tampico, Mexico.

GEOLOGIC RANGE: Pleistocene to Recent.

Family PETRICOLIDAE Deshayes, 1831

Genus *Petricola* Lamarck, 1801

Subgenus *Petricolaria* Stoliczka, 1870

FALSE ANGEL WING

Petricola (Petricolaria) pholadiformis (Lamarck, 1818), *Hist. Natur. Anim. sans Vert.* 5: 505.

L. *petra* stone, *colere* to inhabit, *forma* form or shape of a *Pholas*.

SIZE: Length up to 2 in.

COLOR: White.

SHAPE: Elongate, somewhat cylindrical.

ORNAMENT OR SCULPTURE: Sculpture on the fragile shell consists of numerous radial ribs and fine growth lines, diminishing posteriorly. The anterior ribs are prominently scaled.

HINGE AREA: Umbones low and near anterior end. Small, brown ligament just posterior to the umbones. Three cardinal teeth are long and pointed in left valve, 2 in right.

PALLIAL LINE & INTERIOR: Both adductor muscle scars are somewhat rounded. Pallial sinus fairly narrow and deep with a rounded end.

PERIOSTRACUM: None visible.

REMARKS: The external appearance and shape of this shell make it easily confused with a *Pholas* but a look at the teeth will correct this error. Filter feeder.

Petricola (Petricolaria) pholadiformis

HABITAT: Open-bay margins, inlet-influenced areas, and near shore in clay banks. Infaunal.
LOCALITIES: Entire.
OCCURRENCE: Common.
RANGE: Gulf of St. Lawrence to the Gulf of Mexico and south to Alvarado, Mexico.
GEOLOGIC RANGE: Pleistocene to Recent.

Genus *Pseudoirus* Habe, 1951

ATLANTIC RUPELLARIA
Pseudoirus typica (Jonas, 1844), *Z. für Malak.* 1:185.
Gk. *pseudo* false; L. *typica* type.

Pseudoirus typica

SIZE: Length 1 in.
COLOR: White.
SHAPE: Oblong. Inequilateral. Variable.
ORNAMENT OR SCULPTURE: The rather heavy shell is sculptured with irregularly spaced radial ribs, narrower over anterior area. Valves gape posteriorly.
HINGE AREA: Umbones rounded, not prominent, curved forward. Three cardinal teeth in left valve, 2 in right.
PALLIAL LINE & INTERIOR: Pallial sinus semicircular and somewhat longer than posterior adductor scar.
PERIOSTRACUM: Thin, brownish.
REMARKS: This coral- and rock-boring bivalve is variable in shape due to its rock-dwelling habit. Syn. *Rupellaria typica*. A filter feeder.
HABITAT: Jetties. Infaunal.
LOCALITIES: Central, south.
OCCURRENCE: Fairly common.
RANGE: North Carolina to the southern half of Florida and the West Indies, Texas, and Isla Mujeres, Mexico.

Order MYOIDA Stoliczka, 1870

Suborder MYINA Newell, 1965

Superfamily MYACEA Lamarck, 1818

Family CORBULIDAE Lamarck, 1818

Genus *Corbula* Bruguière, 1797

BARRATT'S CORBULA
Corbula barrattiana C. B. Adams, 1852, *Contr. Conch.* 12:237.
L. *corbula* little basket; dedicated to Dr. A. Barratt, friend of Adams.
SIZE: Length 9 mm.
COLOR: White.
SHAPE: Trigonal. Inequivalve. Inequilateral.
ORNAMENT OR SCULPTURE: Sculptured with small concentric irregular ridges, that are larger on the large valve. Radial ridge on posterior slope.
HINGE AREA: Beaks small, not involute. Umbones with an acute angle posteriorly. Teeth moderately developed.
PALLIAL LINE & INTERIOR: Pallial line only slightly sinuate, connecting 2 muscle scars.

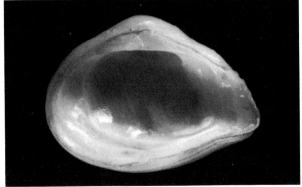

Corbula barrattiana

Corbula contracta

PERIOSTRACUM: Deciduous, brownish.

REMARKS: Adventitious. In bamboo roots on the outer beaches.

HABITAT: In mud. Infaunal.

LOCALITIES: Entire, more south.

OCCURRENCE: Uncommon.

RANGE: North Carolina to both sides of Florida, Texas and the West Indies.

CONTRACTED CORBULA

Corbula contracta Say, 1822, *J. Acad. Natur. Sci. Phila.* 2:312.

L. *corbula* little basket, *contractus* drawn together.

SIZE: Length 9 mm.

COLOR: Dirty gray.

SHAPE: Trigonal. Very inequivalve.

ORNAMENT OR SCULPTURE: Sculpture consists of concentric, elevated threads. Posterior slope rostrate. The ventral margin of the right valve overlaps that of the left.

HINGE AREA: Umbones high, anterior to middle of

shell. Right valve has one prominent cardinal tooth with a corresponding socket in the left valve.

PALLIAL LINE & INTERIOR: Two small muscle scars. Pallial sinus only a slight depression in the pallial line.

PERIOSTRACUM: Thin, yellowish brown.

REMARKS: Screen drift for this tiny bivalve. A filter feeder.

HABITAT: Open sound on sandy bottom. Infaunal.

LOCALITIES: Entire.

OCCURRENCE: Common.

RANGE: Cape Cod to Florida, Texas and the West Indies.

GEOLOGIC RANGE: Pliocene to Recent.

DIETZ'S CORBULA

Corbula dietziana C. B. Adams, 1852, *Contr. Conch.* 12:235.

L. *corbula* little basket; dedicated to a Mr. Dietz.

SIZE: Length 10.5 mm.

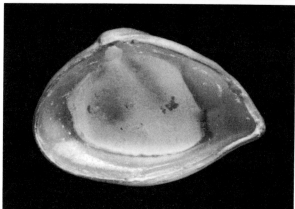

Corbula dietziana

RANGE: North Carolina to southeastern Florida, Texas and the West Indies.
GEOLOGIC RANGE: Pliocene to Recent.

KREBS' CORBULA
Corbula krebsiana C. B. Adams, 1852, *Contr. Conch.* 12:234.
L. *corbula* little basket; dedicated to Henry Krebs of the Virgin Islands, nineteenth-century collector.
SIZE: Length 6 mm.
COLOR: Cream. Left valve pinkish, more colored around margins.
SHAPE: Trigonal. Very inequivalve. Inequilateral.
ORNAMENT OR SCULPTURE: Sculpture consists of strong irregular concentric line. Right valve more

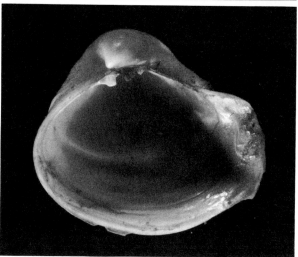

Corbula krebsiana

COLOR: Pink.
SHAPE: Trigonal. Inequivalve. Inequilateral.
ORNAMENT OR SCULPTURE: Sculpture consists of fine, irregular concentric lines. Posterior slope has strong radial ridge. Right valve larger than left.
HINGE AREA: Umbones raised, rounded. Slightly involute. Strong cardinal tooth in right valve, corresponding socket in left valve. No laterals.
PALLIAL LINE & INTERIOR: Two small muscle scars. Posterior scar thickened and elevated. Pallial line barely indented by pallial sinus.
PERIOSTRACUM: Thin, brownish.
REMARKS: This species has 2 distinct stages of growth with an abrupt transition from the first to the second. In the first the valves are nearly equal, in the second the right valve grows nearly 3 times as much as the other. A filter feeder.
HABITAT: Calcareous banks offshore. Infaunal.
LOCALITIES: Entire.
OCCURRENCE: Uncommon.

strongly sculptured, much larger and more pointed posteriorly than left valve.

HINGE AREA: Umbones rounded. Prominent. One large cardinal tooth in right valve with corresponding socket in left.

PALLIAL LINE & INTERIOR: Two small muscle scars. Pallial line barely indented by pallial sinus.

PERIOSTRACUM: Thin, yellowish brown. Heavier on left valve.

REMARKS: In bamboo roots on the outer beaches. A filter feeder.

HABITAT: On banks offshore. Infaunal.

LOCALITIES: Entire, probably.

OCCURRENCE: Uncommon.

RANGE: Jamica and Texas.

SWIFT'S CORBULA

Corbula swiftiana C. B. Adams, 1852, *Contr. Conch.* 12:236.

L. *corbula* little basket; dedicated to Robert Swift (1796–1872), American.

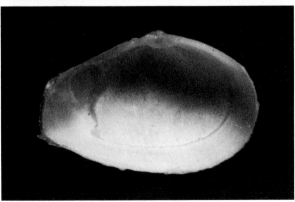

Corbula swiftiana

SIZE: Length 9 mm.

COLOR: White.

SHAPE: Trigonal. Inequivalve. Almost equilateral.

ORNAMENT OR SCULPTURE: Sculpture consists of very weak, irregular concentric lines. Posterior slope pointed and marked with a sharp radial ridge. Right valve deeper and longer than left.

HINGE AREA: Umbones sharp. A single cardinal tooth in right valve with corresponding socket in left.

PALLIAL LINE & INTERIOR: Two small muscle scars. Pallial line is simple.

PERIOSTRACUM: Brownish, deciduous.

REMARKS: Probably the most common *Corbula* on this coast. A filter feeder.

HABITAT: Inlet-influenced areas. Open-bay margins. Infaunal.

LOCALITIES: Entire.

OCCURRENCE: Common.

RANGE: Massachusetts to eastern Florida, Texas and the West Indies.

GEOLOGIC RANGE: Miocene to Recent.

Genus *Varicorbula* Grant & Gale, 1931

OVAL CORBULA

Varicorbula operculata (Philippi, 1849), *Z. für Malak.* 5:13.

L. *varius* various, *corbula* little basket, *operculata* oval.

SIZE: Length 9 mm.

COLOR: Whitish.

SHAPE: Subtrigonal. Inflated. Inequivalve.

ORNAMENT OR SCULPTURE: The sculpture consists of strong, concentric ridges. Right valve larger than left, more inflated and stronger sculpture.

HINGE AREA: Umbones high, almost central, curved in. Right valve has a single prominent cardinal tooth, left valve has corresponding socket. Ligament internal.

PALLIAL LINE & INTERIOR: Two small roundish muscle scars. Pallial line well impressed, pallial sinus is but a slight sinuation in pallial line.

PERIOSTRACUM: None visible.

REMARKS: A common beach *Corbula*, sometimes found in bamboo root clumps. A filter feeder.

HABITAT: Offshore, 20 to 40 fathoms. Infaunal.

LOCALITIES: Central, south.

OCCURRENCE: Fairly common.

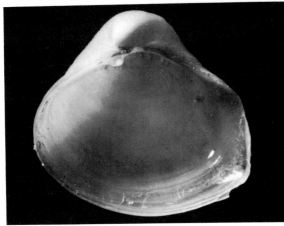

Varicorbula operculata

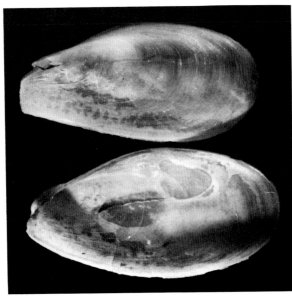

Rocellaria hians

RANGE: North Carolina to Gulf of Mexico and West Indies.
GEOLOGIC RANGE: Pliocene to Recent.

Superfamily GASTROCHAENACEA Gray, 1840

Family GASTROCHAENIDAE Gray, 1840

Genus *Rocellaria* Blainville, 1828

ATLANTIC ROCELLARIA
Rocellaria hians (Gmelin, 1790), *Syst. Natur.*, 12th ed., p. 3217.
N.L. *rocellaria* a little rock; L. *hiatus* opening, to gape;
SIZE: Length 12 to 18 mm.
COLOR: White.
SHAPE: Petal shaped.
ORNAMENT OR SCULPTURE: Sculpture consists of low, indistinct, fine, concentric lines. Posterior end large and rounded. Entire ventral-anterior end is gaping to accommodate the foot.
HINGE AREA: Umbones low. Hinge area weak with 1 large toothlike structure in the umbonal cavity.
PALLIAL LINE & INTERIOR: Posterior adductor scar large, anterior scar degenerate. Pallial sinus deep and angular.
PERIOSTRACUM: Thin, dark brown.
REMARKS: This fragile shell constructs a hollow "cell" of sand and bits of shell glued together in which it lives. According to Pulley (1953) the mollusk is often found inside empty pecten shells. A filter feeder.
HABITAT: Offshore. Infaunal.
LOCALITIES: Entire.
OCCURRENCE: Uncommon.
RANGE: Florida Keys, western Florida, Texas and the West Indies.
GEOLOGIC RANGE: Pliocene to Recent.

Superfamily HIATELLACEA Gray, 1824

Family HIATELLIDAE Gray, 1824

Genus *Hiatella* Daudin, 1801

ARCTIC SAXICAVE
Hiatella arctica (Linné, 1767), *Syst. Natur.*, 12th ed., p. 1113.
L. *hiat* cleft, opening, *ella* dim.; from the Arctic seas.

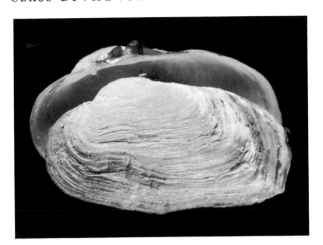

Hiatella arctica

Panopea bitruncata

SIZE: Length 1 in.

COLOR: Chalky white.

SHAPE: Irregularly oblong.

ORNAMENT OR SCULPTURE: Sculpture consists of coarse, irregular growth lines. A weak radial rib on posterior end, may be scaled. Posterior may gape.

HINGE AREA: Umbones close together and about one-third back from anterior end. Teeth are indefinite.

PALLIAL LINE & INTERIOR: Pallial line discontinuous, sinus large but irregular.

PERIOSTRACUM: Thin, gray, deciduous.

REMARKS: The nestling and boring habits of this shell cause its shape and sculpture to vary greatly.

HABITAT: Inlet areas. Hypersaline lagoons. Offshore. Boring infaunal.

LOCALITIES: Entire.

OCCURRENCE: Fairly common.

RANGE: Arctic seas to deep waters in West Indies and Panama. Also Pacific.

GEOLOGIC RANGE: Miocene to Recent.

HINGE AREA: Umbones low and incurved. Almost central. One large single, simple "tooth" and a broad ledge to which the ligament is attached.

PALLIAL LINE & INTERIOR: Anterior adductor muscle elongated, posterior rounded and small. Pallial sinus short and rounded.

PERIOSTRACUM: Thin, yellowish.

REMARKS: This extremely large bivalve is seldom found alive or dead because it lives so deep in the substrate that the valves rarely wash out of the burrows. Live juvenile specimens have been found in the Aransas Pass channel during a very extreme, extended low tide. A filter feeder.

HABITAT: Offshore and inlet areas. Boring infaunal.

LOCALITIES: Port Aransas, south.

OCCURRENCE: Uncommon.

RANGE: North Carolina to Florida and South Texas.

GEOLOGIC RANGE: Miocene to Recent.

Genus *Panopea* Menard de la Groye, 1807

GEODUCK

Panopea bitruncata (Conrad, 1872), *Proc. Acad. Natur. Sci. Phila.* 24:216.

Panope, a sea nymph; L. *bi* two, *truncatus* cut off.

SIZE: Length 5 to 9 in.

COLOR: White.

SHAPE: Quadrate, elongated.

ORNAMENT OR SCULPTURE: Sculpture consists of concentric, wavy growth lines. Valves gape at both ends.

Suborder PHOLADINA Newell, 1965

Superfamily PHOLADACEA Lamarck, 1809

Family PHOLADIDAE Lamarck, 1809

Genus *Barnea* Risso, 1826

FALLEN ANGEL WING

Barnea truncata (Say, 1822), *J. Acad. Natur. Sci. Phila.* 2:321.

Honoring D. H. Barnes, American; L. *truncatus* cut off.

Barnea truncata

SIZE: Length 2¾ in.

COLOR: White.

SHAPE: Elongate elliptical.

ORNAMENT OR SCULPTURE: Rather thin shell is sculptured with radial ribs and concentric ridges. The concentric ridges are laminated anteriorly becoming weaker toward the posterior end. Radial ribs are lacking on the posterior slope. Imbrications are formed where the concentric ridges cross the radial ribs. The shell is beaked anteriorly and truncate posteriorly.

HINGE AREA: Umbones prominent, near anterior third.

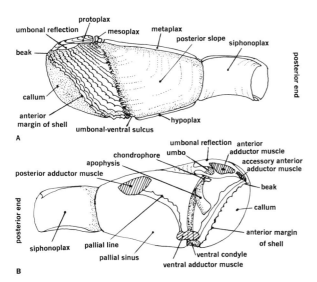

Diagrammatic pholad shell: *A*, external view showing accessory plates; *B*, internal view showing muscle scars and apophysis (after Turner 1954, p. 11).

Apophysis narrow, long, bladelike, and curved. Protoplax lanceolate, with a posterior nucleus and definite growth lines.

PALLIAL LINE & INTERIOR: Pallial line and muscle scars well marked. Pallial sinus almost as wide as shell is high, extending about half-way to the umbones.

PERIOSTRACUM: Thin, yellowish brown.

REMARKS: Siphons capable of extending 10 to 12 times the length of the shell. Animal cannot withdraw into shell and usually lives in deep burrows in mud, but is known to live in wood also. A filter feeder.

HABITAT: In clay bottoms of bays. Boring infaunal.

LOCALITIES: Central.

OCCURRENCE: Fairly common.

RANGE: Massachusetts Bay to southern Florida and Texas.

Genus *Cyrtopleura* Tryon, 1862

Subgenus *Scobinopholas* Grant & Gale, 1931

ANGEL WING

Cyrtopleura (Scobinopholas) costata (Linné, 1758), *Syst. Natur.*, 10th ed., p. 669.

Gk. *cyrtos* curved, *pleuron* side; L. *costa* rib.

SIZE: Length up to 7¼ in.

COLOR: White.

SHAPE: Elongate oval.

ORNAMENT OR SCULPTURE: This rather light shell is sculptured with concentric ridges and strong radial ribs that extend the entire length of the shell. Where the ribs cross the concentric line, imbrications are formed; these are stronger on the anterior and posterior slopes.

Cyrtopleura (Scobinopholas) costata

HINGE AREA: Umbones prominent, located near anterior fourth of shell. The umbones are well separated from the raised umbonal reflections. Protoplax triangular in outline and mostly chitinous. Mesoplax calcareous. Apophyses large, spoon shaped, and hollow at the upper end.

PALLIAL LINE & INTERIOR: Anterior and posterior muscle scars fairly well marked, but pallial sinus is not evident.

PERIOSTRACUM: Thin, yellowish, deciduous.

REMARKS: This beautiful, large bivalve bores into a clay substrate in the bays to a depth of about 18 inches. Its inability to withdraw into its shell makes it very susceptible to changes of salinity. The freshwater inundation following Hurricane Beulah in 1967 has virtually wiped out the colonies but they are reestablishing. A filter feeder.

HABITAT: Open-bay margins. Inlet-influenced areas. Boring infaunal.

LOCALITIES: Entire.

OCCURRENCE: Common.

RANGE: Massachusetts to Florida, Texas, Mexico and the West Indies.

GEOLOGIC RANGE: Pliocene to Recent.

Genus *Pholas* Linné, 1758

Subgenus *Thovana* Gray, 1847

CAMPECHE ANGEL WING

Pholas (Thovana) campechiensis Gmelin, 1792, *Syst. Natur.*, 13th ed., p. 3216.

Gk. *pholas* a rock-boring mollusk, boring in a hole; from Campeche, Mexico.

SIZE: Length up to 4½ in.

COLOR: White.

SHAPE: Subelliptical.

ORNAMENT OR SCULPTURE: This thin, fragile shell is sculptured with laminated, concentric ridges and radial ribs. This sculpture becomes weaker posteriorly. Shells gape. Shells rounded at both ends.

HINGE AREA: Umbones prominent, located near the anterior fourth of shell, covered by double septate umbonal reflections. Apophyses delicate, short, and broad, projecting beneath the umbo at a sharp posterior angle. Three accessory plates: a nearly rectangular protoplax, a transverse mesoplax, and an elongate metaplax.

PALLIAL LINE & INTERIOR: Adductor muscle scars

Pholas (Thovana) campechiensis

and pallial line well marked. Pallial sinus wide, extending anteriorly almost two-thirds the distance to the umbo.

PERIOSTRACUM: Thin.

REMARKS: This shell is common in the beach drift but is seldom found alive. At times it is found in wood on the outer beaches. It is distinguished by the comb-like umbonal reflection. A filter feeder.

HABITAT: Offshore in wood and clay. Boring infaunal.

LOCALITIES: Entire.

OCCURRENCE: Fairly common.

RANGE: North Carolina to the Gulf states, Mexico and Central America.

GEOLOGIC RANGE: Pleistocene to Recent.

Genus *Martesia* G. B. Sowerby I, 1824.

Subgenus *Martesia* s.s.

FRAGILE MARTESIA

Martesia (Martesia) fragilis Verril & Bush, 1890, *Proc. U.S. Nat. Mus.* 20(1139):777.

L. *martes* a marten, a burrowing animal, *fragilis* fragile.

Martesia (Martesia) fragilis

SIZE: Length 18 mm.

COLOR: White.

SHAPE: Pear shaped.

ORNAMENT OR SCULPTURE: The valves are divided into two sections by a shallow umbonal-ventral sulcus. The anterior part sculptured with concentric, denticulated ridges and weak radial ribs. The posterior part sculptured with smooth, rounded ridges. The truncated beaks gape. Shell rounded and closed posteriorly.

HINGE AREA: Umbones prominent, located near anterior end of shell in adults. Mesoplax circular to oval, depressed with strong concentric sculpture. Metaplax long and narrow, pointed anteriorly, wider and rounded posteriorly. Apophyses long, thin, and extending under the umbones anteriorly at an angle.

PALLIAL LINE & INTERIOR: Muscle scars strong, pallial sinus wide and deep, extending almost to the umbonal-ventral ridge.

PERIOSTRACUM: Thin, yellowish.

REMARKS: This species can be easily confused with *M. striata*; however, the mesoplax is depressed, has concentric sculpture, and has sharply keeled edges. A filter feeder.

HABITAT: Pelagic in floating wood. Boring infaunal.

LOCALITIES: Entire.

OCCURRENCE: Common.

RANGE: Virginia, south through the Gulf of Mexico, West Indies to Brazil.

Subgenus *Particoma* Bartsch & Rehder, 1945

WEDGE-SHAPED MARTESIA

Martesia (Particoma) cuneiformis (Say, 1822), *J. Acad. Natur. Sci. Phila.* 2:322.

L. *martes* marten, a burrowing animal, *cuneus* wedge, *forma* form.

SIZE: Length 12 to 21 mm.

COLOR: White.

SHAPE: Pear shaped.

ORNAMENT OR SCULPTURE: Valves divided into parts by narrow, umbonal-ventral sulcus. Anterior part sculptured with concentric, denticulated ridges. Posterior part sculptured with smooth, rounded, concentric ridges and weak growth lines. Shell gapes anteriorly and is rounded and closed posteriorly.

HINGE AREA: Umbones prominent, located near anterior end of shell. The umbonal reflection is small, adpressed, not free. Mesoplax is cuneiform or wedge

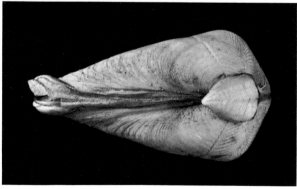

Martesia (Particoma) cuneiformis

shaped, with a median groove and radiating growth
lines. Metaplax long and narrow and divided. Apophyses long, thin, and almost parallel with umbonal-
ventral ridge.

PALLIAL LINE & INTERIOR: Muscle scars well marked.
Posterior adductor scar is long and oval; anterior
scar is kidney shaped. Pallial sinus broad and extending anteriorly beyond the umbonal-ventral ridge.

PERIOSTRACUM: Thin, yellowish.

REMARKS: This chubby woodborer is easily confused
with *Diplothyra smythi* unless the characteristic
wedge-shaped mesoplax is present. The material in
which the animal is boring readily affects the shape
of this family, causing some problem in identification. A filter feeder.

HABITAT: Pelagic in wood. Boring infaunal.

LOCALITIES: Entire.

OCCURRENCE: Fairly common.

RANGE: Southeastern United States, Texas and the
West Indies to Brazil.

GEOLOGIC RANGE: Miocene to Recent.

Genus *Diplothyra* Tryon, 1862

OYSTER PIDDOCK

Diplothyra smythi Tryon, 1862, *Proc. Acad. Natur.
Sci. Phila.* 14:450.

Gk. *diplos* double, *thyreos* large shield; dedicated to
Ed. A. Smith.

SIZE: Length 15 mm.

COLOR: White.

SHAPE: Pear shaped.

ORNAMENT OR SCULPTURE: Divided into two parts
by an umbonal-ventral sulcus. Anterior part triangular in shape, sculptured by fine, close-set, wavy, concentric ridges and numerous weak radial ribs. Posterior part has only growth lines. Widely gaping anteriorly; rounded and closed posteriorly.

HINGE AREA: Umbones prominent and near the anterior fourth of shell. Umbones imbedded in a
callum that extends posteriorly on either side of mesoplax. Apophyses long, thin, and fragile. A large
chondrophore in left valve. Mesoplax divided into
concentrically sculptured posterior section and wrinkled anterior section. Mesoplax and hypoplax forked
posteriorly.

PALLIAL LINE & INTERIOR: Adductor scars large. Pallial sinus broad and deep, extending anteriorly past
the umbonal-ventral ridge.

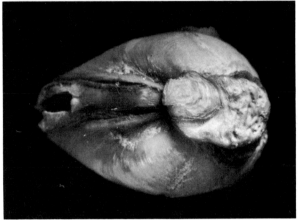

Diplothyra smythi

PERIOSTRACUM: Yellowish and nearly covering pedal
gape.

REMARKS: This is similar in appearance to *M. cuneiformis* but is not found in wood, preferring shell
and coquina rock. It bores into oyster shell but is not
commercially damaging. A filter feeder.

HABITAT: High-salinity oyster reefs. Boring infaunal.

LOCALITIES: Entire.

OCCURRENCE: Common.

RANGE: Massachusetts south to Daytona Beach and
Sanibel Island, Florida, and west to Texas.

Genus *Jouannetia* Des Moulins, 1828

Jouannetia quillingi Turner, 1955, *Johnsonia* 3(34):
139.

Derivation of *Jouannetia* unknown; dedicated to Ben
Quilling, original collector of species.

Jouannetia quillingi

SIZE: Length 21 mm.

COLOR: White to gray.

SHAPE: Globose, pear shaped.

ORNAMENT OR SCULPTURE: Valves divided into two parts by umbonal-ventral sulcus. Anterior part triangular and sculptured with numerous laminated, imbricate, concentric ridges and weak radial ribs. The shell is constricted at the sulcus. Posterior part sculptured with thin, concentric ridges and growth lines. The ridges bear long, curved spines.

HINGE AREA: Umbones prominent and near the anterior third of shell. Umbonal reflections free and raised. The callum extends dorsally between the beaks and on the left valve is enlarged to form the covering for the anterior adductor muscle. Mesoplax small, wedge shaped. There is a chondrophore on left valve.

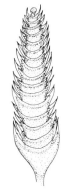

Bankia (Bankiella) gouldi
(after Turner 1966, p. 247)

PALLIAL LINE & INTERIOR: Muscles scars barely visible. Pallial sinus extends anteriorly to the umbonal-ventral ridge.

PERIOSTRACUM: None visible.

REMARKS: This unusual species has only recently been found on the Texas coast in floating wood. A filter feeder.

HABITAT: In wood and calcareous rock off shore. Pelagic. Boring infaunal.

LOCALITIES: Central.

OCCURRENCE: Rare.

RANGE: North Carolina to Lake Worth, Florida, and Texas.

Family TEREDINIDAE Latreille, 1825

Genus *Bankia* Gray, 1840

Subgenus *Bankiella* Bartsch, 1921

GOULD'S SHIPWORM

Bankia (Bankiella) gouldi (Bartsch, 1908), *Proc. Biol. Soc. Wash.* 21:211.

Honoring Gilbert S. Banks, 1906–; dedicated to Dr. A. A. Gould, (1805–1866), American.

SIZE: Length of pallets about ½ in.

COLOR: White.

SHAPE: Convex, equivalve.

ORNAMENT OR SCULPTURE: See *Johnsonia* 2(19):13.

REMARKS: Only the calcium-lined tunnels of these mollusks are found in pieces of wood, pilings, etc. They are not visible from the outside. Dr. Turner (1966, p. 61) states that the shells in this genus are nearly useless for purposes of identification. Identity is made from the pallets located on either side of the siphon. A filter feeder.

HABITAT: In wood. Boring infaunal.

LOCALITIES: Entire.

OCCURRENCE: Common.

RANGE: New Jersey to Florida, Texas and West Indies.

Genus *Teredo* Linné, 1758

BARTSCH'S SHIPWORM

Teredo bartschi Clapp, 1923, *Proc. Boston Soc. Natur. Hist.* 37:33.

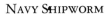

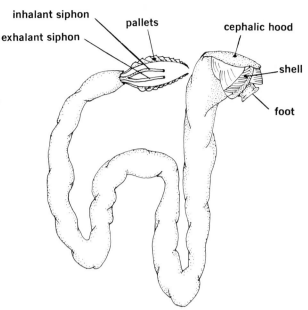

inhalant siphon
exhalant siphon
pallets
cephalic hood
shell
foot

Diagram of an entire teredinid indicating relative position of pallets and shell (after Clench & Turner 1946, p. 4).

L. *teredo* a worm; honoring Paul Bartsch, U.S. Museum.
SIZE: Length of pallets about ½ in.
COLOR: White.
SHAPE: Stalk long, blade short.
ORNAMENT OR SCULPTURE: Blade deeply excavated at top.
REMARKS: The shell is useless in identifying this species. A filter feeder.
HABITAT: A wood borer. Infaunal.
LOCALITIES: Entire.
OCCURRENCE: Fairly common.
RANGE: South Carolina, to northern half of Florida and Texas; introduced to San Diego, California.

NAVY SHIPWORM
Teredo navalis Linné, 1758, *Syst. Natur.*, 10th ed., p. 651.
L. *teredo* a worm, *navis* ship.
SIZE: Length 5 to 6 mm.
COLOR: White.
SHAPE: Stalk short. Blade mammilliform.
ORNAMENT OR SCULPTURE: Blade roundly excavated at top.
REMARKS: The *Teredo* is identified by the pallets located on either side of the siphon. A filter feeder.
HABITAT: A wood borer. Infaunal.
LOCALITIES: Central.
OCCURRENCE: Uncertain.
RANGE: Newfoundland to Daytona Beach, Florida. Eastern Florida to Panama City and Port Aransas, Texas. Also in Puerto Rico.

Genus *Lyrodus* Binney, 1870.

Lyrodus pedicellata Quatrefages, 1849, *Ann. Sci. Natur.* (*Zool.*) 11:26.
L. *lyricus* lyre, *pediculus* dim. of a foot.
SIZE: Length 6 mm.
COLOR: White.
SHAPE: Stalk long. Blade long.
ORNAMENT OR SCULPTURE: Blade very deeply and sharply excavated at top.
REMARKS: Rarely found in drift because they usually stay in the wood. A filter feeder.
HABITAT: In wood. Infaunal.
LOCALITIES: Entire.
OCCURRENCE: Uncertain.
RANGE: New Jersey to West Indies, Gulf of Mexico, Central America to north coast of South America.

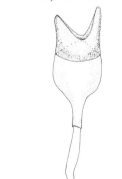

Teredo bartschi
(after Turner 1966, p. 147)

Teredo navalis
(after Turner 1966, p. 159)

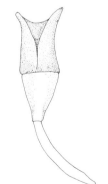

Lyrodus pedicellata
(after Turner 1966, p. 133)

Order PHOLADOMYOIDA Newell, 1965

Suborder PHOLADOMYINA Newell, 1965

Superfamily PANDORACEA Rafinesque, 1815

Family PANDORIDAE Rafinesque, 1815

Genus *Pandora* Bruguière, 1797

Subgenus *Clidophora* Carpenter, 1864

SAY'S PANDORA
Pandora (*Clidophora*) *trilineata* Say, 1822, *J. Acad. Natur. Sci. Phila.* 2:261.
Gk. Pandora, all-gifted mythological person; L. *trilinea* three-lined.
SIZE: Length 18 to 25 mm.
COLOR: White to cream. Interior pearly.
SHAPE: Semilunar. Flat. Inequivalve. Inequilateral.
ORNAMENT OR SCULPTURE: The flat, compressed valves are sculptured with microscopic, concentric growth lines. Left valve has a radial ridge along posterior slope. Strong ridge along hinge margin.
HINGE AREA: Umbones tiny, near rounded anterior end. Hinge is internal just below umbones; lamellar plates replace teeth.
PALLIAL LINE & INTERIOR: Two small, round adductor muscles. Pallial line simple, discontinuous.
PERIOSTRACUM: Thin, light colored.
REMARKS: One wonders how any animal finds space to live between the two flat, half-moon–shaped valves. A filter feeder.

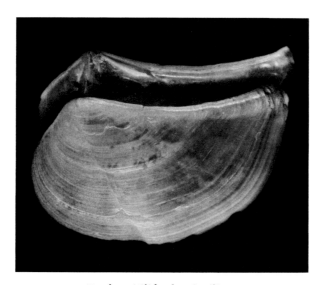

Pandora (*Clidophora*) *trilineata*

HABITAT: Inlet areas. Open-sound and lagoon centers in clayey sediments. Infaunal.
LOCALITIES: Entire.
OCCURRENCE: Common.
RANGE: Cape Hatteras, North Carolina, to Florida and Texas.
GEOLOGIC RANGE: Miocene to Recent.

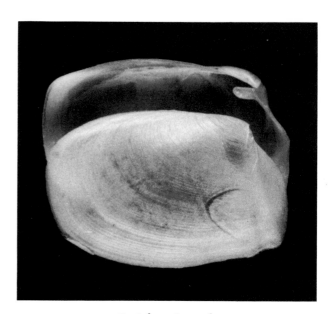

Periploma inequale

Family PERIPLOMATIDAE Dall, 1895

Genus *Periploma* Schumacher, 1816

UNEQUAL SPOON CLAM
Periploma inequale (C. B. Adams, 1842), *Amer. J. Sci.* 43:145.
Gk. *peri* around, about, *ploimos* fit for sailing; L. *in* prefix not, *aequalis* equal.
SIZE: Length 18 to 25 mm.
COLOR: White, translucent.
SHAPE: Subquadrate. Inequivalve. Inequilateral.
ORNAMENT OR SCULPTURE: Sculpture on this rather fragile shell consists of fine, concentric growth lines. Posterior margin very truncated. A low keel extends from umbones to the anterior ventral margin.
HINGE AREA: Umbones small, near posterior end. Ligament internal and located in a spoon-shaped "tooth" in each valve.
PALLIAL LINE & INTERIOR: Two small adductor mus-

cle scars. Pallial sinus broad, short, and strongly defined.

PERIOSTRACUM: None visible.

REMARKS: At times this fragile little shell is on the outer beaches in large numbers. Most of the valves bear the neat, round holes of a predatory gastropod. Originally described from Texas coast. May be syn. *P. margaritaceum* Lamarck, 1801. A filter feeder.

HABITAT: Open-sound, open-lagoon centers, inlet areas in sandy bottoms. Infaunal.

LOCALITIES: Entire.

OCCURRENCE: Common.

RANGE: South Carolina to Florida, Texas to Mexico.

Family LYONSIIDAE Fischer, 1887

Genus *Lyonsia* Turton, 1822

FLORIDA LYONSIA

Lyonsia hyalina floridana Conrad, 1849, *Proc. Acad. Natur. Sci. Phila.* 4:121.

Dedicated to Pierre Lyonett (1706–1789); L. *hyalus* glass; from Florida.

SIZE: Length 16 mm.

COLOR: White, translucent.

SHAPE: Elongate, rather tear shaped. Inequilateral. Inequivalve.

ORNAMENT OR SCULPTURE: The thin shell is sculptured with fine growth lines. Valves gape posteriorly.

HINGE AREA: Umbones tiny, pointed. There are no teeth; a small groove is posterior to umbones to house the ligament.

PALLIAL LINE & INTERIOR: Two adductor muscle scars. Pallial line indistinct.

PERIOSTRACUM: Thin with numerous raised radial lines. Sand grains are usually attached.

REMARKS: This little bivalve has a delicate byssus. A filter feeder.

HABITAT: Open-bay margins, inlet areas, in grass beds. Infaunal.

LOCALITIES: Central.

OCCURRENCE: Fairly common.

RANGE: West coast of Florida, and Texas.

Lyonsia hyalina floridana

Class CEPHALOPODA Cuvier, 1797

Subclass COLEOIDEA Bather, 1888

Order DECAPODA Leach, 1818

Suborder TEUTHOIDEA Naef, 1916

Family LOLLIGINIDAE d'Orbigny, 1835

Genus *Lolliguncula* Steenstrup, 1881

BRIEF SQUID
Lolliguncula brevis (Blainville, 1823), *J. Phys. Chem. Hist. Natur.* 6:133.
L. *lolligo* cuttlefish, *brevis* short.
SIZE: Length up to 3 in.
COLOR: Cream colored with numerous reddish purple spots uniformly distributed over the dorsal and ventral surface of the mantle, head, and arms with the exception of the ventral surface of the fins.
DESCRIPTION: The animal is cylindrical, pointed posteriorly, and the width is less than half the length. The anterior mantle edge comes to a blunt point at the middorsal line. The fins are large, about half the mantle length. They form an elliptical circle and are joined posteriorly by a fleshy ridge that encircles the posterior end of the body. The head is small with moderate eyes. The tentacles are long and slender with clublike end. The arms are short, with the dorsal pair being the shortest.
REMARKS: Commonly used for fish bait. These soft-bodied animals swim backward by means of jet propulsion. Water is expelled forcibly from a siphon beneath and a little back of the head. There is a chitinous structure beneath the mantle on the anterior side that is called the pen. In some non-American species this pen is calcareous and is used in bird cages. This is an economically important animal. When you catch it in your seine be careful not to let it nip your fingers with its sharp beak. The tentacles are used to capture food.
HABITAT: Bays and inlets.
LOCALITIES: Port Aransas, south.

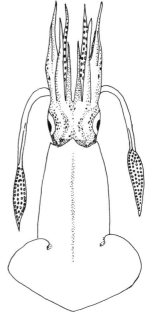

Lolliguncula brevis
(after Voss 1956, p. 113)

OCCURRENCE: Common.
RANGE: Maryland to Florida, Gulf of Mexico, Caribbean Sea; South America to Río de la Plata.

Suborder SEPIOIDEA Zittel, 1895; emend. Naef, 1916

Family SPIRULIDAE Owen, 1836

Genus *Spirula* Lamarck, 1799

COMMON SPIRULA
Spirula spirula (Linné, 1758), *Syst. Natur.* 10th ed., p. 710.
L. *spira* coil.

Spirula spirula

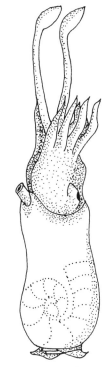

(after Lane 1960, p. 14)

SIZE: Length of mantle 55 mm., diameter of shell up to 25 mm.

COLOR: The shell is porcelaneous white, animal light brown.

DESCRIPTION: The closely coiled and chambered shell is enclosed in the posterior end of the fleshy animal.

The body of the animal is cylindrical with parallel sides. The side walls form two folds that cover the coiled shell, which is visible dorsally and ventrally but is completely enclosed within the mantle. The posterior end of the body is truncated with two lateral fins. Between these fins is a circular disk that emits a light. The large head has two eyes that protrude beyond the mantle edges. The arms are short, stout, and pointed, with four rows of suckers. The tentacles are short with numerous rows of minute suckers.

REMARKS: The first specimen taken intact from the Gulf of Mexico was obtained by Dall (1896) from the mouth of a deep-sea fish. The coiled shell is often found by the thousands on the outer beaches in the spring indicating a large concentration of the species in longshore waters. Normal position for the animal is vertical with arms hanging down.

HABITAT: Offshore in deep waters.

LOCALITIES: Entire.

OCCURRENCE: Common.

RANGE: Worldwide.

Order OCTOPODA Leach, 1818

Suborder INCIRRATA Grimpe, 1922?

Family OCTOPODIDAE d'Orbigny, 1835

Genus *Octopus* Lamarck, 1798

COMMON ATLANTIC OCTOPUS

Octopus vulgaris Lamarck, 1798, *Hist. Natur. Anim. sans Vert.* 11:361.

Gk. *okto* eight, *pous* a foot; L. *vulgaris* common.

SIZE: Length 1 to 3 ft.

COLOR: Reddish brown.

DESCRIPTION: This soft-bodied animal has a saclike body with a small head. The arms are long and tapering.

REMARKS: This small octopus lives among the rocks of the jetties and sea walls. It moves by jet propulsion and exudes a black cloud of ink when frightened. An octopus does not have tentacles as does the squid. It uses its tapering arms to reach into crevices

in search of food. The female broods her eggs, which
are laid in clusters. Larvae are planktonic.

HABITAT: Open-bay center, inlet areas, and the Gulf.

LOCALITIES: Entire.

OCCURRENCE: Fairly common.

RANGE: North and South Atlantic; western Africa;
Mediterranean; Atlantic coast of Europe to Great
Britain; western Atlantic from New York to Brazil,
Gulf of Mexico.

Octopus vulgaris
(after Lane 1960, p. 6)

PART TWO

Fig. 29. The author's daughter, Jinxy, listens to the song of the left-handed whelk amid a group of beachcombing treasures beside the waving sea oats.

4. A Shell Collecting Trip

When planning a shelling trip to the coast of Texas, the first question is—how to get there? Without a boat many of the areas are inaccessible, but the majority can be reached in a passenger car. A map of Texas, then, is the first necessity. The Texas Highway Department in Austin is only too happy to send its official state highway map. This same agency recently published some delightful maps describing the newly designated Texas Trails. The two titled "Independence Trail" and "Tropical Trail" cover most of the coastal area and are more than worth the trouble it takes to order them. Another map of interest to shellers is Number 1967–306–116/21, published by the United States Department of the Interior, which covers the national seashore on Padre Island. All these maps are free.

Unless one of these maps shows a road to the outer beach there just isn't one. Often maps of the immediate locality are available at motels and eating places along the coast or may be purchased at sporting goods stores. Simplified maps in this book show the roads to the beaches and back bay areas and can be used to mark the map you keep in your automobile.

In Chapter 1 the coastal area has been described. This section will only deal with actually getting there. Let's begin at Sabine Pass.

State Highway 87 will take you along the beach for about fifty miles from Sabine Pass to Bolivar Peninsula. As you approach Bolivar Pass an old lighthouse built in 1852 will loom up in front of you—turn to the left short of the light to go to the north jetty. A sign directing you to Buster's Bait Stand is the only marking on this turn. The area is muddy and marsh grasses grow to the water's edge. When the tide is out a very wide flat is exposed east of the jetty. Another spot exposed by the tides is just to the left of the ferry landing.

The free ferry trip from Port Bolivar to Galveston lasts about fifteen minutes. The boat operates on an irregular schedule twenty-four hours a day. Navigators on the ocean-going vessels entering Bolivar Roads and the Houston Ship Channel recognize this passage as one of the most hazardous ship channels in the world and never attempt it in doubtful weather conditions; hence, you often see many large ships anchored nearby waiting to enter the channel.

Galveston is a large city. In 1848 Ferdinand von Roemer, the early Texas-German historian, considered Galveston to be the most important city in Texas. It is still important, but its main attractions are now more historical and recreational. Texas A&M University maintains its marine laboratory and oceanographic research ship here. A visit to the port on the bay side of

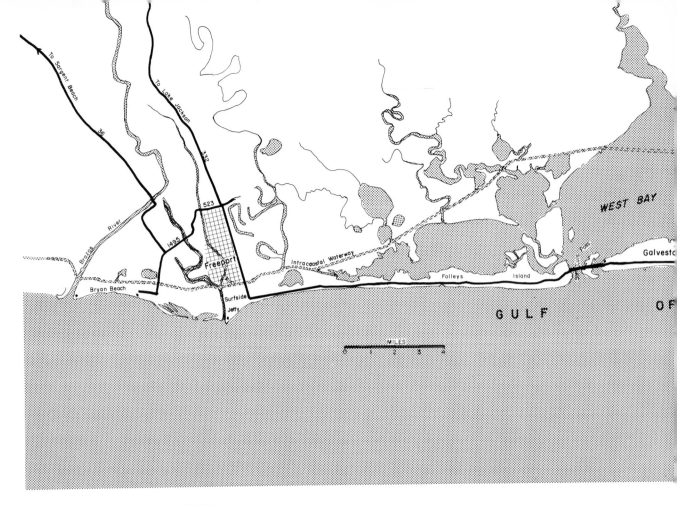

Map 1. GALVESTON TO FREEPORT

the island is an enjoyable experience—mountains of yellow sulphur, bales of cotton, foreign seamen, shrimp boats, and United States Navy ships are but a few of the things that lend color to the scene.

A wide avenue beside the ten-mile-long sea wall is not the least of the attractions to be found in Galveston. Before embarking on this beautiful drive after leaving the ferry, turn left off State Highway 87 onto Seawall Boulevard and drive to the south jetty. Stewart Beach is adjacent to this jetty. The beaches of Galveston are teeming with seashore visitors in the summer but the winter months find them almost as lonely as the shores of Padre Island. Return to Seawall Boulevard and head west.

Near the end of this drive is a large, well-stocked marineland. This is a first-class aquarium display that will interest all who love the sea and its creatures. Leave the sea-wall drive here to get on Stewart Road, which runs behind the marineland, and detour on Ninety-ninth Street to Offatts Bayou in order to sample bay species found there. Then back to Stewart Road.

You can drive through several resort developments to San Luis Pass on this paved road or you may drive the shores of West Beach, which are passable at low tides. The toll bridge spanning San Luis Pass, Vacek Bridge, delivers you to the mainland shores of Brazoria County. The tidal flat at this bridge is the favorite spot for shell collectors in the area.

The Gulf is never out of sight as you drive to Surfside, and frequent access roads permit driving on the beach. If you would like to sample the beach west of Freeport to the mouth of the Brazos River, turn at Surfside on State Highway 332. At the intersection of 332 and Farm Road 523 turn left on 523 and go to Farm Road 1495. This road will take you over the harbor, past the Freeport Inn, and out to Bryan Beach. There you can drive the beach to the river and return by the same route. The Freeport area is being enclosed with fifty miles of levee as a means of hurricane protection.

On leaving Freeport follow State Highway 332 to Lake Jackson and spend a little time at the Brazosport

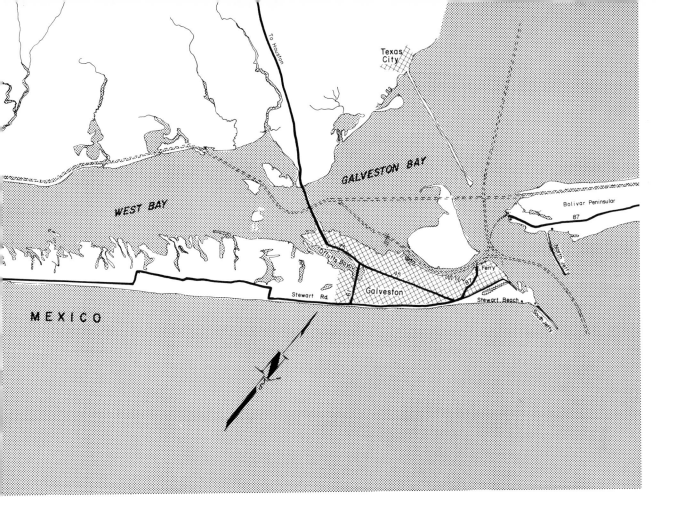

Museum of Natural Science. This small museum has the most impressive collection of shells from the Gulf of Mexico to be found in Texas. You will receive a warm welcome from the volunteers who staff it during the days it is open—Tuesday, Thursday, Saturday, and Sunday.

From Lake Jackson continue on 332 to Brazoria. Here you will have a choice of routes. The first is a new road to the beach that as yet is not shown on the maps. If you leave Brazoria on State Highway 36 you can turn right onto Farm Road 2611, go across the Fisherman's High Bridge, and take a left turn onto Farm Road 2918. This new road will carry you to a point where the San Bernard River intersects the Intracoastal Waterway. At present the Gulf may only be reached by a small boat from this point but a bridge is proposed to close the gap in the near future. Return to Farm Road 2611 and follow it until it meets Farm Road 457, which will take you to Sargent Beach. Farm roads in Texas are well-maintained, paved roads.

The second choice of routes from Brazoria is to go to Sargent Beach via Farm Road 521. At Cedar Lane crossroads turn onto Farm Road 457 to reach Sargent Beach, a mainland shore near Caney Creek. There is a small resort area here, but, as tourist facilities are very modest, it is best not to plan to spend the night. Walk the beach. Return by the same road to Cedar Lane and get back on 521 until you enter Wadsworth. Turn at this village onto State Highway 60 for Matagorda and go beyond the town on Farm Road 2031 to the water.

This beach is at the mouth of the Colorado River and can be a bonanza to the collector at low tides. It is possible to drive a four-wheeled vehicle north on the sand at Matagorda Peninsula for about fifteen miles, but at Brown Cedar Cut you must turn around and come back. Unless you are seeking driftwood it is a hard drive for the results; pickings are better nearer the Colorado. A one-day trip is all that is recommended for this spot, as there are no overnight facilities. Be sure to pack a lunch!

Back again over 60 to 521. At the intersection of 521 and State Highway 35 turn left toward Palacios.

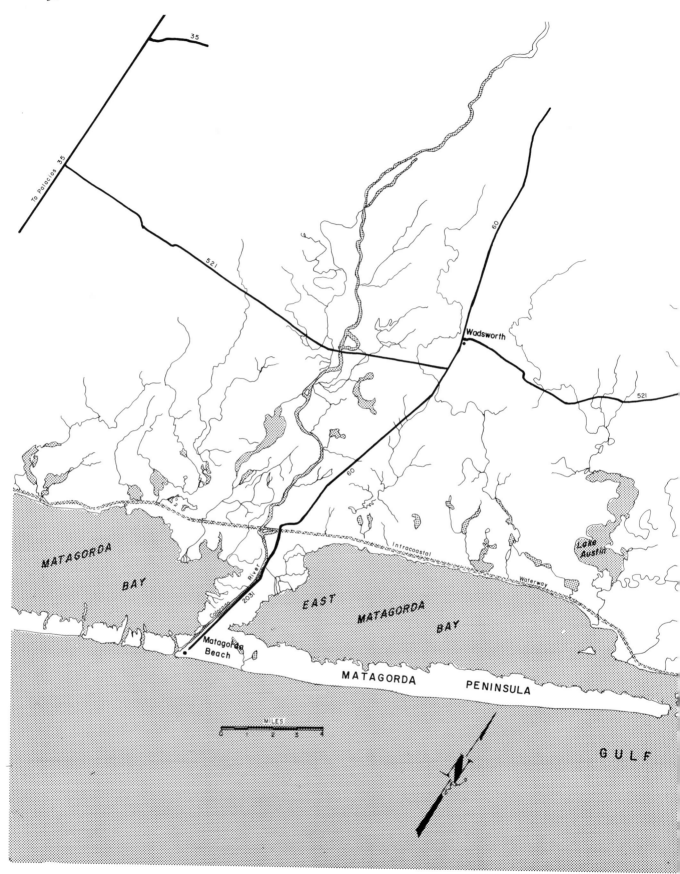

Map 2. FREEPORT TO THE COLORADO RIVER

238

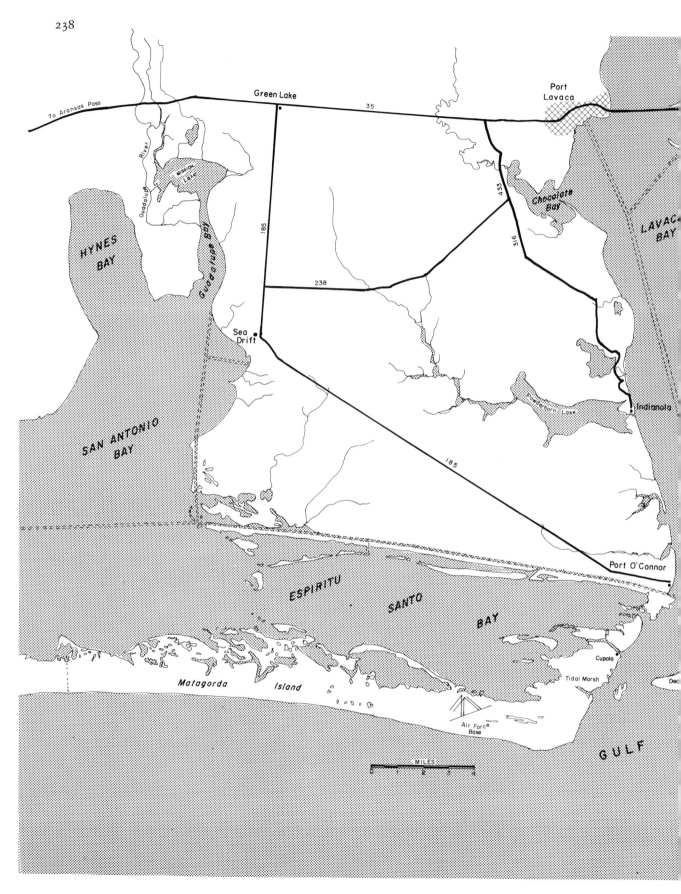

Map 3. COLORADO RIVER TO MATAGORDA ISLAND

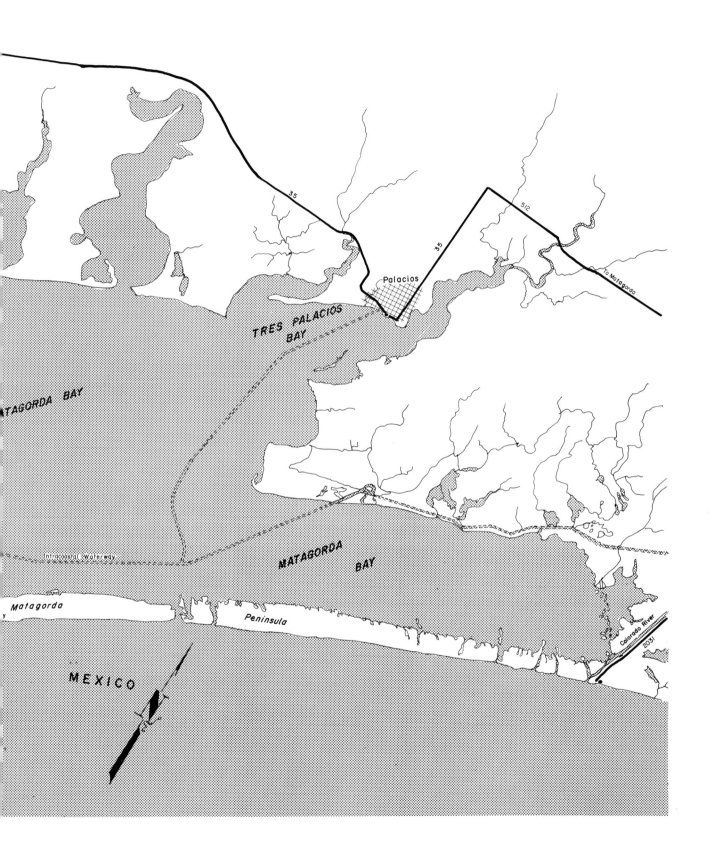

The beaches in Palacios may not look very promising to the collector, but a person interested in minute shells will find the stop rewarding. Take First Street to Bay Boulevard, past the old Luther Hotel, built in 1903; beyond a big pier at the corner of Sixth Street and Duson Avenue is a little cove whose beach is covered with rock and old oystershell. Stop here. The drift from this spot is worth sifting when you return home and the detour through this little town is an adventure in itself. Return to 35 and travel on to Port Lavaca, over the flat coastal prairie and picturesque bayous.

From Port Lavaca you can make a jaunt to the former site of Indianola (see Chapter 1) by going south on 35 and turning left on Farm Road 2433, which intersects State Highway 316. Brackish-water species may be collected from the bay shores along this drive, but Indianola itself yields only some old foundations and a few tombstones in the abandoned cemetery. A statue in honor of La Salle stands in the bay that he named La Vaca. After returning to Port Lavaca, continue on 35 to Green Lake, keeping a lookout for the turn onto State Highway 185 to Port O'Connor.

The collector without a boat must content himself with the little jetties along the Intracoastal, but the flats around them may produce specimens at low tide. It takes a boat and a good knowledge of the waters to reach the Gulf beaches adjoining Pass Cavallo. Shoaling in the pass makes boat passage very tricky for a stranger to the waters, but boats, comfortable lodging, and good food can be had in this fisherman's paradise. Nothing is fancy, but all is new, since Hurricane Carla almost completely destroyed the town in 1961. Waterfowl abound in the extensive marsh areas of this tidal delta. It is not difficult in your boat to reach Decros Point on Matagorda Peninsula, but you are afoot once you are there. A new jettied ship channel a few miles north of this point permits exit from the bay onto the open Gulf. The flats at Saluria Bayou on the south side of the pass are good at low tide for small species. An abandoned Coast Guard station called the Cupola and an old lighthouse constructed in 1852 mark this spot. Beyond this point, toward the pass, the going gets rough.

Matagorda Island is not open to the public, for this northernmost island in the chain of barrier islands is privately owned. The state law covering beach usage (Article 5415d—State Beaches) permits the use of all seaward beaches by the people if there is access by public road or ferry. This island has ferry service but it is operated by the United States Air Force, which maintains a bombing range at the north end of the island and strictly prohibits entry by unauthorized personnel. The major part of the remainder of the island is a large ranch owned by Toddie Lee Wynne of Dallas and can be visited only by invitation. Three thousand head of cattle are being raised at the Wynne Star Brand ranch in an operation maintained by air and sea that is a marvel in itself.

The shores of both passes that border either end of the island can be visited by an enterprising collector interested in a lot of boat riding, hard work, and walking. Unless you want to haul a great amount of camping gear overland on your back, these excursions should be planned as long one-day trips. Since so much work and expense is involved to get to these spots, plan your trip carefully to coincide with the lowest tides of the year. Do watch the tide, for shoals can be very hard to see in these murky waters.

Highway 35 will be the route to a large marina, formerly Mill's Wharf, on the north shore of Copano Bay, at the site of the old village of Lamar. A lovely lodge, the Sea Gun (owned by Mr. Wynne), welcomes the traveler just short of the big causeway. Here, you can launch a boat for the voyage to Mesquite Bay. Follow a well-marked channel for four miles to canal marker number 13 in the Intracoastal Canal and then follow the canal north. This part of the canal traverses the Aransas National Wildlife Refuge where whooping cranes feed along the shores, completely unconcerned over the constant flow of tugs and barges. At a small barrel marker, also number 13, turn right into Mesquite Bay. The channel is not well identified, so hold close to the south shore or you will run aground. Once inside, the channel is well marked and maintained all the way to Cedar Bayou because the state Parks and Wildlife Department has established a small experimental station in Cedar Bayou. The Star Brand ranch headquarters are visible to the north at the opening of the pass. A Parks and Wildlife Department sign prohibiting the passage of boats under penalty of $100.00 fine leaves the collector a fairly long hard walk to the beach—but to follow a low tide behind a norther is worth all the work entailed, including the sore muscles that will follow. The south shore is St. Joseph Island, and it is the most rewarding to the shell enthusiast.

St. Joseph Island is another privately owned island. The northern end may be explored from Cedar Bayou and the southern end from Port Aransas. We will return to the southern end of the island after we arrive in

Port Aransas; in the meantime, back to the Sea Gun and your dependable automobile.

After crossing the causeway on 35 take the first left turn opposite Farm Road 1781 onto an unmarked, tree-bordered road into the town of Fulton. On one side of the drive is the shallow bay and on the other wind-sculptured live oak trees and charming vacation homes. The historic old Fulton home still stands along this route. Motels and restaurants are plentiful in this popular resort area. Fulton and Rockport are almost continuous, separated only by bird-filled marsh flats; however, these havens for birds are being filled to make additional building sites.

The state of Texas operates a Marine Biological Laboratory in Rockport where welcome visitors will find aquaria filled with local fish and some Texas shells on display. The laboratory owns a notable library on mollusks that is available to the serious collector by prior arrangement. After this stop, return to Business 35 and drive to the Post Office. Turn south on 35 at the Post Office, continue for two miles, and turn left onto an unmarked street just after crossing the railroad track. This road will wind you to the cove, which is a good spot to look for bay shells at low tide. Return to Highway 35 by the same route and enter the shrimp fishing port of Aransas Pass. The large shrimp boat fleet in the harbor is always a picturesque sight. Before taking the causeway to Port Aransas try some bay collecting in the Aransas Pass–Ingleside area.

Aransas Pass has Beasley Street as its southern city limit. Winter low tides can make the bay end of this street most profitable for collectors. The city fathers would do well to prohibit dumping along this street, for dumpers are spoiling one of the town's most attractive locations. To be reached by boat (a very short trip) from this spot is Ransom Island, one of the most productive shell sites in Redfish Bay. A hand dredge would be a help here, since the collecting is in the very soft mud along the grassy shores of this little island. A hurricane in the nineteen-thirties destroyed the causeway to this former resort site, but the foundations of buildings may still be seen. You will need your own boat.

Go back to the entrance to Beasley Street and turn south on Farm Road 2725. Take samples from the drift at Harbor City. (Look for oil storage tanks at a turn in the road.) Retrace your steps and turn off 2725 on to Farm Road 1069. The grass flats at Bahia Azul Marina, a well-marked spot farther on this road, once the home of the emerald nerite and the spiny pen shell, have re-

cently been dredged into a channel so that the boats of vacationers living in the adjoining marina may reach the Intracoastal Canal. A large steel company and a chemical firm plan to build on the peninsula where the storage tanks are, so it appears that the curtain is rapidly going down on bay collecting. A turn south, off Farm Road 632 between Ingleside and Gregory, will take you along hideous piles of red sludge from bauxite ore, deposited by the nearby aluminum plant, onto a very bad road to a spot called La Quinta. Many years ago this was the headquarters of the Taft ranch and President Taft was sometimes a guest at his brother's home there. Now, the mud flats along the shore are the home of the beautiful white angel wing clam. Because of a gate placed across this road in 1968, if you want to reach the angel wing flats you will have to hike to the beach. Let's return to Aransas Pass.

Along the causeway (State Highway 361) several spots afford good shelling, especially for those interested in the minute mollusks in the drift. One is opposite a bait stand–restaurant, Fin and Feather, and the other is on the north side of the road at the last bait stand (the name of this one changes frequently).

After crossing the ferry, we come to the flats on either side of the Port Aransas landing that offer good "drift" collecting; Cline's Point, north of the Coast Guard Station, is excellent at low tide, and the jetties are always interesting.

From Port Aransas on the northern tip of Mustang Island it is possible to explore the southern tip of St. Joseph Island on foot. In Port Aransas at the marine service station adjoining The University of Texas Marine Science Institute one can obtain information about a water taxi service to the island. This is a new service and may not be regular. At one time it was possible to rent one of the old trucks owned by the commercial fishermen who worked on St. Joseph. If you were lucky enough to obtain a truck the entire island could be covered in a day's combing. However, getting a car was something else because these local fishermen have little trust in the strangers who have invaded their once-sleepy village. At unpredictable intervals the owners of St. Joseph have invoked Article 5415d and caused the old cars and fishing huts to be removed from the island. Peace reigned between the two camps for more than ten years but in late 1968 the owners had the collection of rusting vehicles hauled away and the shacks burned down, which put the seiners out of business for the time being and virtually eliminated the possibility of beachcombing except at either tip of the island.

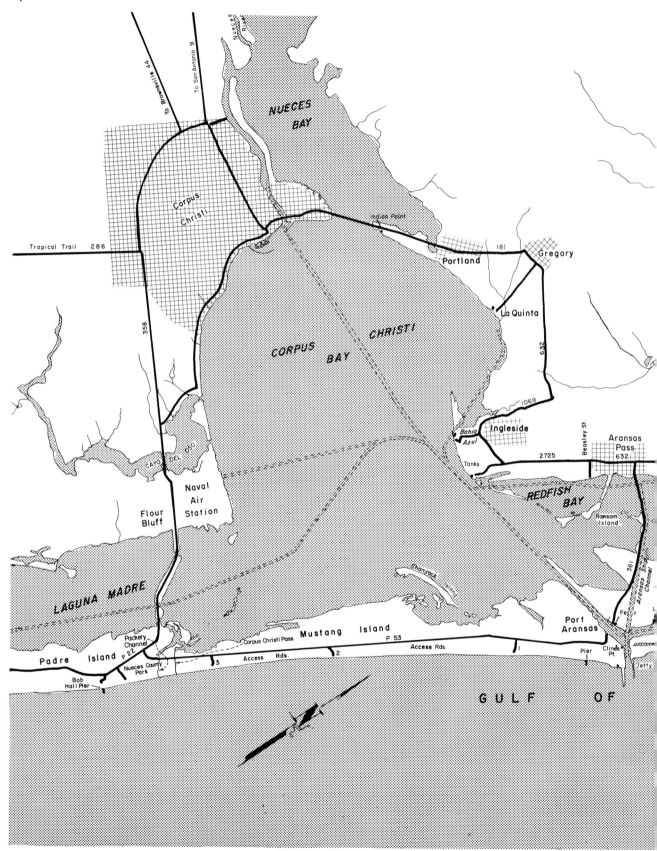

Map 4. CEDAR BAYOU TO PADRE ISLAND

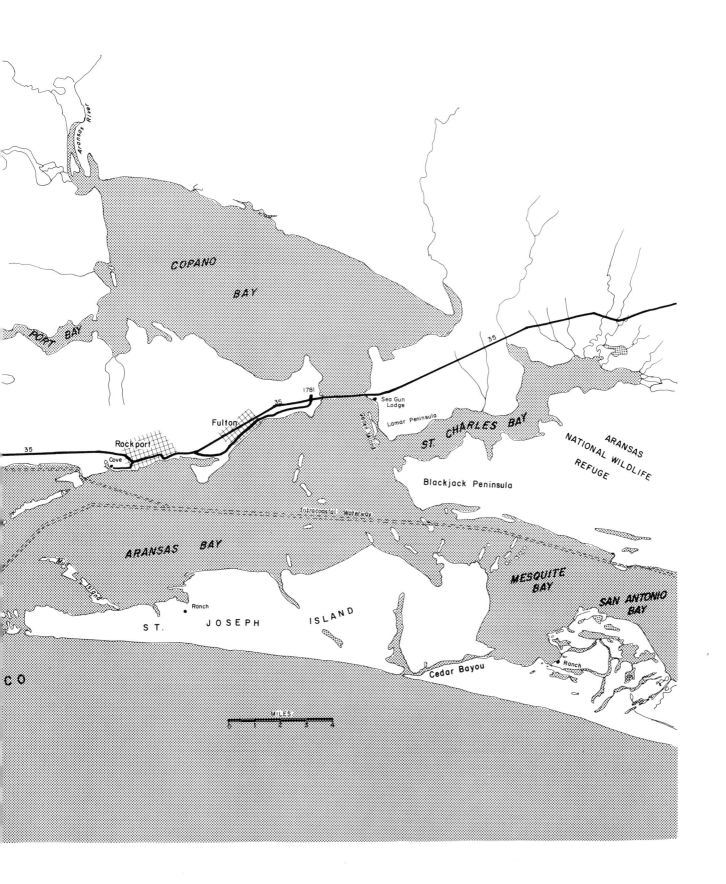

COPANO BAY

Aransas River

PORT BAY

35

1781

35

Fulton

Rockport

Cove

Sea Gun Lodge

Lamar Peninsula

ST. CHARLES BAY

ARANSAS NATIONAL WILDLIFE REFUGE

Blackjack Peninsula

Intracoastal Waterway

ARANSAS BAY

Ranch

ST. JOSEPH ISLAND

MESQUITE BAY

SAN ANTONIO BAY

Cedar Bayou

Ranch

CO

MILES
0 1 2 3 4

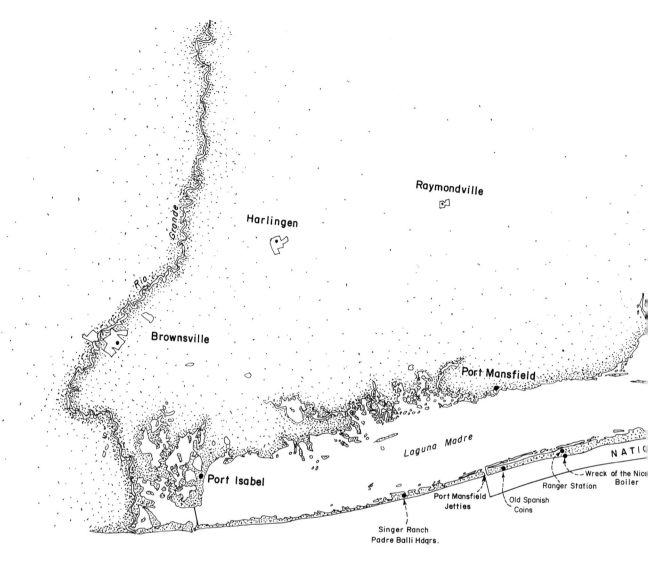

Map 5. PADRE ISLAND

Mustang Island and the northern twenty miles of Padre Island are completely accessible to the collector in a passenger car. A paved road (Park Road 53) with numbered access roads runs down the center of Mustang Island. The beach is passable at most times, but extremely high tides and water from offshore disturbances occasionally flood the usually firm-packed beach. Natural, wind-formed cusps (ridges of sand perpendicular to the beach road) make excessive speed hazardous. Besides this natural speed hazard an alert police beach patrol also works to this end.

On spring and summer weekends after 3 P.M. long lines of automobiles form a creeping procession to the ferry and the Kennedy Causeway as weary, salty, sand-coated visitors return to the mainland. It is hoped wider roads and bridges will ease this traffic jam in the future but, meanwhile, it is well to try to plan your trip to avoid this miserable situation.

Eighteen miles south of Port Aransas is fabulous Padre Island. Between Packery Channel and the Nueces County Park on the north end of the island much new development is taking place. Luxury hotels and

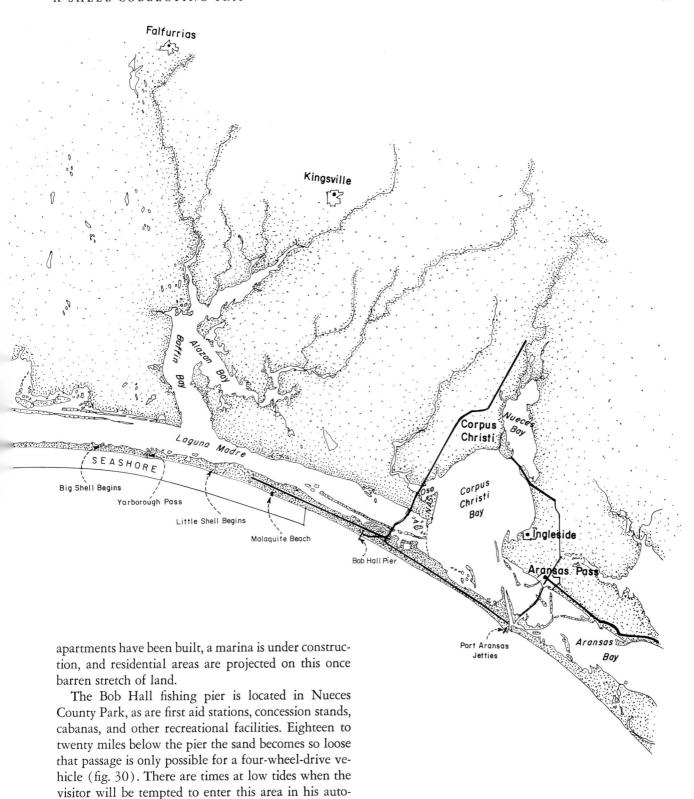

Falfurrias

Kingsville

Corpus Christi

Nueces Bay

Baffin Bay

Alazan Bay

Laguna Madre

SEASHORE

Big Shell Begins

Yarborough Pass

Little Shell Begins

Malaquite Beach

Bob Hall Pier

Oso

NAS

Corpus Christi Bay

Ingleside

Aransas Pass

Port Aransas Jetties

Aransas Bay

apartments have been built, a marina is under construction, and residential areas are projected on this once barren stretch of land.

The Bob Hall fishing pier is located in Nueces County Park, as are first aid stations, concession stands, cabanas, and other recreational facilities. Eighteen to twenty miles below the pier the sand becomes so loose that passage is only possible for a four-wheel-drive vehicle (fig. 30). There are times at low tides when the visitor will be tempted to enter this area in his automobile but it is a foolhardy endeavor that will be re-

Fig. 30. The author's four-wheel-drive vehicle in the dunes near the Port Mansfield jetty on Padre Island.

warded with hours of backbreaking digging under the sand-stranded car. The public-use buildings of the national seashore located at Malaquite Beach are easily reached by a paved road that terminates there. Below this area you are able to return to the shore and resume the drive over the sandy beach to the Port Mansfield jetties.

The section of Padre Island that makes up the main portion of the National Seashore Area, Little Shell to Port Mansfield jetty, can be visited only with a four-wheel-drive vehicle. It is seventy-eight miles from Bob Hall Pier to Port Mansfield and seventy-eight miles back over areas of loose shell and deep sand that have been deeply cut by the beach cars of adventurous anglers and other shellers. The vehicle chosen for this jaunt must be in good repair and well equipped with spare parts, gasoline, and water. Current status of conditions in this area may be investigated through the

Padre Island National Seashore Area office, Box 8560, Corpus Christi, Texas 78412.

I have made the trip in as little as two and one-half hours when I was the first to travel it after a hurricane and the beach was firmly packed, but on another occasion when it had been much in use by the many beach cars in the area during dry weather, my loaded vehicle required thirty-five gallons of gasoline for the round trip (Corpus Christi to the Port Mansfield jetty, about 98 miles) because I had to grind in the deep sand so much of the distance using the "grandma" gear.

A word of caution about travel on these islands: Matagorda, St. Joseph, and the middle of Padre are virtually uninhabited. If the beachcomber gets in trouble, he should not panic. Set up some sort of signal—a fire, mirror, flag, writing in the sand—and remember that fresh-water wells for cattle run the length of the islands. The Park Rangers patrol the national seashore between its entrance and Little Shell from 8:00 A.M. until 10:00 P.M. daily with three weekly scheduled trips to the Mansfield Ship Channel, when Gulf tide conditions permit. After the facilities are in operation these rescue services will be expanded. A paved road down the center of the island is planned, and twelve miles of it have been completed.

Preparations for the trip should include an ample supply of water and enough food for your party, also, a jack, shovel, and tow rope. Never visit these islands alone. On one occasion, after being injured when sixty-five miles down the beach, I was fortunate in being able to flag down an airplane, which landed on the beach and flew a message to a rescue party; but you might not be so lucky, especially if the tricky Texas weather finds you stranded there.

A doctor keeps regular hours at Port Aransas and there are twenty-four–hour first aid stations at Port Aransas and the county and national parks on Padre, but travel to them is slow at best. In summer months law officers, with first aid equipment and radio-controlled cars patrol the area of beach that is passable by automobile. Ambulance service comes from Corpus Christi and Aransas Pass and can take hours. A helicopter evacuation plan is being tried for serious emergencies.

During the winter when there are few people on the islands, the Coast Guard and Park Rangers have only praise for the visitor to middle Padre who calls their station before he leaves, to give his destination and an estimated time of return, and then checks in when he returns. Don't forget to call back in, for a search party will be on its way within the hour. This may seem overcautious but on the barrier islands of Texas, history has proved that anything can happen and often does.

The Laguna Madre can be traveled by boat down the Intracoastal Canal but the salinity conditions of these enclosed waters render the search for shells fruitless. The dredging of the Intracoastal Canal caused spoil islands to be built along the canal from the vast amounts of sediments removed to deepen the waterway. Fishermen have thrown up ingenious squatter's shacks on these barren little islands.

To reach the southern part of Padre Island you may leave Corpus Christi by several roads that connect with U.S. 77. Try the Tropical Trail. At Raymondville a drive over State Highway 186 and Farm Road 497 will lead you to Port Mansfield where you can go by boat to the Mansfield jetties. There are modest motels and restaurants at Port Mansfield.

Continue on 77, but turn on State Highway 100 before reaching the border city of Brownsville and take a side trip to Port Isabel. Everything in this well-developed area can be reached by car except the beach drive north to the Mansfield jetties; for this a four-wheel-drive vehicle is necessary. To get to the last of the islands, Brazos, one must leave Port Isabel on Farm Road 1792 and travel almost to Brownsville. Short of Brownsville turn south onto State Road 4 and go to Boca Chica Beach.

While you are in the area you may find it worthwhile to cross the border at Brownsville into Matamoros for a trip to the Mexican beach south of the Rio Grande. Before going into the town ask for directions to Washington Beach—an appropriate name for the beach nearest to the United States. Things change so fast in this border town I don't dare direct you to the beach myself.

Check all this information against the most recent road map. We live in a world of constant change.

EQUIPMENT

Elaborate equipment is not necessary for the type of collecting to be done in Texas. The seasoned collector usually has his pet essentials, but the beginner will find the following to be handy:

Sun glasses: The glare on the white sand and the water is blinding.

Sun hat: The Texas sun can produce painful burns even on a cloudy day.

Canvas shoes: Broken shell, bottles, and cans buried in the sand can be a hazard to bare feet.

Long-sleeved shirt and pants: Even if a tan is desired the skin can only stand so much.

Pockets: Either built in or added apron fashion.

Plastic bucket: A light-weight carry-all.

Pill bottles, vials, small cotton bags: For small and special specimens.

Plastic bags: For collecting drift or wrapping messy material.

Trowel: For digging in the sand and mud or lifting the layer of drift from the sand.

Small strainer or wire-bottomed boxes: For sifting shells from sand and mud.

Tweezer or toothpicks: To pick up minute shells (wet the end of the toothpick).

Notebook and pencil: To record data.

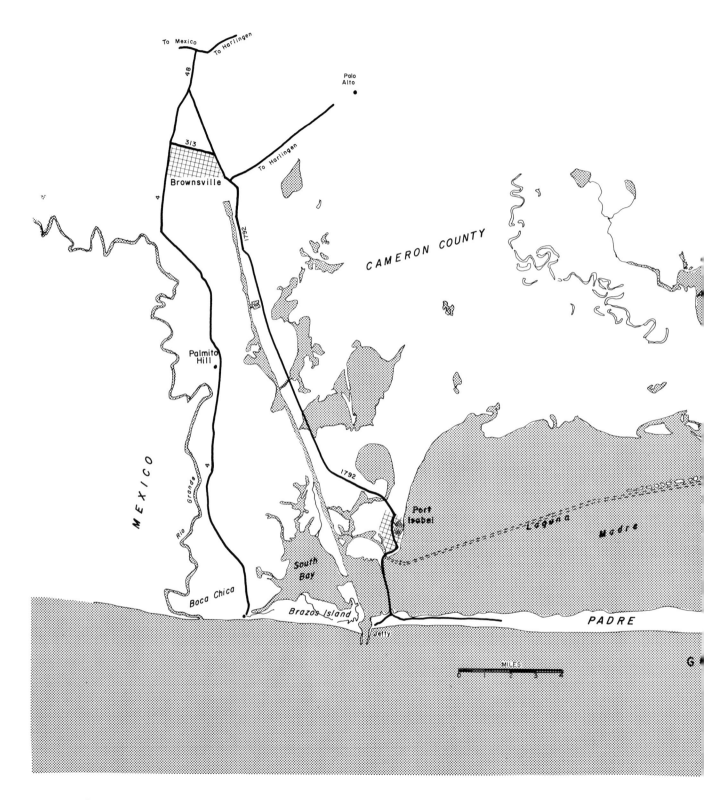

Map 6. PORT MANSFIELD TO THE RIO GRANDE

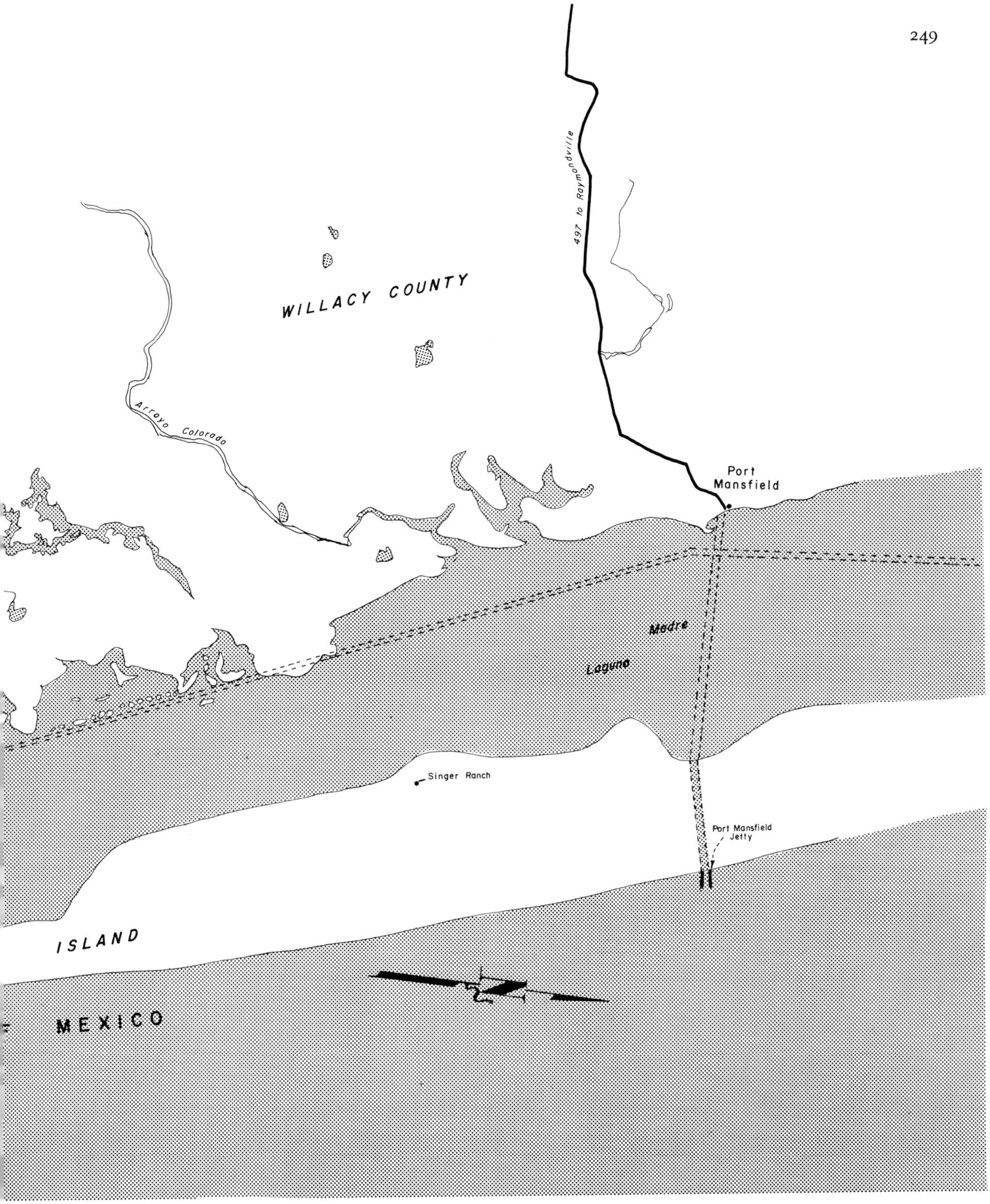

WILLACY COUNTY

497 to Raymondville

Arroyo Colorado

Port
Mansfield

Madre

Laguna

Singer Ranch

Port Mansfield
Jetty

ISLAND

MEXICO

Fig. 31. A beached mass of the sea whip coral, *Leptogorgia setacea*, the home of the little yellow and red *Neosimnia uniplicata*.

Thread or string: To tie valves together.

Waders, rubber boots, and thermal underwear: Since winter brings the best collecting, these items will make the northers bearable.

Mask, snorkle, and diving gear: To look along the jetties.

When and Where to Look

Like fishermen, shell collectors have their favorite haunts and many are reluctant to reveal their "trade" secrets. However, a few fundamentals can act as a guide to obtaining a representative sample of the shells in Texas waters. Abbott (1958a) has said that there are four ingredients to securing good shells:

1. A knowledge of the habits of mollusks
2. A familiarization with the physical conditions of the ocean and seashore

3. A sensible choice of collecting equipment
4. A large proportion of perseverance

A familiarity with the first three of these may be acquired from this book or a similar reference, but the last will come only after hours in the field or at the microscope. The more common forms of shells may be easy to come by for any beachcomber, but the true collector seeks the living shell in its natural habitat.

If one is fortunate enough to visit the open Gulf beach following a storm or hard freeze, the shore may be strewn with shells that normally live offshore. On an ordinary day, these sparsely inhabited shores may seem unproductive to casual beachcombers. The knowing collector, however, does not pass up a waterlogged piece of wood, for here dwell many boring shells. The tangle of yellow and red sea whip coral will reward the person who stops to shake it (fig. 31)—a tiny yellow

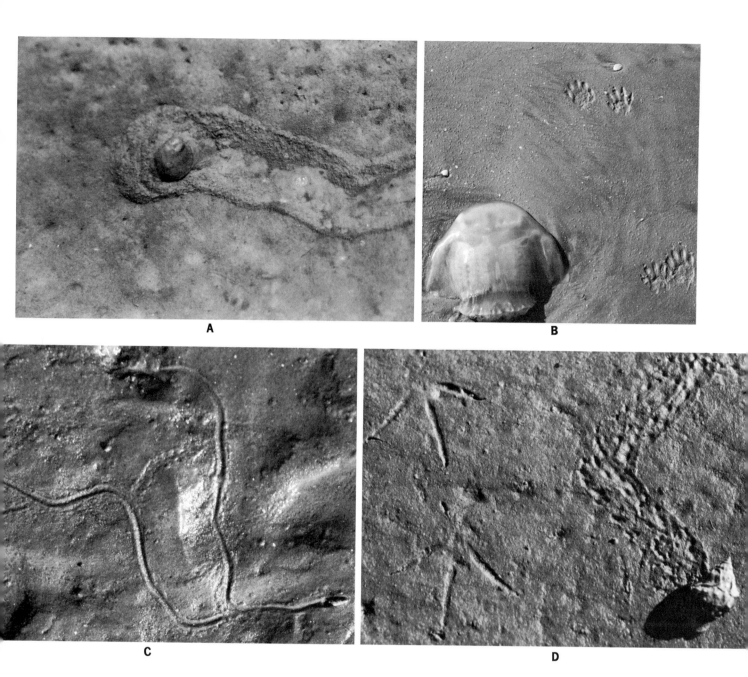

Fig. 32. Trails along a shore: *A*, the wide trail produced by the plowing motion of *Polinices duplicatus*; *B*, a raccoon leaves his tracks after a meal of clams; *C*, a cerith worms its way across the wet sand; *D*, a hermit crab in a *Thais* shell leaves its punctured trail beside the footprints of a shore bird.

Neosimnia and other such gifts the sea may drop out. That barnacle-covered coconut or clump of bamboo roots may have a rare tropical species attached. Shore collecting is rewarding in direct proportion to the regularity and persistence of the search. That happy coincidence of time and tide which brings a rare treasure is unpredictable.

The bays and back waters will provide more reward-

ing collecting any time of the year. On these mud and sand flats the trail of a snail may be followed or the siphon hole of a buried clam investigated. One soon learns to distinguish what animal he is tracking by the type of trail left in the sand (fig. 32). For example, the unbroken line left by the dragging shell of a hermit crab is bordered on either side by rows of dots made by the crab's legs. A snail's track will not have the pattern

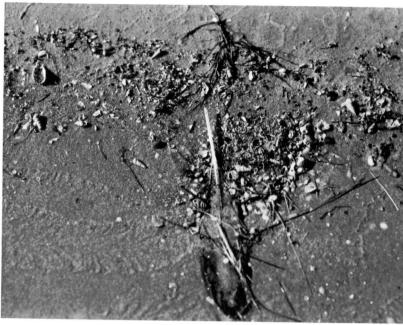

Fig. 33. Beach drift: *A*, the author examining a line of beach drift, that accumulation of rubble deposited along the high-tide line; *B*, a close-up of beach drift along an inlet area.

of dots. Many species are more active on a cloudy day or at night. Many inhabitants of the tidal flats are more in evidence about half an hour after the tide has begun to rise.

Pay attention to the winds, currents, seasons, and tides. Normally, in Texas the winter months are the most productive on the open beaches, but a hurricane during the early fall can bring a windfall of shells. The windy days of March and April drive the pelagic (floating) forms, such as the purple sea snail, to shore. After a quiet tide many small specimens may be encountered in the drift line along the edges of the ship channels and the bays. Wading along the bay edges and raising and examining submerged broken shell, bottles, rocks, and timber often produces fascinating results. Examine any attached algae. Always remember to return the rock or shell to its former position in order not to disturb the ecology of the area.

Another conservation note—don't take more than you need from a site where specimens are living. On the open beach where shells have been stranded by the waves, take all you want; the waves will only carry them back into the sea after they are dead and these are already doomed. But when you find an exposed flat

where *Busycon* and *Murex* are popping up, leave some to repopulate the area. It is hard to understand why some people want a bucket of live whelks. A beautiful specimen or two for your collection should be enough. When you start cleaning all those you lugged home you will probably wish you had left them in the bay or inlet in the first place. It is very sad to watch a shell club on a field trip to a new area practically denude it in one day's hunt. Vast beds of clams like those found in other areas are not the rule in Texas. Leave a few for seed.

Use of the tide tables published annually by the Coast and Geodetic Survey is most beneficial in determining the lowest tides of the year. These tables may be purchased from the United States Department of Commerce or from a marine supply company. Local newspapers also publish daily tide times and moon phases.

Certain species associate with definite types of environment. For example, the lovely angel wing lives in hard mud flats. The habitat of each species is given with its description in Chapter 3 of this book.

More than half the shells found in Texas are minute; thus good eyes, a hand lens, or a microscope are needed

Fig. 34. A collector using a trowel to collect fine drift from the high-tide line along the open Gulf beach to be screened later under a microscope.

Fig. 35. Equipment needed to classify minute mollusks: *left to right*, plastic boxes, vials, capsules, slides, labels, fine brush, technical pen, dissecting needle, microscope with tray of screened material, and slide storage box.

to identify them. Drift may be taken home to be picked over when you have more time or when the T.V. is broken; but be careful, you may become an addict. Trying to identify the tiny shells can be more challenging and frustrating than solving a Russian crossword puzzle or tracing your family tree, and the satisfaction more rewarding when you identify a confusing little fellow. First, however, you must gather beach drift.

Beach drift is that unassuming-looking fine rubble or trash deposited by the waves along the high-water line (fig. 33). Usually the last high-water line will hold the richest assortment. Scrape the drift up with your hand or a trowel and put it in plastic bags or a plastic bucket (fig. 34). When you get home wash it with fresh water to remove all traces of salt. Salt left on these tiny shells will eventually cause complete deterioration. Screen the drift through strainers, either homemade or laboratory models. Mine are a nest of simple redwood boxes with rust-resistant wire bottoms in three sizes—hardware cloth, window-screen wire, and marine filter mesh. The washing and screening operations can be done simultaneously.

After your drift is screened and washed, spread it to dry on an old sheet or other piece of fabric—paper comes apart when wet. The drift must be completely dry before it can be sorted. Do not attempt to dry it in the oven because small particles of asphalt that are present on all Texas beaches (see Chapter 5) will melt and ruin your material. A magnifying glass or microscope (dissecting type) is needed for the sorting (fig. 35). The specimens can be picked up with a wet finger or a damp, fine (No. 000), sable brush. You will soon develop your own method for handling this operation.

In recent years skin diving or scuba diving along the jetties and wharves has become a popular pastime; however, this can be a very dangerous undertaking unless the diver is properly trained, well equipped, and accompanied by another diver. Courses are given at regular intervals at the Corpus Christi Y.M.C.A. for those interested in learning this sport. Skin diving has proved very beneficial to science, and marine biology students are using diving as a method for collecting and observing. With the invention of the aqualung much new territory has been opened to the diving scientist. Many species of mollusks once considered to be rare are now more easily obtained by divers.

A mask, fins, and snorkel would be of great help to the collector on a Texas jetty. Choose only simple, rugged, compact equipment that is properly fitted and comfortable. Jetty diving is not a sport for a person alone or the untrained, for the jetties are very treacherous at times as a result of tidal currents sweeping through the passes and currents set up by passing ocean-going steamers. A study of the currents and tides must be made before a diving trip is planned. It is, however, worth all the trouble to train and equip one's self to see the brightly colored soft corals, the spiny sea urchins, the glowing tunicates, an olive shell plowing under the sand at the base of the jetties, or a shy octopus in a crevice in the rocks.

To me dredging is unrewarding in proportion to the work and expense involved and it will not be discussed here as a form of collecting. Those interested in attempting dredging can learn from a discussion of the methods and equipment in *American Seashells* by R. Tucker Abbott or at greater length by reading the benthic sampling methods detailed by Holmes (1964). This is a difficult and backbreaking procedure; I leave it to the organized research group.

Cleaning

Cleaning your shell is of great importance, since a beautiful specimen, improperly preserved, can quickly become an unlovely thing. First consider the mollusk that is taken alive.

Many types of shells may be cleaned by simply boiling them and then removing the animal. Start them in cold water with a folded piece of cloth on the bottom of the pan to reduce breakage. Boil slowly until the meat is firm or the clam has opened. Cool to lukewarm before washing. Boil only a few shells at a time and remove the bodies immediately when cool enough to handle because cooling causes the body to contract and shrink up into the shell. A corkscrew motion will usually get all the animal from the winding snail shell. If the visceral portion of the body breaks off, a few drops of 5 percent formalin (formaldehyde gas dissolved in water) placed in the shell will eliminate bad odors. The formalin should then be removed; no shell should be stored in a solution of formaldehyde, because of the destructive action it has on the shell itself. Dental tools, nut picks, and wires are handy for the body-removing operation.

Some highly polished shells like *Strombus*, *Oliva*, or *Cypraea* will craze if boiled. The animals in porcelaneous shells can be killed by placing them in fresh water. Most can then be picked clean, but some are stubborn. These may be dehydrated in the refrigerator.

Wrap each shell in a paper towel with more thickness next to the aperture and place with the aperture down in the refrigerator for a week. At this time the animal should come out easily. If it does not, don't soak it in anything, for this type of shell waterspots easily. Dry it carefully and allow it to sit with the aperture up until the remaining animal has dried up and can be shaken out.

The true collector wants to preserve the operculum or "trap door" of the operculate shell. Not all snails have an operculum, but they are an interesting feature of many. Mark them in some way so that you keep the correct operculum with its own shell. When these have been dried they may be glued in their natural position after the thoroughly clean shell has been stuffed with cotton.

Tiny snails are best preserved by allowing them to remain out of water until they die (fresh water will make them withdraw far into the shell) and then placing them in a solution of 70 percent ethyl or grain alcohol or 50 percent methyl or isopropyl alcohol. If they are placed directly in the alcohol the animal will contract and draw the operculum into a position where it will not be visible. After several days in the solution the small snails may be drained and spread to dry before storing.

The bivalves or clams are cleaned in the same way, taking care not to break the hinges that hold the two sections together. Many collectors prefer that the shell be dried closed in its natural position. After the animal is removed, cotton string is wrapped around the closed valves until the ligament dries. Rubber bands, adhesive tape, and other tapes may prove disastrous. Heat causes the rubber to melt and the adhesive coating on the tape comes off, leaving a hard-to-remove residue on the once-beautiful shell. Other collectors prefer that the bivalve be dried in a butterfly position so that the inside is visible. If one is fortunate enough to have several specimens it is desirable to dry the clam in both ways. William Clench, the editor of *Johnsonia*, recommends preserving bivalves in a solution of four parts alcohol and one part glycerine to keep the periostracum and ligament soft and pliable. An occasional dab of glycerine on the ligament will prevent it from becoming brittle.

The fastidious collector may want to clean off all the foreign matter that is attached to the outside of the shell. This again is a matter of taste, as many desire their shells in the natural state with the exterior uncleaned. It is a good idea to have an example of a cleaned and an uncleaned shell in your collection be-cause cleaning often reveals hidden beauty.

A stiff brush is usually enough to do the job of clean-ing. Growth that cannot be removed with the brush can often be removed with a sharp pointed instrument, such as an awl, a knife, or a dental tool. The limey de-posits can prove quite stubborn and often one must resort to muriatic acid. Experience has proved that this job must be done with running water at hand. The shell is held with forceps and *quickly* dipped into full strength acid, then immediately under running water. Extreme caution must be used with this method. The acid is a poison and can burn your skin. Use it only as a last resort. Apertures and delicate parts of the shell can be coated with melted paraffin to protect them from the acid. Shells overcleaned with acid take on an unnatural sheen and can be spotted a yard away in a shell shop. The composition of shells varies; some will withstand the acid and others, like the lion's paw, will dissolve immediately when put in contact with the powerful caustic.

Often the shell is covered with vegetation and sponge that can be removed by the use of household chlorine compounds. The chlorine will not dissolve the calcium of the shell as do muriatic acid and formalin, but it does have a tendency to bleach the shell if contact is maintained over too long a period. (A note from experience: It will also dissolve your fingernails.)

The following method is not recommended as an enjoyable pastime but in some cases it may prove nec-essary. Delicate shells like the purple sea snail may be cleaned by "watering." Place them overnight in fresh water and then rinse away the fleshy parts. This pro-cedure may need to be repeated several times. A word of warning—change the water daily or more often. The decomposition of the animal's body forms corro-sives that dull the surface of the shell if it is allowed to remain in the same water for too long a period.

After the shell is cleaned its natural lustre may be restored by rubbing a thin coating of some type of greasy substance on it. Vegetable fats become sticky with age; it is best to use mineral oil, baby oil, or vase-line. Store oiled shells in closed cabinets to avoid the accumulation of dust on the coated surfaces.

Most collectors shudder at the thought of a shell that has been coated with varnish or shellac or over-cleaned with acids. Don't kill the shell with kindness. A shell carefully cleaned to preserve its natural beauty can be a joy forever.

A word on traveling with a collection of live caught specimens; you may develop your own method but un-til you do let me recommend a method learned the hard way. Leave the animal in the shell and wrap the indi-vidual shell with newspaper, protecting the lip with extra thickness and stuffing the aperture firmly. If pa-per towels or tissue are available, they work well for the stuffing. Place the wrapped shell in a plastic bag and tie tightly. Plastic garbage-can liners are good for putting several wrapped specimens together and are stronger than smaller bags. Small specimens may be packed in plastic jars in a 30 percent alcohol solution. Carry the whole mess in a garbage can with a tightly fitting lid or cover the top lightly with plastic before putting the lid on. Plan to bring the materials you will use with you from home. Not every place has news-paper. Decomposition of sea life, especially in the Texas climate, is rapid and odorous. In other words, it will stink quickly.

One trip without proper preparation will be enough. A return trip from the west coast of Mexico was ghast-ly; even the customs agents passed our car without coming near to examine it, waving us on as quickly as possible.

STORING

The storing of a sea shell collection presents special problems. Although properly cleaned sea shells do not deteriorate, they do tend to fade in direct sunlight and extreme heat. This feature, combined with their va-riety of size and shape, makes the shell collection hard to store.

There are several types of collections—the whatnot shelf, the display cabinet, the study collection, the aquarium, and the photograph collection. The first is usually the outcome of a summer's beachcombing by the beginner or it may be an auxiliary to the study col-lection. The second is generally found in museums, clubs, or libraries and even the home. It is planned for visual appeal, is not overcrowded, is well labeled, has good lighting, and contains choice specimens. The study collection is housed more compactly with the catalog number, scientific name, date found, and loca-tion found recorded for each specimen.

In a study collection it is essential to record care-fully the depth of water, special conditions, occurrence, and date when the specimen was secured. A shell that cannot be given a definite locality adds no value to a collection. (Not just "Texas" but "beach drift, near jet-ties, Port Aransas, Texas.")

Small plastic pill bottles and gelatine capsules are

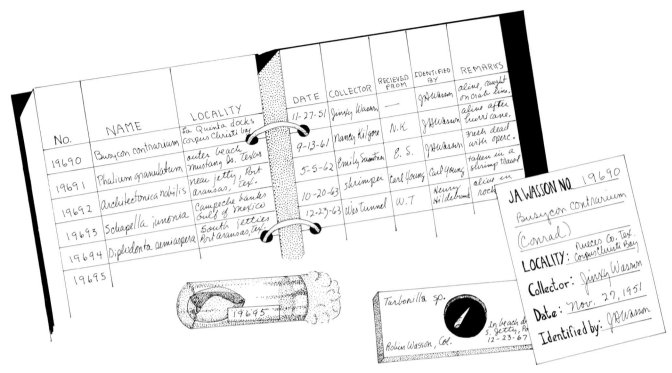

Fig. 36. Cataloging a shell collection: *A*, catalog; *B*, vial for small shells with label inside; *C*, slide for minute shells; *D*, label to accompany larger specimens.

handy for storing the smaller specimens and may be obtained in a variety of sizes from your pharmacist. Modular plastic boxes with a plastic container to hold a group of them can be ordered from Althor Products, 2301 Benson Avenue, Brooklyn, New York 11214. This company's catalog illustrates a great many types of plastic containers. The tiny shells are best housed in slides. I prefer the micropaleontological slides with glass covers and metal slide holders that can be ordered from W. H. Curtin & Co., P.O. Box 1546, Houston, Texas 77001.

Larger specimens can be stored in drawers with small boxes for dividers. Avoid the use of cotton because it clings to rough surfaces and is difficult to remove. A little more trouble but worthwhile is lining the drawers with a thick (dependent on drawer depth and shell thickness) layer of foam rubber or styrofoam. Using an X-acto knife or razor blade cut a berth to house each individual shell. The drawer is beautiful to look at and chipping is kept to a minimum because the shells never touch each other or jiggle around.

A catalog list in a notebook or a card file is a necessity to prevent the loss of the valuable locality data

(fig. 36). You will want to record the precise location, the date, and many other data. The method of collection is important too. All this information cannot be attached to each shell, hence a numbering system must be used for ease in locating the recorded data. Whatever system is used, you will assign to each shell a specific number. This number can be written in an inconspicuous place on the larger shells with India ink and a fine pen. The Rapidograph pen is excellent for this job and also for recording in your catalog or notebook. Water will not smear dry India ink. Smaller shells can have a number affixed to their container or written on the slide.

I prefer the double-entry type of catalog, but it is only one of many systems. With it, acquisitions are first entered as received, the number is placed on the shell or its container. Separate localities may be indicated by letters that are added to the numbers. The second entry is in an alphabetical listing of the genera. Care must be taken to allow space between each name for future additions. The numbers are entered opposite the individual species but are not consecutive. Use a loose leaf notebook for this catalog.

The aquarium collectors are a growing group and are adding much to our knowledge of the habits of mollusks by their patient observations. The animals' natural protection against drying will usually permit you to carry them home safely in a small amount of sea water. Unless you live near the seashore a marine aquarium may become an expensive proposition due to the high cost of specimens and sea water, but it is well worth a try.

The basic problems to be solved are pollution, feeding, disease, and rust and corrosion. The correct selection of a tank will solve some of these problems. An ideal marine aquarium is all glass with a glass cover to slow evaporation, but those constructed of plywood coated with fiberglass and having a glass front or a plastic or lucite vat are acceptable. All are preferable to the stainless-steel-and-glass aquarium. The cement of this type of container reacts with the salt water to form a poison, and the salt corrodes the metal creating another toxin to seep into the water. The metal can be coated with an epoxy paint but it will corrode regardless of the manufacturer's claims.

Once you have the proper tank the next step is a sub-sand filter, which acts as a false bottom under the sand. It aerates, filters, removes food particles that might decompose, and permits the use of sand without the danger of toxic gases forming. Choose a filter with holes small enough to prevent sand from passing through them.

Due to the nature of the creatures you will keep in your aquarium an air pump is necessary to provide oxygen and a moderate current. They will also need some sun, but not too much. Place the tank in a draft-free location that will receive some sun every day but is not in strong or direct sunlight, for sunlight encourages the growth of algae. You must also remember that most marine animals live in crevices, reefs, or under the sand and are light sensitive.

All equipment used in a marine aquarium must be sterilized. To do this, soak the article (pump, shell, coral, etc.) for several days in fresh water to which three tablespoons of table salt have been added. This removes surface oils. Then rinse thoroughly in fresh water. If you collect shells or coral to decorate your tank you should select only shallow, easily cleaned examples, then soak them in fresh water for several weeks and dry thoroughly in the sun before adding them to the aquarium.

Beach sand is too full of organisms to be used in your aquarium. Choose a builder's sand and wash it several times with fresh water in an enamel bucket. Do not have it over one and one-half inches deep and make certain all air bubbles are removed before adding specimens.

The temperature in the tank should be kept rather constant between seventy and eighty degrees. If it happens to get over ninety degrees cool the water with ice and if it drops under sixty degrees place a heater nearby. Marine animals are able to stand changes in temperature if they are gradual. During cold weather increased aeration is indicated.

A cover glass will prevent your specimens from leaving the tank and will retard evaporation. Extra sea water may be stored for long periods in a glass container in a darkened location. Collect, transport, and store the natural sea water in a nonmetallic container. Filter the water as thoroughly as possible to remove all organic material that might cause the water to foul during storage. Artificial sea water is used with success by many. Also, a fairly constant density (specific gravity) should be maintained. Sea water is heavier than distilled water and evaporation tends to cause the ratio to increase. The ideal specific gravity is 1.020. Your pet shop will have special hydrometers with which you can measure the density of the water.

When you add a specimen to the aquarium try not to excite it. Keep it in a small amount of the water in which it was transported and add a cup of water from the tank every fifteen minutes until it is acclimated. Transfer only a few at a time and do so very slowly. If you have too many specimens you may have to add an air stone (purchase from aquarium dealer). Artificial salt water requires even more care when adding specimens.

Feeding is not the problem it once was; however, overfeeding with failure to remove the excess food particles can present a serious situation. If available, feed the snails a small, live clam every month. You may keep ground clam, oyster, or snail in your freezer to feed to your carnivorous snails. When a supply is abundant, chop it in a blender and place it in an ice cube tray. These frozen cubes are easily added to the tank. A snail does not require frequent feeding and often after a big meal it will bury itself for days only to emerge with a new growth of shell. Some creatures require algae for food. Small rocks covered with algae or pieces of the seaweed *Ulva* will give herbivorous species something to feed on. Fish eat three to four times a day but if there is algae in the tank for them to browse on you will only need to feed them in the morning and

Fig. 37. Aquarium observation: This series of the gastropod *Fasciolaria hunteria* was raised by Mannette Wilson of Corpus Christi in a twenty-gallon salt-water aquarium. The larger female (next to the rock) was found stranded on the beach of Mustang Island. Later the smaller male (next) was found and added to the aquarium. Following mating, the female worked at constructing tulip-shaped egg cases on a rock over a period of three months. In fifty-eight days the young began to crawl out of the gaping egg cases. By a month later 125 snails had been hatched. Those that survived reached a length of 2½ inches within a year.

evening. Do not attempt to grow algae without a sub-sand filter.

A variety of species adds interest to your aquarium but if you have live clams or scallops in your tank they may become the dinner of a starfish, snail, or crab. Anemones devour any hapless little mollusk, such as a *Cyphoma*, that might come within their reach.

The main causes of aquarium failure are:
1. Specimens introduced too quickly
2. Soiled hands in contact with the specimens or water
3. Insufficient aeration
4. Specimens in poor condition when added
5. Painting or spraying in the home
6. Insecticides

However, if you get all this working you can have many pleasurable hours observing the antics of your aquarium's inhabitants. Remember to make notes and sketches of the behavior of these creatures; you may be observing an activity never before recorded by man.

An example of the type of study to be made with an aquarium is illustrated in the growth series shown in Figure 37.

New developments in 35 mm. camera lenses make it possible for the camera bug to photograph his collection. This hobby is expensive because the initial equipment is costly; however, collectors so equipped could add greatly to our store of information concerning the various characteristics of the mollusk. These high-quality lenses are often more sensitive than the human eye peering through a microscope and the photograph allows several people to examine the same feature of a specimen under great magnification. The opportunity to photographically record actions of the mollusk living in an aquarium should not be overlooked. Very technical books are written on macro- and microphotography for those interested, but unless you have the patience of a saint don't bother to equip yourself for this type of photography.

The photographs in this book were made by Jean

Bowers Gates and the author. Mrs. Gates used an Ex-acta VX II B single lens reflex camera with an F/2 Zeiss panacolor 50 mm. and an F/3.5 Leitz summaron 35 mm. lens to produce the individual shell shots. Others were made by the author using a Nikon F. with Nikon Auto 50 mm. F 1.4 lens for the situation shots and adding custom-made lenses for the micromollusks (fig. 38); these were 32 and 16 mm. macrotessar Bausch and Lomb lenses. Both photographers added bellows and mounted their cameras on stationary copy stands except in the field. Kodak Panatomic X film developed with Rodinal and printed on Polycontrast paper was found to give most satisfactory results.

In taking pictures of extremely high magnification the two main problems encountered are vibration and limited depth of field. The first is more easily overcome than the second. High magnification reduces the depth of field to practically zero and with a three-dimensional subject like a shell this can lead to total failure. Lighting can also be difficult and time consuming. For these photographs two side lights were used with reflectors and diffusers when needed; each shell was then individually lighted to bring out its significant characteristics.

Many illustrated references are available for use in identifying your collection (see Bibliography). The Corpus Christi Museum maintains a labeled display of Texas shells and has a study collection of the smaller species that may be viewed on request. The national seashore office has a display and will have more when the facilities are completed. There is a study collection at The University of Texas Marine Science Institute at Port Aransas and another at the Texas Parks and Wildlife Marine Laboratory in Rockport. Special arrangements must be made in advance to use both these collections. The Brazosport Museum of Natural Science in Lake Jackson also has a beautiful collection. The Houston Museum of Natural History has recently added an excellent library on mollusks. Some museums will identify specimens if you are unable to do so after serious effort. First ask permission of the museum; then send only specimens for which you have exact locality data. If possible, include the animal with the shell to facilitate identification. Wrap the shell in cotton saturated in 30 percent denatured alcohol and place in a vial. Sometimes a living specimen will arrive in good condition if wrapped in damp seaweed and layers of loosely wadded newspaper in a cardboard box.

Members of the various Texas shell clubs are also willing to assist the beginner. Contact any of the fol-

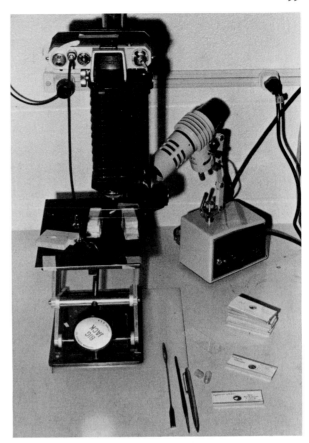

Fig. 38. One of the cameras used to make the photographs of minute shells for this book: a Nikon F. with bellows and special lenses mounted on a built-in copy stand.

lowing: Coastal Bend Shell Club, Corpus Christi Museum, Corpus Christi, Texas 78401; Conchology Group of the Outdoor Nature Club, 130 Hickory Ridge, Houston, Texas 77024; Galveston Shell Club, 1417 Market Street, Galveston, Texas 77550; Gulf Coast Shell Club, 1760 Wexford Avenue, Vidor, Texas 77662; San Antonio Shell Club, 102 East Hermosa, San Antonio, Texas 78212; and South Padre Island Shell Club, P.O. Box 2110, South Padre Island, Texas 78478.

By exchanging specimens a collection may be expanded and new friends made. A list of the many people interested in exchanging is published by Richard E. Petit, P.O. Box 133, Ocean Drive Beach, South Carolina 29582. The beginner would do well to "feel out" his prospective exchanger, since many advanced collectors become very choosy. Great care must be given to packaging the material to be shipped. Nothing is more disappointing than to anxiously await an exchange from a distant land only to receive a box full of chipped and broken shells. Wrap the shell in tissues, then in cotton, and pack with loosely wadded newspapers in a stout box.

The earliest recorded shell collection was found in the ruins of Pompeii, which was buried by lava and volcanic ash in A.D. 79. Today the United States National Museum, under the Smithsonian Institution in Washington, D.C. 20560, contains the largest mollusk collection in the world. Others of note to Gulf collectors are Department of Mollusks, Academy of Natural Sciences of Philadelphia, Nineteenth and the Parkway, Philadelphia, Pennsylvania 19103; Department of Mollusks, Museum of Comparative Zoology, Harvard University, Cambridge, Massachusetts 02138; Department of Living Invertebrates, American Museum of Natural History, Central Park West at Seventy-ninth Street, New York, New York 10024; Beal-Malthie Museum, Rollins College, Winter Park, Florida 32789; Division of Mollusks, Museum of Zoology, University of Michigan, Ann Arbor, Michigan 48104; Lowe Invertebrates, Department of Zoology, Field Museum of Natural History, Roosevelt Road and Lake Shore Drive, Chicago, Illinois 60605; and Department of Mollusks, Carnegie Museum, Pittsburgh, Pennsylvania 15213. The splendid collections on the West Coast are of less interest to the collector in this area.

EATING YOUR CATCH

The oyster is the only mollusk regularly found along the Gulf in quantities sufficient to eat but others can sometimes be found, especially after a norther. These shells that wash up on the beach in the winter are usually alive and edible. Do not risk cooking those that are dead when found. Check to see that they are living by attempting to pry the valves apart; if the muscles contract, they are alive and therefore safe to eat. In the case of the snails, punch them with your finger; if they draw up into the shell you can eat them.

Oysters may be eaten raw with lemon juice or a cocktail sauce or fried to a golden brown, but have you tried an oyster stew or better yet what I call "panned oysters"? Both are simple to prepare.

Oyster Stew

6 oysters per serving, well drained
1 cup of half-and-half cream per serving
1 teaspoon of butter per serving
Salt and freshly ground pepper to taste

Melt the butter in a heavy stew pan but do not brown. Add a few carefully drained oysters at a time and cook only until the edges curl. Stir in the cream (milk if weight conscious) and simmer until the mixture is thoroughly heated but be careful not to boil. Serve steaming hot with buttered French bread and a tossed green salad.

Panned Oysters

At least 6 oysters per person, but better have more
1 stick of butter, not margarine
Freshly ground pepper

In a chafing dish or electric skillet at the table, melt the butter. When ready to eat add about 6 or 8 well drained oysters at a time and sauté until edges curl, turn once. Serve on Ritz-type crackers. Sprinkle with lemon juice if desired. Everyone does his own.

Clam Chowder

Dinocardium, *Tellina*, *Dosinia*, *Mercenaria*, or *Atrina* are sometimes found in enough quantity to make this dish. A mixture of clams may be used.

1 large onion
4 slices bacon, cut and fry together until onion is golden and translucent. Then add
2 cups of diced potatoes with enough water to cover, boil until potatoes are tender. While this is cooking, steam
2 dozen clams, remove from shell, carefully wash out sand, and grind in a food chopper or blender. Strain the broth through a clean cloth and let settle to remove sand. Then combine both mixtures.
1 quart milk, add and heat to a simmer. Never boil. Blend
¼ cup milk and
1 tablespoon flour and stir into chowder to thicken. Salt and pepper to taste. Serve hot with crackers and a green salad. Serves 6.

Stuffed Clams

Any kind of clam or a mixture of clams can be used. Open shells by inserting a thin knife between the valves and severing the adductor muscles. Remove the meat and wash carefully to remove sand. If using *Dinocardium*, cut the foot open and remove visceral material before placing in food chopper.

12 large cockles or a proportionate number of smaller clams, finely chopped. To this add
1 egg, beaten

2 tablespoons tomato catsup
1 tablespoon minced onion
1 tablespoon minced celery
1 cup bread crumbs
Salt and freshly ground pepper to taste

Mix all ingredients together, and stuff 4 of the large cockle shells. Cover with half a slice of bacon and bake at 350 degrees until bacon is crisp. Serves 4.

Coquina Chowder

Wash two quarts of *Donax* or coquina clams thoroughly to remove sand. Steam with two cups of water for 20 minutes. Strain the broth through a fine cloth and let settle. Discard the clams and shells. This chowder can be prepared the same way as the clam chowder or as Euell Gibbons, author of *Stalking the Blue-eyed Scallop*, favors, with wine. Substitute 1 cup of dry white wine for 1 cup of water when steaming the clams.

4 slices of bacon, fried crisp and drained, crumble
1 cup diced boiled potato
½ cup minced white part of green onions

Cook the last two items together in the bacon fat until the potato is a delicate brown and the onion clear and yellow. Drain to remove as much fat as possible.

6 ears of fresh sweet corn, score, scrape on side of mixing bowl to produce corn cream.

Add the corn cream, bacon, potato, and onion to the broth; reserve ½ cup of broth to mix for thickening. Bring to a boil, reduce heat, and simmer for 20 minutes. Heat 2 cups of half-and-half cream. Make a paste with the reserved broth and 4 tablespoons of flour, stir into chowder until thickened. Add the hot cream, stir again. Season to taste. Serve hot to 4.

Seviche

I have eaten this in Mexico made of every combination of seafood imaginable. Anything can be substituted for a fish or shellfish mentioned. Great for cocktail snacks on crackers; let your guests dip it from the bowl themselves.

1½ pounds red fish, snapper, trout, or a mixture
2 cups mixed clams, oysters, shrimp, snails, chopped
3 cups lime or lemon juice (Mexican limes preferred)
2 mild, green *jalapeño* peppers, chopped
1 large fresh tomato, chopped
1 large onion, chopped fine
Fresh cilentro or Chinese parsley if available.

Filet the fish and cut in very narrow thin strips; then chop the shellfish and place both in a large glass or stainless steel bowl. Cover with half the lemon juice and marinate in the refrigerator several hours or overnight. Drain and wash in cold water. Drain again. Return to bowl. Pour in remainder of lemon juice and other ingredients. Season to taste. Serves 8 to 10.

5. Beachcombing

The beachcomber will find every day on a Texas beach rewarding even if the shells are not "coming in" that day. Certain items occur with regularity that might arouse the curiosity of the visitor, even though it often is difficult to ascertain just what the new-found "treasure" is, and the fishermen or bait-stand operators whom you might ask for information are not always reliable in their replies. Much of the information received will be "old wives' tales" and "fishermen's stories"—more fiction than fact. To remove some of the mystery surrounding these objects the most common ones will be listed and identified; often, the truth about them is stranger than the fiction.

Treasure hunting is a popular pastime on these isolated beaches. However, the Antiquities Act of 1906 limits the extent to which treasure hunters can enjoy the fruits of their labors on federally owned property. In other words, the removal of items of historic or prehistoric nature or items of antiquity is not permitted within the Padre Island National Seashore Area, which is federally owned and thus falls under the Antiquities Act. Permission to remove such items as pieces of eight, old ships, Civil War items, or Indian artifacts must come from the Secretary of the Interior. However, materials not of a scientific or historic nature that are washed in by recurring tides may be taken as beachcomber items. Driftwood, shells, bottles, floats, or other contemporary objects may be lugged home.

For those interested in this romantic pastime, a copy of the Antiquities Act is available at the offices of Padre Island National Seashore Area, 10235 South Padre Island Drive, Corpus Christi, Texas 78418.

Common law designates all goods lost at sea as *wreck*. *Jetsam* is goods that are cast into the sea and there sink and remain under water; *flotsam* is goods that continue floating on the surface; *lagan* (ligan) is goods that are deliberately sunk at sea, but tied to a cork or buoy in order to be found at a later date. In the absence of the true owner flotsam, jetsam, and lagan belong to the sovereign. Wreck (which is stranded on the beach) also belongs to the sovereign because it is said that by loss of the ship the owner loses all right to his property.

The federal government and the state of Texas say you can have all the desirable wreck that does not come under the Antiquities Act. Comb to your heart's content. But if you come across a shrimper or other boat run aground you should check on who has salvage rights before you start stripping her; that comes under the heading of felony and penalties can be steep. Happy hunting!

FROM THE ANIMAL KINGDOM

Portuguese Man-of-War.
Physalia physalia (Linné), a hydroid coelenterate, is commonly mistaken for a true jellyfish (*Scyphozoa*) but it is really a colony of animals that sails the sea with its iridescent blue air bubble (pneumatophore) floating on the waves at the mercy of the winds. Stretched out, invisible in the water beneath this colorful gas-filled float, are stinging tentacles up to twenty feet long. The man-of-war is an inhabitant of the Atlantic from the Bay of Fundy to the Hebrides. Strong offshore winds drive these creatures up on the beach in vast numbers from time to time. It is great sport to pop the air bladders with the wheels of a car but the temptation to touch the balloonlike float should be resisted, as the result can be a painful sting.

If contact is made with a man-of-war, wash the affected area with clean sand and water immediately. Morphine is effective in relieving the pain and oral histaminics and topical cream are useful in treating the rash. Dilute ammonium hydroxide or sodium bicarbonate might be helpful. Artificial respiration, cardiac and respiratory stimulants, and other forms of supportive measures may be required if the victim has allergic tendencies. There are no known specific antidotes.

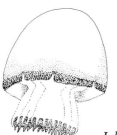

Jellyfish. Several species of these translucent, colorless to whitish, jellylike coelenterates occur in Texas waters. However, they are *Scyphozoa* and no relation to fish. They may be seen by the thousands moving smoothly in the quiet waters along wharves and docks or in a melting lump on the open beach. Give these delicate organisms wide berth in the water, for they have stinging tentacles trailing them for several feet. On the beach, apparently dead, they are quite capable of inflicting a serious sting; however, they are not as painful as the man-of-war, from which they are easily distinguished by the absence of blue color.

Three of the most common varieties to be encountered are the cabbagehead, *Stomolophus meleagris* Agassiz; the moon jelly, *Aurelia aurita* Linné, which has a maroon border design; and the beautiful sea nettle, *Dactylmetra quinquecirrha* Desor. After storms remnants of tentacles may be found in the waters, making swimming very unpleasant. If stung, wash the affected area with sand and water first, then apply ammonia, soda, oil, or alcohol.

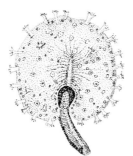

By-the-Wind-Sailor. Velella *mutica* Bosc, is a coelenterate. When it is alive it is deep blue green in color but when dead it is transparent and colorless. This colonial animal is common in the spring when its chitinous, elongate body, about 3–4 inches long with keel-like crest, may be encountered in great numbers on the open beach.

Porpita. Porpita linneana
Lesson, occurs with the by-the-wind-sailor on the open beach after high winds in the spring. When this colonial coelenterate is living the flat disclike body is a bright blue, shading to green with a fringe of short tentacles; but dead, its body becomes a colorless, transparent piece of chitin about the size of a quarter. Its sting is not harmful.

Sea Pansy. Renilla mulleri
Gunter is a fleshy, stemmed, rose-colored, pansy-

shaped, flat animal about 1.5–2 inches in diameter. The stinging cells of this alcyonarian coelenterate are not powerful enough to affect man. The phosphorescent sea pansy makes its home just offshore on the sandy bottom. Children love to put one in a glass of sea water in a dark room and stir it to produce a glow. The sea pansy will live in a salt-water aquarium or may be preserved by drying it slowly in a warm oven.

Sea Whip Coral. Eugorgia setacea Pallas, a coelenterate, is a soft coral that looks like long yellow or red cords. Large, tangled rolls of this animal are found on the open beach where it lives offshore on the sandy bottom. The dried cores of these colorful strings look like wire. Interesting objects often become entangled in a mass of sea whip, rewarding the person who takes time to stop and shake a pile of it. The sting does not bother man. Some people braid the wiry cores into bracelets or hatbands.

The shorter, more treelike variety found living on the jetties is *Eugorgia virgulata* Lamarck.

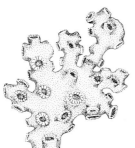

Stony Coral. Astrangia cf. *A. astreiformis* Milne-Edwards & Haime is one of the few calcareous corals found along the Texas coast. This solitary coelenterate of the order Madreporaria can be found growing on the jetties. Occasional pieces of this strange animal are found in beach drift.

Starfish. Three relatively small species of starfish occur on the Texas

coast. *Luidia clathrata* Say, *Luidia alternata* Say, and *Astropecten duplicatus* Gray are echinoderms. They may be prepared for the collection by slowly drying them on a cookie sheet in a warm oven.

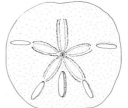

Sand Dollar. Mellita quinquiesperforata Leske (the keyhole urchin), an echinoderm, is a favorite among beachcombers. The chalky, brittle disc has five perforations in its flat body. When living, the pliable, rose-colored animal may be found in great numbers along the third trough just off the barrier islands. Post cards bearing the legend of the sand dollar are in many shops for those interested in the tale.

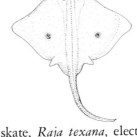

Sting Ray and Skate. The skate, *Raja texana,* electric ray, *Narcine brasiliensis,* stingaree, *Dosyatis sabina,* and the cownose ray, *Rhinoptera bonasus,* are bat-shaped fish with long barbed tails that hide just under the sand in the bays and open Gulf waters where they feed on mollusks. These fish may inflict a painful wound in the ankle or foot of the unwary wader. To avoid an injury from stepping on a ray, wade by shuffling your feet along the bottom. The electric ray will shock you if you step on it with bare feet.

Worm Tube. Diopatra cuprea Claparede is a worm that builds a curious tube covered with bits and pieces of small shell. The animal is seldom found, but the odd piece of tube is often found in the drift along bays or open beach. Examine these for living *Epitonium* and *Vitrinella* shells.

Gooseneck Barnacle. This curious arthropod, *Lepas* several species, has a flattened body, enclosed in a white shell mounted on a more or less slender, fleshy stalk. This animal with its bright orange and blue colorings may be found on wood, floats, and ships that have floated onto the beach from tropical waters. The calcareous plates of this animal are found in the drift, often confusing the shell collector.

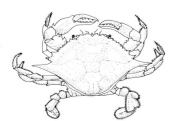

Ghost Crab. Ocypode albicans Bosc is a nocturnal crustacean that feeds on carrion. It can run thirty miles per hour at times. This species is colored like the sand with eyes standing on protruding stalks. When it runs on the beach and suddenly stands still, it seems to disappear; hence, the name "ghost crab." It burrows about two to four feet deep and can live without water for as long as forty-eight hours but must regularly return to the water's edge to moisten its gills.

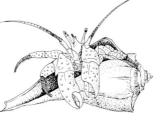

Crabs. Descriptions of the species common to the Texas coast may be found in a bulletin, *Crabs of Texas,* published by the Texas Parks and Wildlife Service. The species include *Arenaeus cribarius* Lamarck, speckled crab; *Callinectes sapidus* Rathbun, blue crab; *Hepatus epheliticus* Linné,

calico crab; *Persephona punctata aquilonaris* Rathbun, purse crab; *Menippe mercenaria* Say, stone crab; and *Libinia emaiginata* Leach, spider crab. Crabs are members of the large invertebrate phylum Arthropoda—animals with jointed legs—and, more specifically, they are the most numerous group of crustaceans representing the order Decapoda, along with the lobster, crayfish, and shrimp. The blue crab is the only one of commercial importance found on the Texas coast.

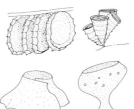

Hermit Crab. Pagurus floridanus Benedict, in the bays, and *Clibinarius vittatus* Bosc are probably the most common species of the Anomuran family Paguiridae occuring here. These funny little animals live in the cast-off shell of various marine gastropods and are often mistaken for the original inhabitant of the borrowed shell they carry on their backs.

Luminescence of the Waters. The phenomenon of luminescence occurs in salt water but not in fresh water. It is produced by many kinds of sea creatures, such as *Noctiluca,* copepods, bacteria, ctenophores, and other living organisms that make up plankton. On a warm, quiet night this phosphorescence produces a glowing surf that can be an unforgettable sight.

Egg Cases. The egg cases of mollusks take many curious forms to puzzle the finder (see fig. 18). Perhaps the most curious are the long strings of disc-shaped, leathery capsules that are all attached to a cartilaginous band up to three feet in length. Miniature left-handed whelks will be found inside these protective cases formed by the mother *Busycon.* The tulip shells will attach their tulip-shaped, bladdery capsules in clusters to old shells or rocks. The *Murex* egg cases look like a wad of dried sponge but examination shows the individual capsules. Collars of sand are produced by the *Polinices*; groups of purple capsules are formed on old cans or rock by the rock shell; other families lay their eggs in jellylike masses or strings. Another egg case of interest is not that of a mollusk but of the skate. It is a shiny, black rectangular purse with pointed corners.

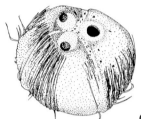

Land and Fresh-Water Snails. These pulmonates are often the most common shells found in the beach drift, especially with small driftwood. Stream and river drainage is the main source of the thousands of land snails commonly found on Texas beaches. Pilsbry and Hubricht (1956) list the following as occurring on the beach:
Anguispira strongylodes (Pfeiffer), *Bulimulus alternatus mariae* (Albers), *Cecilioides acicula* (Müll), *Euconulus chersinus trochulus* (Reinh), *Gastrocopta contracta* (Say), *G. cristata* (Pilsbry & Vanatta), *G. pellucida hordeacella* (Pilsbry), *G. tappaniana* (C. B. Adams), *G. riograndensis* (Pilsbry & Vanatta), *Guppya gundlachi* (Pfeiffer), *Hawaiia minuscula* (Binn), *Helicina chrysocheila* Binney, *H. fragilis elata* Shuttleworth, *H. orbiculata* (Say), *Helicodiscus eigenmanni* Pilsbry, *H. parellelus* (Say), *H. singleyanus* (Pilsbry), *Holospira montivaga* Pilsbry, *H. roemeri* Pfeiffer, *Lamellaxis mexicanus* (Pfeiffer), *Lucidella lirata* (Pfeiffer), *Polygyra ariadnae* (Pfeiffer), *P. auriformis* (Bland), *P. cereolus* (Mühlfeld), *P. cereolus febigeri* (Bland), *P. dorfeuilliana* Lea, *P. dorfeuilliana sampsoni* Wetherby, *P. implicata* (Martens), *P. leporina* (Gould), *P. mooreana* (W. G. Binney), *P. oppilata* (Morelet), *P. rhoadsi* Pilsbry, *P. scintilla* Pilsbry & Hubricht, *P. texasensis* Pilsbry, *P. texasiana* (Moricand), *P. texasiana tamaulipasensis* Lea, *P. tholus* (W. G. Binney), *Praticolella berlandieriana* (Moricand), *P. griseola* (Pfeiffer), *P. pachyloma* ('Menke' Pfeiffer), *Pupoides albilabris* (C. B. Adams), *Retinella indentata paucilirata* (Morelet), *Stenotrema leai aliciae* (Pilsbry), *Strobilops Labyrinthica* (Say), *S. texasiana* (Pilsbry & Ferriss), *Succinea luteola* Gould, *Synopeas beckianum* (Pfeiffer), and *Thysanophora horni* (Gabb).

The most attractive of these snails is the large, glo-

bose apple snail, *Pomacea* sp. from the rivers of Cuba, Mexico, Florida, and Georgia. It can be up to two inches in diameter, and its grayish white surface is spirally striped with purplish brown lines on the body whorl.

FROM THE PLANT WORLD

Coconuts. The fruit of the palm tree, coconuts have floated onto the Texas shores from Mexico and the Indies. They are found with and without the thick husk.

Bamboo. The jointed stems of the bamboo plant that wash ashore from tropical areas are easily recognized and may be utilized in various ways, but the roots are a different matter. The twisted, ribbed roots of the giant bamboo often take the form of weird animals, and the imaginative beachcomber may carry home an entire menagerie to decorate his garden.

Sea Beans. Entada gigas (syn. *E. scardens*) is the largest and most common of the several types of seeds to be found on Gulf shores. This large, red, kidney-shaped bean that grows on a one-hundred-foot-long woody vine principally in Jamaica is found washed up on the beaches from the West Indies to Norway. The seeds occur in pods five feet long. Medicinal properties are attributed to it and to the several smaller species (*Mucuna sloanei*, brown with dark black band; *Guilandia bonducella*, W. Indies and Cartagena) by primitive peoples. The uncooked kernel is toxic to man.

The journal of the explorer Cabeza de Vaca, who lived among the Karankawas from 1528 to 1536, acting as a trader with tribes in the interior, reports that from the coastal region he took sea shells, "conches used for cutting, and fruit like a bean of the highest value among them, which they use as a medicine and employ in their dances and festivities" (Newcomb 1961, p. 70). In return for this mysterious sea bean, he brought back to the coastal tribes skins, red ocher for painting the face and body, canes for arrows, sinew, flint for weapons and tools, "cement" with which to affix flint tools to handles and shafts, and deer hair, dyed and fixed into tassels. I have seen this distinctive bean in the herb sections of Indian markets throughout Mexico where its magical powers are still respected.

Nuts and Seeds. Lofty pecan, black walnut, and other nut-bearing trees line the shady Texas rivers to have their dropped fruits carried to the barren, sunsoaked beaches. The fastidious coon carries peaches and other fruits to the banks of an inland stream in order to wash them for his dinner. Floods carry these and many other strange seeds to the Texas coast.

Among the most unusual looking is a large, wrinkled, black, hard, oval seed about three and one-half inches in length, which is called cabbage bark. This strange seed is probably *Andira inermis* HBK described in Strandley's book on the flora of British Honduras as from a tree that is widespread in tropical America.

Driftwood. Most of the wood found on the beaches north of Big Shell comes from the banks of Texas rivers and may be cottonwood,

pecan, cypress, or walnut. The tropical woods, such as bamboo, mahogany, and cedar, are more common south of Big Shell, although both types are mingled in the two areas. Ocean currents bring the tropical woods from southern Mexico and Central America.

Sargassum. *Sargassum nutans* Meyer is the dominant, large floating seaweed. These floating "weeds" are brown algae that support an entire community of animal life. At times it comes to the shore in enormous quantities, two and three feet deep for miles, obstructing travel along the lower beach.

MISCELLANEOUS WRECK

Pumice. The rounded pieces of this volcanic rock float on the waves to the Texas shores after having been carried to sea by rivers in Africa, Central America, and the region around Tuxla in southern Mexico.

Naval Flares. These flares are often thought to be practice bombs and at times are numerous on the beaches of Padre, Mustang, and St. Joseph islands. The eighteen-inch-long aluminum cylinder is a U.S. Navy Marine Location Marker, MK-Mod.3. The Navy uses it to simulate the outline of a carrier when landing approaches are practiced in the Gulf. During times of intensive training they are found by the hundreds. It is best to look at them and leave them on the beach; they may be harmless, but that would depend on the use to which they were put when you got them home.

Mud Balls. Strange round balls of hard mud are often found along ship channels and near inlets, in large numbers. They are about the size of a baseball and usually covered with broken shell. These balls began as chunks of clay that were eroded by wave action from spoil banks piled up when the channel was dug. Rolling back and forth along the bottom has formed them into neat balls.

Floats. The highly prized glass floats along with the plastic, cork, metal, styrofoam, and wooden ones come ashore after breaking away from nets and lines at sea. Contrary to popular opinion the glass floats are not the invention of the Japanese, although most of those in use today are made in Japanese factories. Christopher Faye, a Norwegian, invented them more than 125 years ago to replace the clumsy wooden ones used on cod nets. Today they are used on tuna long lines and for other fishing operations in the Canaries, Puerto Rico, Portugal, North Africa, and North, Central, and South America, being carried to the Texas coast by the North Equatorial Current through the Yucatán Channel to finally become a prized possession of a lucky beachcomber. These hollow, glass balls vary in size dependent on the use to which they are put. The more bouyant larger floats are used on tuna long lines that may stretch for as far as fifty miles in the open sea. Medium-sized ones are pressure resistant, hence they are employed in bottom tangle-net operations. The smaller floats are from gill nets. The colors range from clear, brown, green, and aqua to lavender due to the inexpensive glass used. They are less expensive than the other types of floats but they are not compatible with the mechanized winch and are consequently being replaced by floats of other materials. Some may be rarities, such as those partially filled with water. There is a high percentage of loss in fishing gear annually and it is thought that the average glass float picked up on the beach is easily ten or more years old. The spring currents bring in the greatest number. All types of netting from very fine to very makeshift may be used to wrap the float so that it is more easily attached to the net.

The rarest float on the beach is hardly recognizable as such, for it is the crudely hand-carved piece of wood

used by the poorer fishermen in the West Indies. They range from one foot to four feet in length, at least those I have seen are this size, and may be riddled with teredo holes. A piece of wood about six inches wide and of various lengths is rounded at one end, while the other is whittled so that it has a narrow neck where the rope attaching it to the net is tied.

Bottles. Since the sixteenth century, when shipping first took place along the Gulf shores, bottles have been collecting in the sands and collectors have been collecting bottles. They were highly prized by the Indians as containers, while the broken pieces were used for arrowheads and scrapers. When early colonists denied the trade of bottles to the Indians, beach bottles became even more highly prized by the natives. Many ancient bottles have been discovered in the dunes of the barrier islands, but it is very difficult to determine the age of a glass bottle from the beach.

This difficulty arises for several reasons, the primary one being that age and deterioration do not change the composition of glass; carbon-14 and other such tests are therefore useless. Beach bottles made in faraway lands and tossed overboard by foreign seamen are difficult to date. Foreign bottles were often made by certain families who passed their secrets of glassmaking from one generation to another so that bottles made hundreds of years ago may have been made by the same formulae and methods as those made yesterday. Molds receive little wear and the same ones may be used for years.

Very primitive methods were used to make bottles until about fifty years ago when great technical strides were made during World War I. Prior to that time all bottles had hand-wrapped necks that were added after the bottle had been formed. The bottlemaking machine formed the neck and lip leaving a mold mark up the sides of the bottle and through the lip. In 1900 the metal bottle cap was invented and the lips were redesigned to accomodate caps instead of corks in certain types of bottles. Prior to the repeal of the prohibition law in 1933, no American bottles bore the mark "Federal law prohibits reuse of this bottle."

To produce a clear glass, bottlemakers would add

manganese oxide, which counteracted the natural green color. Glass containing manganese turns a purple color when exposed to the sun, the length of time that it takes to become purple depending on the amount of manganese, the amount of exposure, and the color of the background against which the bottle is placed. When conditions are favorable the change may take place in less than a month. The invention of the bottle-making machine in the early part of this century made it possible to use materials with fewer impurities, with the result that glass made after this date seldom turns the beautiful purple color. Early wine bottles are green or black because that was the cheapest glass available.

The iridescence of bottles comes from excessive alkalis like sodium in the mix. The rust is an accumulation of iron from the soil as a result of long burial. Hand-blown bottles usually have air bubbles in the glass and some have brown specks caused by iron in the sands used in their manufacture. These bottles will often have a pontil, or punty, mark on the bottom where the glass blower's iron rod was attached.

Asphalt and Petroleum Tars. The beachcomber often finds his feet covered with gooey tar and usually blames ocean-going ships or offshore oil wells for the mess he is in. They are not the culprits. The Karankawa Indians found this same asphalt and used it to waterproof and decorate pottery. The survivors of De Soto's expedition left us the earliest reference to asphalt on the beaches. Oil slicks in the Gulf of Mexico have been re-ported long before ships began using oil for fuel. It is said that pieces of tar weighing several hundred pounds have been found and I have seen hunks three feet in diameter, but most pieces are small. (The area of Padre Island near the wreckage of the *Nicaragua* is the site of the greatest stranding of tars on the Texas coast.) Hildebrand and Gunter (1955) suggest that these deposits float to our coast from natural asphalt seeps from the Tampico-Tuxpan and the Isthmus of Tehauntepec areas of Mexico. Kerosene or any petroleum solvent will remove the accumulation from your feet and your treasures.

Plastic Pellets. The beach drift often yields tiny polyethylene pellets ranging from 3.5 to 1 mm. in diameter, with colors from dull white to amber. These pellets come from the chemical plants located at Port Arthur, Orange, Houston, Baytown, Texas City, Corpus Christi, Seadrift, and Freeport and are carried by winds and currents along the coast of Texas and into Mexico. Polyethylene, according to the Dow Company, is produced by the polymerization of ethylene, usually at a high temperature and pressure. The hot polyethylene is passed through an extruder where it is cut into pellets and bagged for sale. The purchaser finds it easy to remelt the pellets and mold a finished product. Some of the plants are installing skimming devices to prevent pellets from leaving the plant site; however, they are not known to cause an aquatic loss-of-life problem at this time.

GLOSSARY

ABAPICAL: away from shell apex toward base along axis or slightly oblique to it.

ABERRANT: deviating from the usual type of its group; abnormal, straying, different.

ACCESSORY: aiding the principal design; contributory, supplemental, additional.

ACUMINATE: sharply pointed.

ACUTE: sharp; a spire with an angle of less than ninety degrees.

ADAPICAL: toward shell apex along axis or slightly oblique to it.

ADDUCTOR MUSCLES: muscles that hold the valves of a bivalve together.

ADPRESSED: overlapping whorls with their outer surfaces very gradually converging; preferred to the term *appressed*.

ADVENTITIOUS: not inherent; accidental, casual.

ANAL: pertaining to or near the anus, or posterior opening, of the alimentary canal.

ANGULATE: having angles, sharp corners, or an edge where two surfaces meet at an angle.

ANNULATED: marked with rings.

ANOMPHALOUS: lacking umbilicus.

ANTERIOR: the forward end of a bivalve shell.

APERTURE: an opening; the "mouth" of the snail.

APEX: the tip of the spire of a snail shell.

APOPHYSIS: a bony protuberance; a fingerlike structure; spoonshaped, calcareous.

ARCHETYPE: ancestral type established hypothetically by eliminating specialized characters of known later forms.

ARCUATE: curved, as a bow, or arched, as the ventral edge in some pelecypods.

AURICULATE: having ear-shaped projections.

AURIFORM: shaped like a human ear.

AXIAL: in the same direction as the axis; from apex to the base of a snail shell. Axial ribs are those parallel to the edge of the the outer lip. *Transverse* is a preferred term.

BACKSHORE: the part of the shore lying between foreshore and the coastline; covered with water only by exceptional storms or tides.

BASE: in snails, the extremity opposite the apex, the bottom of. In clams, the part of the margin opposite the beaks.

BEAK: the earliest-formed part of a bivalve shell; the tip near the hinge; the umbo.

BENTHOS: the whole assemblage of plants or animals living in or upon the sea bottom.

BICONICAL: similar in form to a double cone; the spire about the same shape and size as the body whorl.

BIFID: divided into two branches, arms, or prongs, or into two equal parts by a cleft; separated down the middle by a slit; divided by a groove into two parts.

BIVALVE: a shell with two valves; pelecypod, such as oyster, scallop, or mussel.

BLADE: the broad flat portion of the pallet of all teredinids; the blade and the stalk are the two parts of the pallet.

BODY WHORL: the last and usually the largest turn in a snail's shell.

BUCCAL: pertaining to the mouth or cheek.

BYSSAL GAPE: an opening or a gaping on the ventral margin of bivalves for the passage of the byssus.

BYSSUS: a series or clump of threadlike fibers that serve to anchor the bivalve to some support.

CALCAREOUS: shelly; of hard calcium carbonate.

CALLUM: a sheet of shelly material filling in the anterior gape in adult shells.

CALLUS: a deposit of calcareous, or enamel-like, material, mostly around the aperture in snails.

CANAL: a tubular prolongation; e.g., in snails, the siphonal canal is at the base of the shell and contains the siphon.

CANCELLATE: sculpture lines intersecting at right angles.

CARDINAL TEETH: the main, or largest, teeth in a bivalve hinge located just below the beaks or umbones.

CARINATE: with a keel-like, elevated ridge, or carina.

CARNIVOROUS: feeding on living animal matter.

CERATA (singular CERAS): external outgrowths, presumably respiratory in nature, along each side of the dorsal surface.

CHITIN: a hard colorless compound that forms the main substance of the hard covering of insects and crustaceans.

CHONDROPHORE: a pit, or spoonlike shelf, in the hinge of a bivalve, such as *Mactra*, into which fits a chitinous cushion, or resilium.

CILIA: hairlike processes on the surface of a cell or organ, shorter and more numerous than flagella.

CIRCUMTROPICAL: throughout the tropics.

COLONIAL: a kind of animal that is organized into associations (colonies) of incompletely separated individuals; e.g., *Physalia*, sponges, and corals.

COLUMELLA: a pillar or column around which the whorls of a snail form their spiral circuit.

COMMENSAL: two or more different species found living with one another, but not as parasite and host; e.g., an anemone attached to an olive shell.

COMPRESSED: flattened or "squashed."

CONCAVE: hollow or dished, as opposed to convex.

CONCENTRIC: as in circles or lines of sculpture, one within the other.

CONCHOLOGY: the branch of zoology that embraces the arrangement and description of mollusks based upon a study of the hard parts.

CONFLUENT: flowing together as to form one.

CONIC: cone shaped, conical, peaked.

CONSPECIFIC: of or pertaining to the same species; members of the same species.

CONSTRICT: draw tight or compress at some point; bind, cramp.

CONVEX: curving outward like a segment of a circle.

COPPICE DUNE: a mound formed by wind in conflict with bunch vegetation.

CORD: coarse, rounded spiral or transverse linear sculpture on the shell surface; smaller than costae.

CORNEOUS: consisting of horn, of a hornlike texture, as the opercula of some mollusks, such as *Busycon*.

CORONATE: encircled by a row of spines or prominent nodes, especially at the shoulder of the last whorl in snails.

COSTAE: ribs.

COSTATE: having ribs.

CRENULATE: notched or scalloped around the margin.

CUSP: a prominence, or point, especially on the crown of a tooth; denticle.

CYLINDRICAL: round, like a cylinder with parallel sides; having the form of a cylinder.

DECIDUOUS: having the tendency to fall off early or before maturity, as the periostracum of most *Cymatiums*.

DECK: a septum, or small sheet of shelly substance in the umbonal region connecting the anterior and posterior ends of a valve; also used to describe the diaphragm of *Crepidula*.

DENTICLES: small projections resembling teeth, around the margin of the gastropod aperture or the pelecypod valve.

DENTICULATED: toothed.

DETRITUS: disintegrated material.

DEXTRAL: turning from left to right; right-handed.

DIAMETER: the greatest width of the shell at right angles to the shell axis.

DIMORPHISM: occurrence of two distinct morphological types in a single population; in sexual dimorphism male and female forms are different.

DISC: the space between the umbo and the margin of a bivalve shell.

DISCOID: disc shaped; whorls coiled in one plane.

DORSAL: belonging to the dorsum, or back; the edge of a bivalve in the region of the hinge; the back of a gastropod opposite the aperture.

ECOLOGY. the study of the relationship between organisms and their environment; both animate and inanimate.

ECTOPARASITE: a parasitic animal, such as the snail *Pyramidella*, which infests the outsides of some bivalves, piercing the shell with a buccal stylet and feeding upon the soft parts.

ELONGATE: lengthened; longer than wide.

ENDEMIC: native, not introduced; having the habitat in a certain region or country.

ENDENTULOUS: without teeth.

ENTIRE: smoothly arched, without a reentrant curve, sinus, or crenulation.

EPIDERMIS: skin of the soft part of a mollusk.

EPIFAUNA: animals that normally live exposed, above the substrate surface; may be with or without attachment.

EQUILATERAL: the anterior and posterior end of each valve being of equal size.

EQUIVALVE: the two valves of a bivalve being of the same shape and size.

ERODED: appearing as if eaten or gnawed away.

ESCUTCHEON: a long, somewhat depressed area of the dorsal area just posterior to the beaks of a bivalve.

FASCIOLE: a small band; a distinct band of color; a spiral band formed in gastropods by the successive growth lines on the edges of a canal.

FATHOM: a nautical unit of measure (six feet), used principally for measuring cables and the depth of the ocean by means of a sounding line.

FLAGELLA: whiplike processes.

FLAMMULATIONS: small flame-shaped spots of color.

FOLIACEOUS: leaflike, flattened, projecting like tiles.

FOOT: muscular extension of body used in locomotion.

FORESHORE: the intertidal zone.

FOSSIL: any hardened remains or traces of plant or animal life of some previous geological period, preserved in rock formations in the earth's crust.

FUSIFORM: spindle shaped with a long canal and an equally long spire, tapering from the middle toward each end; applied to univalves.

GAMETE: either male or female mature reproductive cells; an ovum or a sperm.

GAPING: having the valves not meet; leaving a space.

GASTROPOD: a univalve; snail, conch, whelk, etc.

GENUS: a group of species, distinguished from all other groups by certain permanent features called generic characters.

GIBBOSE or GIBBOUS: swollen.

GLABROUS: smooth.

GLAZED: having a shiny surface.

GLOBULAR: globe or sphere shaped, like a ball.

GRANULATED: having a rough surface of grainlike elevations.

GROWTH LINES: lines on the shell surface indicating rest periods during growth, denoting a former position of the outer lip.

HABITAT: the kind of place where an organism normally lives.

HEIGHT: in gastropods, the greatest length parallel to the shell axis through the columella; in pelecypods, the greatest vertical dimension through the beak at right angles to a line bisecting the adductor scars.

HERBIVOROUS: feeding on vegetable matter.

HERMAPHRODITE: having the sexes united in the same individual.

HETEROSTROPHIC: having the protoconch coiled in a reverse, or left-hand, direction.

HINGE: where the valves of a bivalve are joined.

HOLOSTOMATE: having the mouth of the shell rounded or entire, without a canal, notch, or any extension.

HOLOTYPE: the original type; a single specimen upon which a species is based.

HYPOPLAX: an accessory shell piece between the valves ventrally on some burrowing clams.

IMBRICATED: overlapping one another at the margins, shinglelike; to lay or arrange regularly so as to overlap one another.

IMPERFORATE: not perforated or umbilicated; when the spire is quite flat, the umbilicus vanishes entirely; when the whorls are so compactly coiled on an ascending spiral that there is no umbilicus, the shells are termed imperforate. *Anomphalous* is a preferred term.

INCISED: sculptured with one or more sharply cut grooves.

INDIGENOUS: native to the country, originating in a specified place or country.

INDUCTURA: smooth shelly layer secreted by general surface of mantle, commonly extending from inner side of aperture over parietal region, columellar lip, and (in some genera) part or all of shell exterior. Preferred to term often used, *parietal callus.*

INEQUILATERAL: having the anterior and posterior sides of the valves unequal; the umbones nearer one end than the other.

INEQUIVALVE: one valve larger or more convex than the other.

INFAUNA: sessile and mobile animals that spend part or all of their lives buried beneath the substrate.

INFLATED: applied to rotund shells of thin structure; swollen, increased unduly, distended.

INTERCOSTAL: placed or occurring between the ribs.

INTERSPACES: channels between ribs.

INVOLUTE: rolled inward from each side, as in *Cypraea.*

IRIDESCENT: displaying the colors of the rainbow.

KEEL: the longitudinal ridge; a carina, a prominent spiral ridge usually marking a change of slope in the outline of the shell.

LAMELLA: a thin plate or ridge.

LANCEOLATE: long and spearhead shaped.

LATERAL: to the side of the midline of the body.

LENGTH: in gastropods, the distance from the apex to the anterior end of the shell, same as height; in bivalves, the greatest horizontal dimension at right angles to the height.

LENTICULAR: having the shape of a double convex lens.

LIGAMENT: a cartilage that connects the valves.

LIPS: the margins of the aperture of a snail.

LITTORAL: pertaining to the shore.

LUNULE: a heart-shaped area, set off by a difference of sculpture, in front of the beaks.

MACULATED: irregularly spotted.

MALACOLOGY: the study of molluscous, or soft, animals; the branch of zoology that deals with mollusks, the animal within the shell.

MAMILLIFORM: in the form of a breast.

MANTLES: a fleshy layer or cape that secretes the shell of the mollusk.

MEDIAN: middle.

MESOPLAX: an accessory plate in Pholadacea.

METAPLAX: accessory plate behind the umbones of some Pholadacea.

MICROSCOPIC: exceedingly minute, visible only under a microscope; opposed to macroscopic.

MILLIMETER: one-thousandth of a meter, 0.03937 of an inch; 25.4 millimeters are equal to one inch.

MORPHOLOGICAL: the structure or form; the morphological features of a spiral shell are the aperture, body-whorl, columella, outer lip, and spire.

MUCOCILIARY: a method of feeding associated with sedentary habits. Food particles are entangled with mucus in the gill leaflets and stored in a food groove to be eaten later.

MUCUS: a sticky, slimy, watery secretion (adjective: MUCOUS).

NACRE: iridescent layer of shell, sometimes called "mother of pearl."

NAUTILOID: resembling the nautilus in shape.

NEOTYPE: a type of a species collected later, or selected to replace the original type if it is lost or destroyed.

NODOSE: having knoblike projections or nodules.

NUCLEAR WHORLS: the first whorls formed in the apex of a snail during the egg or veliger stages.

NUCLEUS: the initial whorl, the protoconch.

OBLIQUE: to deviate from the perpendicular; slanting, as the aperture of some shells.

OPERCULUM: cover or lid; in snails, a shelly or horny plate attached to the foot and used to close the aperture of the shell.

ORBICULAR: round or circular.

ORIFICE: a small opening into a cavity.

OUTER LIP: the apertural margin of the last part of the body whorl of a snail shell.

OVATE: egg shaped in outline.

PALLET: one of the two lance-shaped plates forming the closing apparatus at the siphonal end of some wood-boring mollusks.

PALLIAL LINE: a groove or channel near the inner margin of a bivalve shell, where the mantle is made fast to the lower part of the shell. When this line is continuous and not marked with a pallial sinus it is said to be simple.

PALLIAL SINUS: a U-shaped indentation in the pallial line produced by the siphon.

PAPILLA: small nipplelike processes.

PARAPODIA: lateral expansions of the food of certain opisthobranchs that can be used for swimming or to envelop the shell.

PARATYPE: a specimen collected at the same place and at the same time as the holotype and used in the description of a species.

PARIETAL WALL: the area on the whorl of a snail near the columella and opposite the outer lip.

PAUCISPIRAL: only slightly spiral, as some of the opercula.

PELAGIC: pertaining to or inhabiting the open sea far from land; animals living at the surface of the water in mid-ocean.

PELECYPOD: a bivalve; clam, mussel, or scallop.

PENULTIMATE WHORL: the next to the last whorl.

PERIOSTRACUM: the outer covering of a shell, composed of a form of sclerotized protein, or conchin.

PERIPHERY: the perimeter of the external surface.

PHYLUM: a primary division of the animal or plant kingdom.

PLANKTON: life floating or drifting in the sea.

PLANORBID: flat, whorls in one plane, discoid.

PLATE: chitinous or calcareous accessory of the complicated pholad shell, somewhat flat in shape and held in place by chitinous folds; e.g., protoplax, mesoplax, metaplax, hypoplax.

PLICATE: folded or with plaits, or plicae.

PLICATION: a fold, especially on the columella of a gastropod.

PLUG: the apical closing formed when the Caecidae discard the hindering juvenile portion of the shell.

PORCELLANEOUS: having the quality of porcelain; hard and shiny.

POSTERIOR: situated away from the anterior part of the shell.

PROBOSCIS: a long flexible snout.

PROPODIUM: the foremost division of the foot of a gastropod used to push aside sediment as the animal crawls.

PROTOCONCH: the embryonic shell of a univalve, frequently different in design, texture, or color from the adult shell; the rudimentary or embryonic shell of a bivalve mollusk is called a prodissoconch.

PROTOPLAX: one of the accessory plates in Pholadacea.

PUSTULOSE: marked with pustules or pimplelike projections.

PYRIFORM: pear shaped.

QUADRATE: squarish or rectangular in general outline.

RADIAL: lines of color or sculpture fanning out from the beaks to the margins of a valve.

RADULA: a rasplike organ, or lingual ribbon, armed with toothlike processes, found in nearly all mollusks except clams.

RECURVED: bent downward.

REFLECTED: bent backward.

RESILIUM: a triangular ligament structure; a tough chitinous pad, residing in a chondrophore, or pit, along the inner hinge margin of a bivalve which causes the shell to spring open when the muscles relax.

RETICULATE: crossed, like a network.

REVOLVING: turning with the whorls, or spirally.

RHINOPHORES: the posterior pair of tentacles on many opisthobranchs, especially nudibranchs.

RIB: a long and narrow ridge, strip; a firm riblike elevation.

ROSTRATE: drawn out, like a bird's beak.

ROSTRUM: a beaklike process or part; usually used in describing the anterior end of bivalves.

RUFOSE: reddish.

RUGOSE: rough or wrinkled.

SCULPTURE: a pattern of raised or depressed markings on the shell's surface.

SEPTUM: a calcareous plate or partition; one of the transverse partitions of a chambered shell; a dividing wall.

SHOULDER: the top, or largest part, of the outline of a whorl.

SINISTRAL: turning counterclockwise; "left-handed."

SINUOUS: undulating, winding and turning in an irregular course.

SINUS: a bend, or embayment, either in growth lines or in the attachment scar of the mantle.

SIPHON: a tubular structure through which water enters or leaves the mantle cavity.

SLOPE: refers to the face of a bivalve shell; e.g., central, anterior, or posterior slope.

SOCKET: a cavity in the hinge of a bivalve to receive the tooth of the opposite valve.

SPECIES: the subdivision of a genus, distinguished from all others of the genus by certain permanent features called specific characters.

SPINES: a pointed process or outgrowth, stiff, sharp, pointed.

SPINOSE: with spines.

SPIRAL: revolving; as lines going in the direction of the turning of the whorls.

SPIRE: the upper whorls, from the apex to the body whorl, but not including the body whorl.

SPOIL BANKS: banks and islands formed as a result of dredging operations.

STRIATED: marked with rows of fine grooves or threads usually microscopic in size.

SUB-: a prefix indicating "somewhat" or "almost"; as subglobular—almost round.

SUBSTRATE: the sea floor.

SUFFUSED: overspreading; to overspread as with color, to cover the surface.

SULCUS: slit or fissure.

SUTURE: a spiral line, or groove, where one whorl touches the other.

SYMBIOTIC: dissimilar organisms living together in a mutually advantageous situation.

SYNONYM: having the same meaning; or being a different name for the same species.

SYNTYPE: one of several specimens of equal rank upon which a species is based.

TAXONOMY: the laws, or principles, of the systematic classification of organisms.

TEETH: in a bivalve, the shelly protuberances on the dorsal margin of a valve that fit into corresponding sockets of the opposite valve.

TELEOCONCH: the entire gastropod shell except the protoconch.

THREAD: a slender linear surface elevation; the silky fibers of the byssus.

TOPOTYPE: a species collected at the same locality where the original type was obtained.

TORSION: a twisting around; twisted spirally, as a gastropod.

TRANSLUCENT: allowing light to pass through, but not transparent.

TRIGONAL: somewhat triangular in shape.

TRUNCATE: having the end cut off squarely.

TUBERCULES: small, raised projections.

TUMID: swollen, enlarged.

TURBINATE: conical with a round base.

TURRICULATE: having the form of a turret; tower shaped; spire whorls regularly stepped in outline forming a long spire with somewhat shouldered whorls.

TYPE: a fundamental structure common to a number of individuals, having the essential characteristics of its group; a specimen or specimens upon which a description of a species is based.

UMBILICUS: a small hole, or depression, in the base of the body whorl of a snail shell.

UMBO (plural UMBONES): the upper, or early, part of the bivalve shell as seen from the outside, opposite the hinge.

UNDULATE: having a wavy surface.

UNGUICULATE: resembling a claw, or talon.

UNIVALVE: a snail shell; gastropod.

VALVE: one of the shelly halves of a clam shell.

VARIX (plural VARICES): a prominent raised rib on the surface of a snail shell, caused by a periodic thickening of the lip during rest periods in the shell's growth.

VELIGER: a mollusk in the larval stage, in which it has a ciliated swimming membrane or membranes; free-swimming young.

VENTRAL: the lower side, opposite the dorsal area.

VENTRICOSE: swollen.

VERMICULATION: surface sculpture of irregular wavy lines.

VITREOUS: glassy.

VIVIPAROUS: producing live young.

WHORL: a complete turn, or volution, of a snail shell.

WIDTH: the maximum dimension measured at right angles to the length or height of the shell.

WING: a more-or-less triangular projection or expansion of the shell of a bivalve, either in the plane of the hinge-line or extending above it.

BIBLIOGRAPHY

Abbott, R. Tucker
    1958a.   *American seashells.* Princeton: D. Van No-
            strand Co., Inc.
    1958b.   The marine mollusks of Grand Cayman Is-
            land. *Monogr. Acad. Natur. Sci. Phila.,* no. 11.
    1963.   The janthinid genus *Recluzia* in the Western
            Atlantic. *Nautilus* 76(4): 151.
    1964.   *Littorina ziczac* (Gmelin) and *L. lineolata*
            d'Orbigny. *Nautilus* 78(2): 65–66.
    1968.   *A guide to field identification of the seashells
            of North America.* New York: Golden Press.
Abbott, R. Tucker, and Ladd, H. S.
    1951.   A new brackish-water gastropod from Texas
            (Amnicolidae; *Littoridina*). *J. Wash. Acad.
            Sci.* 41(10): 335–338.
Akin, R. M., and Humm, H. J.
    1960.   *Macrocallista nimbosa* at Alligator Harbor.
            J. Florida Acad. Sci. 22(4): 226–228.
Allen, J. Frances
    1958.   Feeding habits of two species of *Odostomia.*
            Nautilus 72(1): 11–15.
Amos, Wm. H.
    1966.   *The life of the seashore.* New York: McGraw-
            Hill Book Co.
Anderson, Arvid A.
    1960.   *Marine resources of the Corpus Christi area.*
            Bureau of Business Research, University of
            Texas, Monograph 21.
Arnold, Winifred H.
    1965.   A glossary of a thousand and one terms used
            in conchology. *Veliger,* vol. 7, supplement.
Bartsch, Paul
    1909.   Pyramidellidae of New England and adjacent
            region. *Proc. Boston Soc. Natur. Hist.* 34(4):
            67–117.
    1934.   *Mollusks: Shelled invertebrates of the past
            and present.* Smithsonian Institution Series,
            vol. 10(3).
    1955.   *The pyramidellid mollusks of the Pliocene de-
            posits of north St. Petersburg, Florida.* Smith-
            sonian Miscellaneous Collections, vol. 125(2).

Bartsch, Paul, and Rehder, Harald A.
    1939.   New turritid mollusks from Florida. *Proc.
            U.S. Nat. Mus.* 87(3070): 127–138.
Bedichek, Roy
    1950.   *Karankaway country.* Garden City, N.Y.:
            Doubleday and Co., Inc.
Behrens, E. Wm.
    1967.   Sorting of bivalves—an indication of long-
            shore drift. Paper read at University of Texas
            Institute of Marine Science, Port Aransas.
Berrill, N. J.
    1964.   *The living tide.* Greenwich, Conn.: Fawcett
            Pub., Inc.
Bird, Samuel O.
    1965.   Upper Tertiary Arcacea of the mid-Atlantic
            coastal plain. *Palaeontogr. Amer.* 5(34): 1–62.
    1970.   Shallow-marine and estuarine benthic mollus-
            can communities from area of Beaufort, North
            Carolina. *Bull. Amer. Ass. Pet. Geol.* 54(9):
            1651–1676.
Borradaile, L. A., and Potts, F. A.
    1961.   *The Invertebrata.* 4th ed. London: Cambridge
            Univ. Press.
Boss, Kenneth J.
    1966.   The subfamily Tellininae in the western At-
            lantic: The genus *Tellina* (part I). *Johnsonia*
            4(45): 217–272.
    1968.   The subfamily Tellininae in the western At-
            lantic: The genera *Tellina* (part II) and *Tel-
            lidora. Johnsonia* 4(46): 273–344.
Boss Kenneth J.; Rosewater, Joseph; and Ruhoff, Flor-
            ence S.
    1968.   The zoological taxa of Wm. Healey Dall. *Bull.
            U.S. Nat. Mus.,* no 287.
Bowditch, Nathanial
    1958.   *American practical navigator.* U.S. Navy
            Oceanographic Office. Washington, D.C.:
            Government Printing Office.
Breuer, E. G.
    1957.   An ecological survey of the Baffin and Alacan
            bays, Texas. *Pub. Inst. Marine Sci.* 4: 134–155.

1962.   An ecological survey of the lower Laguna Madre of Texas. *Pub. Inst. Marine Sci.* 8: 153–183.

Brown, Roland Wilbur
1954.   *Composition of scientific words.* Privately printed.

Burch, R. D.
1965.   New terebrid species from the Indo-Pacific Ocean and from the Gulf of Mexico. *Veliger* 7(4): 241–253.

Burkenrod, Martin
1833.   Pteropoda from Louisiana. *Nautilus* 47(2): 54–57.

Bush, Katharine J.
1885.   Additions to the shallow-water Mollusca of Cape Hatteras, N. C. dredged by the U.S. Fish Commission steamer *Albatross* in 1883 and 1884. *Trans. Conn. Acad. Sci.* 6(2).

1897.   Revisions of the marine gastropods referred to the Vitrinellids. *Trans. Conn. Acad. Sci.* 10(1): 97–144.

1899.   Descriptions of new species of *Turbonilla* of the western Atlantic fauna, with notes on those previously known. *Proc. Acad. Natur. Sci. Phila.* 51: 145–177.

Bushbaum, Ralph
1967.   *Animals without backbones.* Chicago: Univ. of Chicago Press.

Cameron, Roderick William
1961.   *Shells.* New York: G. P. Putnam's Sons.

Carr, John T., Jr.
1967.   *Hurricanes affecting the Texas Gulf coast.* Texas Water Development Board Report 49.

Carson, Rachel L.
1950.   *The sea around us.* New York: Oxford Univ. Press.

1955.   *The edge of the sea.* Boston: Houghton Mifflin Co.

Castañeda, Carlos E.
1936.   *Our Catholic heritage in Texas.* Austin: Von Boeckman-Jones Co.

Catlow, Agnes
1845.   *The conchologist's nomenclator.* London: Reeve Bros.

Ciampi, Elgin
1960.   *The skin diver.* New York: Ronald Press.

Clarke, Arthur H., Jr.
1965.   The scallop superspecies *Aequipecten irradians* (Lamarck). *Malacologia* 2(2): 161–188.

Clench, Wm. J., and Pulley, T. E.
1952.   Notes on some marine shells from the Gulf of Mexico with a description of a new species of *Conus. Texas J. Sci.* 4: 59–62.

Clench, Wm. J., and Turner, Ruth D.
1946.   The genus *Bankia* in the western Atlantic. *Johnsonia* 2(19): 1–28.

1950a. The western Atlantic marine mollusks described by C. B. Adams. *Occas. Papers on Mollusks* 1(15): 233–403.

1950b. The genera *Sthenorytis, Cirsotema, Acirsa, Opalia* and *Amaea* in the western Atlantic. *Johnsonia* 2(29): 221–248.

1951.   The genus *Epitonium* in the western Atlantic, part I. *Johnsonia* 2(40): 249–288.

1952.   The genus *Epitonium*, part II, *Johnsonia* 2(41): 289–356.

1962.   *Names introduced by H. A. Pilsbry in the Mollusca and Crustacea.* Academy of Natural Sciences of Philadelphia special publication.

Coker, R. E.
1954.   *This great and wide sea.* Chapel Hill: Univ. of North Carolina Press.

Collier, Albert, and Hedgpeth, Joel W.
1950.   An introduction to the hydrography of tidal waters of Texas. *Publ. Inst. Marine Sci.* 1(2): 121–194.

Cooke, A. H.
1959.   *Molluscs.* Codicote, Engl.: Wheldon and Wesley Ltd.

Coomans, H. E.
1963a. The marine Mollusca of St. Martin, Lesser Antilles, collected by H. J. Krebs. *Stud. Fauna Curaçao & Other Caribbean Islands* 16: 59–87.

1963b. Systematics and distribution of *Siphocypraea mus* and *Propustularia surinamensis. Stud. Fauna Curaçao & Other Caribbean Islands* 15: 51–71.

Copeland, B. J.
1965.   Fauna of the Aransas Pass inlet, Texas: Emigrations as shown by tide trap collection. *Publ. Inst. Marine Sci.* 10: 9–21.

Corgan, James X.
1969.   Review of *Parodostomia* Laseron, 1959, *Telloda* Hertlein and Strong, 1951, and *Eulimastoma* Bartsch, 1916 (Gastropoda: Pyramidellacea). *Nautilus.* In press.

Corpus Christi Caller Times
1959.   75th Anniversary Edition, January 18.

Cox, L. R.
1960.   Gastropoda, general characteristics of Gastropoda. In *Treatise on invertebrate paleontology*, ed. R. C. Moore. Part I, Mollusca 1, pp. 184–269. Boulder, Colo.: Geological Society of America, Inc.

Cox, L. R.; Nuttall, C. P.; and Trueman, E. R.
1969.   General features of Bivalvia. In *Treatise on*

invertebrate paleontology, ed. R. C. Moore. Part N, Mollusca 6, pp. 2–128. Boulder, Colo.: Geological Society of America, Inc.

Cross, E. R.
1956.   Underwater safety. *Healthways*. Los Angeles, Calif.

Dall, Wm. H.
1889*a*.   A preliminary catalog of the shell-bearing marine mollusks and brachiopods of the southwestern coast of the United States. *Bull. U.S. Nat. Mus.*, no. 37.

1889*b*.   Report on the Mollusca. *Bull. Mus. Comp. Zool.*, vol. 18.

1892.   Tertiary fauna of Florida. *Wagner Free Inst. Sci. Phila.* 3(1–6).

1894.   Monograph of the genus *Gnathodon* Gray. *Proc. U.S. Nat. Mus.* 17: 89–106.

1902.   Illustrations & descriptions of new, unfigured, or imperfectly known shells, chiefly American in the U.S. National Museum. *Proc. U.S. Nat. Mus.* 24(1264): 199–566.

Dall, Wm. H., and Bartsch, Paul
1906.   Notes on Japanese, Indo-Pacific, and American Pyramidellidae. *Proc. U.S. Nat. Mus.* 30(1452): 321–326.

Dall, Wm. H., and Simpson, Charles T.
1902.   The Mollusca of Porto Rico. *Bull. U.S. Fish Comm.* 20(1): 351–524.

Dance, S. Peter
1966.   *Shell collecting: An illustrated history.* Berkeley: Univ. of California Press.

Fitch, John E.
1953.   *Common marine bivalves of California.* Fish Bulletin 90. Department of Fish and Game Marine Fisheries Branch, State of California.

Frisch, Karl Von
1965.   *Man and the living world.* New York: Time Inc.

García-Cubas, Antonio, Jr.
1963.   *Sistemática y distribución de los micromoluscos recientes de la Laguna de Términos, Campeche, México.* Instituto de Geología, Universidad Nacional Autónoma de México, Bulletin 67(4).

Gibbons, Euell
1964.   *Stalking the blue-eyed scallop.* New York: David McKay Co., Inc.

1967.   *Beachcomber's handbook.* New York: David McKay Co., Inc.

Gilbert, Samuel A.
1860.   *Report of the superintendent of the coast survey showing the progress of the survey during the year 1860.* Appendix 34. Washington, D.C.: Government Printing Office.

Gilbert, W. S.
1860.   *Report of the superintendent of the coast survey showing the progress of the survey during the year 1860.* Appendix 33. Washington, D.C.: Government Printing Office.

Goode, Brown
1876.   Animal resources of the United States. *Bull. U.S. Nat. Mus.*, no. 6.

Grassé, Pierre P.
1968.   *"Traité de zoologie" mollusques gastéropodes et scaphopodes.* Vol. 5, fasc. 3. Paris: Masson et Cie.

Guest, Wm. C.
1958.   *The Texas shrimp fishery.* Marine Laboratory Bulletin no. 36(5). Austin: Texas Game and Fish Commission.

Gunter, Gordon
1950.   Seasonal population changes of certain invertebrates of the Texas coast, including the commercial shrimp. *Pub. Inst. Marine Sci.* 1(2): 7–51.

1952.   Historical changes in the Mississippi River and adjacent marine environment. *Pub. Inst. Marine Sci.* 2(2): 119–139.

Gunter, Gordon, and Hildebrand, H. H.
1951.   Destruction of fishes and other organisms on the south Texas coast by the cold wave of January 28–February 3, 1951. *Ecology* 32(4): 731–736.

1954.   The relation of total rainfall of the state and catch of the marine shrimp (*Penaeus setiferus*) in Texas waters. *Bull. Marine Sci. Gulf & Caribbean* 4(2): 95–103.

Halstead, Bruce W.
1959.   *Dangerous marine animals.* Cambridge, Md.: Cornell Maritime Press.

Halter, R. E.
1958.   Records of the Office of Civil Assistants and of Superintendents relating to R. E. Halter and his activities in and near Corpus Christi and Padre Island, 1873–1881. (Microfilm). General Services Administration, National Archives and Record Service, Record Group 23.

Harry, Harold W.
1966.   Studies on bivalve molluscs of the genus *Crassinella*, in the northwestern Gulf of Mexico: Anatomy, ecology, and systematics. *Pub. Inst. Marine Sci.* 11: 65–69.

1967.   A review of the living tectibranch snails of the genus *Volvulella*, with descriptions of a new subgenus and species from Texas. *Veliger* 10 (2): 133–147.

1969a.   A review of the living leptonacean bivalves of the genus *Aligena*. *Veliger* 11(3):164–181.

1969b.   Anatomical notes on the mactrid bivalve *Raeta plicatella* Lamarck, 1818 with a review of the genus *Raeta* and related genera. *Veliger* 12(1): 1–23.

1969c.   Cuttlebones on the beach at Galveston. *Veliger* 12(1): 89–94.

Hedgpeth, Joel
1950.   Marine invertebrate fauna of salt flat areas in Aransas National Wildlife Refuge, Texas. *Pub. Inst. Marine Sci.* 1(2): 103–120.

1953.   An introduction to zoogeography of the northwestern Gulf of Mexico with reference to the invertebrate fauna. *Pub. Inst. Marine Sci.* 3(1): 111–211.

Hedley, Charles
1899.   The Mollusca of Funafuti. Part 1. Gastropoda. *Mem. Aust. Mus.* 3(7): 398–488.

Henry, W. S.
1847.   *Campaign sketches of the war with Mexico.* New York: Harper and Bros.

Hertlein, L. G.
1951.   Descriptions of a new pelecypod of the genus *Anadara* from the Gulf of Mexico. *Texas J. Sci.* 3(3): 487–490.

Hildebrand, Henry H.
1954.   Fauna of the brown shrimp grounds in the western Gulf. *Pub. Inst. Marine Sci.* 3(2): 233–366.

1955.   A study of the fauna of the pink shrimp (*Penaeus duorarum* Burkenroad) grounds in the Gulf of Campeche. *Pub. Inst. Marine Sci.* 4(1): 171–231.

Hildebrand, Henry H., and Gunter, Gordon
1955.   A report on the deposition of petroleum tars and asphalts on the beaches of the northern Gulf of Mexico, with notes on the beach conditions and the associated biota. Report at Institute of Marine Science, University of Texas.

Hofstetter, Robert P.
1959.   *The Texas oyster fishery.* Texas Game and Fish Commission Bulletin no. 40, Series no. 6.

Hollister, S. C.
1957.   On the status of *Fasciolaria distans* Lamarck. *Nautilus* 70(3): 73–84.

1958.   Illustrated contributions to the paleontology of America: A review of the genus *Busycon* and its allies. *Palaeontogr. Amer.* 4(28).

Holmes, N. A.
1964.   Methods of sampling the benthos. In *Advances in marine biology*, 2:171–260. London: Academic Press.

How to collect shells.

1955.   Publication of the American Malacological Union. Buffalo: Buffalo Museum of Science.

Hubricht, Leslie
1960.   Beach drift land snails from southern Texas. *Nautilus* 74(2): 82–83.

Hulings, Neil C.
1955.   An investigation of the benthic invertebrate fauna from the shallow waters of the Texas coast. Masters thesis, Texas Christian University, Fort Worth.

Hunt, Chas. B.
1959.   Dating of mining camps with tin cans and bottles. *Geotimes* 3(8): 8–9.

Hyman, Libbie H.
1967.   *The invertebrates: Volume VI, Mollusca I.* New York: McGraw-Hill Book Co.

Jaeger, Edmund C.
1960.   *The biologist's handbook of pronunciations.* Springfield, Ill.: Charles C. Thomas.

Johnson, Charles W.
1934.   List of marine Mollusca of the Atlantic coast from Labrador to Texas. *Proc. Boston Soc. Natur. Hist.* 40(1):1–104.

Johnson, Richard I.
1961.   The recent Mollusca of Augustus Addison Gould. *Bull. U.S. Nat. Mus.,* no 239.

Jones Fred B.; Rowell, Chester M.; and Johnston, Marshall C.
1961.   *Flowering plants and ferns of the Texas coastal bend counties.* Sinton, Tex.: Rob and Bessie Welder Wildlife Foundation.

Jordan, David Starr
1905.   *Guide to the study of fishes.* New York: Columbia Univ. Press.

Kauffman, Erle G.
1969.   Form, function, and evolution. In *Treatise on invertebrate paleontology,* ed. R. C. Moore. Part N, Mollusca 6, pp. 129–203. Boulder, Colo.: Geological Society of America, Inc.

Keen, A. Myra
1960.   *Sea shells of tropical West America.* Palo Alto: Stanford University Press.

1961.   What is *Anatina anatina? Veliger.* 4(1): 9–12.

1963.   *Marine molluscan genera of western North America.* Palo Alto: Stanford University Press.

Keith, Don, and Hulings, Neil
1965.   A quantitative study of selected nearshore infauna. *Pub. Inst. Marine Sci.* 10: 22–40.

Kennedy, E. A., Jr.
1959.   A comparison of the molluscan fauna along a transect extending from the shoreline to a point near the edge of the continental shelf of the Texas coast. Masters thesis, Texas Chris-

tian University, Fort Worth. (Excellent plates.)

Kidd, D. A.
1964. *Collins Latin gem dictionary*. London: Collins.

Korniker, Louis; Oppenheimer, Carl; and Conover, John T.
1958. Artificially formed mud balls. *Pub. Inst. Marine Sci.* 5: 148–150.

Kuykendall, J. H.
1903. Reminiscences of early Texans. *Quart. Texas Hist. Ass.* 6: 236–253.

Ladd, H. S.
1951. Brackish water and marine assemblages of the Texas coast. *Pub. Inst. Marine Sci.* 2(2): 125–163.

Lane, Frank W.
1960. *Kingdom of the octopus*. New York: Sheridan House.

Laursen, Dan
1953. *The genus "Ianthina": A monograph*. Carlsberg Foundation's "Dana" Report no. 38, pp. 5–44, Copenhagen, Denmark.

Lee, Anderson Smith
1958. Ecological catalog and bibliography of the invertebrate fauna and environments of the Gulf with special reference to the molluscs, vol. 1. Masters thesis, Texas Christian University, Fort Worth.

Leipper, Dale F.
1954. Physical oceanography of the Gulf of Mexico. In *The Gulf of Mexico*. Fishery Bulletin of the U.S. Fish and Wildlife Service 55(89): 119–137.

Loesch, Harold
1957. Studies on the ecology of two species of *Donax* on Mustang Island, Texas. *Pub. Inst. Marine Sci.* 4(2): 201–227.

McKenna, Verna Jackson
1956. *Old Point Isabel lighthouse: Beacon of Brazos Santiago*. Harlingen, Tex. Privately printed.

McLean, Richard A.
1951. The Pelecypoda or bivalve mollusks of Puerto Rico and the Virgin Islands. *Scientific survey of Puerto Rico and the Virgin Islands*. New York: New York Academy of Sciences.

MacGintie, G. C., and MacGintie, Nettie
1949. *Natural history of marine animals*. New York: McGraw-Hill Book Co., Inc.

Marland, Frederick Charles
1958. An ecological study of the benthic macrofauna of Matagorda Bay, Texas. Masters thesis in oceanography, Texas A.&M. University, Bryan.

Marmer, H. A.
1954. The tides and sea level in the Gulf of Mexico. In *The Gulf of Mexico*. Fishery Bulletin of the U.S. Fish and Wildlife Service 55(89): 101–114.

Maury, Carlotta J.
1920. Recent mollusks of the Gulf of Mexico and Pleistocene and Pliocene species from the Gulf states, Part 1, *Pelecypoda*. *Bull. Amer. Paleontol.* 8(34): 3–115.
1922. Ibid., Part 2, *Scaphopoda, Gastropoda, Amphineura and Cephalopoda*. 9(36): 1–140.

Mayr, Ernst
1942. *Systematics on origin of species*. New York: Columbia University Press.
1969. *Principles of systematic zoology*. New York: McGraw-Hill Book Co.

Melvin, R. Gordon
1966. *Sea shells of the world*. Rutland, Vt.: Charles E. Tuttle Co.

Menzel, R. W.
1955. Some phases of the biology of *Ostrea equestris* Say and a comparison with *Crassostrea virginica*. *Pub. Inst. Marine Sci.* 4(1): 68–153.

Merrill, Arthur S., and Boss, Kenneth J.
1964. Reactions of hosts to proboscis penetration by *Odostomia seminuda* (Pyramidellidae). *Nautilus* 78(2): 42–45.

Miner, Roy Waldo
1950. *Field book of seashore life*. New York: G. P. Putman's Sons.

Mitchell, J. D.
1894. *A list of Texas Mollusca collected by J. D. Mitchell, Victoria, Texas*. Victoria: Times Steam Print.

Moore, Donald R.
1958. Additions to Texas marine Mollusca. *Nautilus* 71(4): 124–128.
1961. The marine and brackish water Mollusca of the state of Mississippi. *Gulf Coast Res. Lab. Bull.* 1(1).
1962. The systematic position of the family Caecidae. *Bull. Marine Sci. Gulf & Caribbean* 12: 695–701.
1964. The Vitrinellidae. Dissertation for Doctor of Philosophy in biology, University of Miami.

Morris, Percy A.
1958. *A field guide to the shells*. Boston: Houghton Mifflin Co.

Morrison, Joseph P. E.
1939. Two new species of *Sayella* with notes on the genus. *Nautilus* 53: 127–130.
1965. New brackish water mollusks from Louisiana. *Proc. Biol. Soc. Wash.* 78: 217–224.

Morton, J. E.
    1960.   *Molluscs.* New York: Harper & Brothers.
Newcomb, W. W., Jr.
    1961.   *The Indians of Texas.* Austin: Univ. of Texas Press.
Nicol, David
    1952.   Revision of the Pelecypod *Echinochama. J. Paleontol.* 26(5): 803–817.
    1953.   The scientific role of the amateur malacologist. *Nautilus* 67(2): 41–44.
Olsson, Axel A., and Harbison, Anne
    1953.   *Pliocene Mollusca of southern Florida, with special reference to those from St. Petersburg.* The Academy of Natural Sciences of Philadelphia, Monograph no. 8.
Palmer, Katherine Van Winkle
    1966.   Who were the Sowerbys? *Sterkiana,* no. 23, pp. 1–6.
Papp, Charles S.
    1968.   *Scientific illustration, theory and practice.* Dubuque: Wm. C. Grown, Co.
Parker, Robert H.
    1955.   Changes in the invertebrate fauna, apparently attributable to salinity changes, in the bays of central Texas. *J. Paleontol.* 29: 193–211.
    1956.   Macro-invertebrate assemblages as indicators of sedimentary environments in east Mississippi delta region. *Bull. Amer. Ass. Pet. Geol.* 40(2): 295–376.
    1959.   Macro-invertebrate assemblages of central Texas, coastal bays and Laguna Madre. *Bull. Amer. Ass. Pet. Geol.* 43(9): 2100–2166.
    1960.   Ecology and distributional patterns of marine macro-invertebrates, northern Gulf of Mexico. In *Recent sediments, northwest Gulf of Mexico,* pp. 302–381. Tulsa: American Association of Petroleum Geologists.
    1964.   Zoogeography and ecology of macro-invertebrates in the Gulf of California and continental slope off Mexico. *Vidensk. Medd. dansk. naturh. Foren.* 126: 1–178.
Parker, Robert H., and Curray, Joseph R.
    1956.   Fauna and bathymetry of banks on continental shelf, northwest Gulf of Mexico. *Bull. Amer. Ass. Pet. Geol.* 40(10):2428–2439.
Patch, Joseph Dorst
    1962.   *The concentration of general Zachary Taylor's army at Corpus Christi, Texas.* Corpus Christi: Privately printed.
Perry, Louise M., and Schwengel, Jeanne S.
    1955.   *Marine shells of the western coast of Florida.* Ithaca: Paleontological Research Inst.

Peterson, Roger Tory
    1960.   *Field guide to the birds of Texas and adjacent states.* Boston: Houghton Mifflin Co.
Pew, Patricia, and Staff of Marine Laboratory.
    1954.   *Food and game fishes of the Texas coast.* Texas Fish and Game Commission, Bulletin 33(4).
Pierce, Wendell E.
    1967.   Flower garden reef. *Texas Parks & Wildlife* 25(12): 6–9.
Pilsbry, Henry A., and Hubricht, Leslie
    1956.   Beach drift Polygyridae from southern Texas. *Nautilus* 69(3): 93–97.
Poirier, Henry Pierre
    1954.   *An up-to-date systematic list of 3200 seashells from Greenland to Texas.* Hudson View Gardens, New York: Villedieu Co.
Pounds, Sandra Gail
    1961.   *The crabs of Texas.* Texas Fish and Game Commission, Bulletin 43(7).
Powell, A. W. B.
    1961.   *Shells of New Zealand.* Christchurch, N. Z.: Whitcombe and Tombs Ltd.
Price, W. Armstrong
    1954.   Shorelines and coasts of the Gulf of Mexico. In *The Gulf of Mexico.* Fishery Bulletin of the U.S. Fish and Wildlife Service 55(89): 39–98.
Puffer, Elton L., and Emerson, William K.
    1954.   Catalogue and notes on gastropod genus *Busycon. Proc. Biol. Soc. Wash.* 67: 115–150.
Pulley, Thomas E.
    1952.   An illustrated check list of the marine mollusks of Texas. *Texas J. Sci.* 4(2): 167–199.
    1953.   A zoogeographic study based on the bivalves of the Gulf of Mexico. Dissertation for Doctor of Philosophy in biology, Harvard University.
Rayburn, John C., and Rayburn, Virginia Kemp
    1966.   *Century of Conflict: 1821–1913.* Waco: Texian Press.
Reese, Pauline
    1938.   The history of Padre Island. Masters thesis, Texas College of Arts and Industries, Kingsville.
Reed, Clyde T.
    1941.   *Marine life in Texas waters.* Texas Academy Publications in Natural History. Houston: The Anson Jones Press.
Rehder, Harald A.
    1935.   New Caribbean marine shells. *Nautilus* 48(4): 127–130.
    1943.   New marine mollusks from the Antillean region. *Proc. U.S. Nat. Mus.* 13(3161): 187–203.

1954. Mollusks. In *The Gulf of Mexico*. Fishery Bulletin of the U.S. Fish and Wildlife Service 55(89): 469–478.

Reinhart, Philip W.
1943. *Mesozoic and Cenozoic Arcidae from the Pacific slope of North America*. Geological Society of America Special Papers, no. 47.

Rice, Winnie H.
1960. *A preliminary checklist of the mollusks of Texas*. Port Aransas: Institute of Marine Science.

Rice, Winnie H., and Kornicker, Louis S.
1962. Mollusks of Alacaran Reef, Campeche Bank, Mexico. *Pub. Inst. Marine Sci.* 8: 366–393.
1965. Mollusks from the deeper waters of the northwestern Campeche Bank, Mexico. *Pub. Inst. Marine Sci.* 10: 108–172.

Ricketts, Edward F., and Calvin, Jack.
1960. *Between Pacific tides*. Rev. Joel W. Hedgpeth. Palo Alto: Stanford Univ. Press.

Robertson, Robert
1957. Gastropod host of an *Odostomia*. *Nautilus* 70(3): 96–97.
1963. Bathymetric and geographic distribution of *Panopea bitruncata*. *Nautilus* 76(3): 75–82.

Robertson, Robert, and Orr, Virginia
1961. Review of Pyramidellid hosts, with notes on an *Odostomia* parasitic on a chiton. *Nautilus* 74(3): 85–87.

Robson, G. C.
1929. *A monograph of the recent Cephalopods*. British Museum of Natural History. Oxford: Oxford Univ. Press.

Roemer, Ferdinand
1935. *Texas*. San Antonio: Standard Printing Co.

Rothschild, Nathaniel Mayer
1961. *A classification of living animals*. New York: John Wiley and Sons, Inc.

Rudloe, Jack
1968. *The sea brings forth*. New York: Alfred A. Knopf, Inc.

Scheltema, Amelie H.
1965. Two gastropod hosts of the pyramidellid gastropod *Odostomia bisuturalis*. *Nautilus* 79(1): 7–9.
1968. Redescriptions of *Anachis avara* (Say) and *Anachis translirata* (Ravenel) with notes on some related species. *Breviora*, no. 304, pp. 1–19.

Schenck, Edward T., and McMasters, John H.
1935. *Procedure in taxonomy*. Palo Alto: Stanford Univ. Press.

Schilder, Franz A.
1966. Personal views on taxonomy. *Veliger* 8(3): 181–188.

Sheldon, Pearl G.
1916. Atlantic slope arcas. In *Palaeontographica America*. Ithaca, N.Y.: Harris Co.

Shepard, Francis, and Rusnak, Gene A.
1957. Texas bay sediments. *Pub. Inst. Marine Sci.* 4(2): 5–13.

Sherborn, Carolo Davies
1902–1933. *Index Animalium*. Part I (1758–1800) and Part II (1800–1850). London: British Museum.

Simmons, E. G.
1957. An ecological survey of the upper Laguna Madre of Texas. *Pub. Inst. Marine Sci.* 4:156–200.

Singley, J. A.
1893. Contributions to the natural history of Texas. *Geological Survey of Texas Fourth Annual Report 1892*. Austin: Ben Jones and Co. State Printers.

Smith, Maxwell
1951. *East coast marine shells*. Ann Arbor, Mich.: Edwards Bros., Inc.

Smylie, Vernon
1964. *The secrets of Padre Island*. Corpus Christi: Texas News Syndicate Press.

Sowerby, G. B., Jr.
1842. *A conchological manual*. London: Henry G. Bohn.

Sowerby, James
1894. Genera of recent and fossil shells. *Ann. Mag. Natur. Hist.* series 6, 13(76): 370–371.

Sowerby, James, and Sowerby, George Brettingham
1820–1825. *The genera of recent and fossil shells*. London: Bernard Quaritch.

Stenzel, H. B.
1940. Mollusks from Point Isabel in Texas. *Nautilus* 69(3): 20–21.

Stohler, Rudolf
1966. Ethics for collectors. *Veliger*. 8(3): 226.

Straughn, Robert P. L.
1964. *The salt-water aquarium in the home*. New York: A. S. Barnes and Co.

Taylor, D. W., and Sohl, N. F.
1962. An outline of gastropod classification. *Malacologia* 1: 7–32.

Taylor, Wm. R.
1954. Sketch of the marine algal vegetation of the shores of the Gulf of Mexico. In *The Gulf of Mexico*. Fishery Bulletin of the U.S. Fish and Wildlife Service 55(89): 177–192.

Tesch, J. J.
1964.   *Thecosomatous pteropods.* Carlsberg Foundation's "Dana" Report no. 28, Copenhagen, Denmark.

Texas Almanac.
1965.   Dallas: A. H. Belo Corporation.

Thiele, Dr. Johannes
1963.   *Handbuch der systematischen weichtierkunde.* Vols. I and II. Amsterdam: A. Asher & Co.

Thompson, Bob
1966.   *Sunset beachcombers' guide to the Pacific coast.* Menlo Park, Calif.: Lane Magazine and Book Co.

Thorne, Robert F.
1954.   Flowering plants of the waters and shores of the Gulf of Mexico. In *The Gulf of Mexico.* Fishery Bulletin of the U. S. Fish and Wildlife Service 55(89): 193–202.

Tinker, Spencer Wilkie
1967.   *Pacific sea shells.* Rutland, Vt.: Charles E. Tuttle Co., Publishers.

Tolbert, Frank X.
1961.   *An informal history of Texas.* New York: Harper & Brothers Publishers.

Tryon, George W., Jr.
1895.   *Manual of conchology: Structural and systematic.* Philadelphia: Academy of Natural Sciences.

Tunnell, John W., Jr.
1969.   The mollusks of seven and one-half fathom reef. Masters Thesis, Texas A & I University, Kingsville.

Turner, Ruth
1954.   The family Pholadidae in the western Atlantic and the eastern Pacific. Part 1. Pholadinae. *Johnsonia* 3(33): 1–64.
1955.   The family Pholadidae in the western Atlantic and the eastern Pacific. Part 2. Martesiinae, Jouannetiinae, and Xylophaginae. *Johnsonia* 3(34): 65–160.
1966.   *A survey and illustrated catalogue of the Teredinidae.* Cambridge: Harvard U. Museum of Comparative Zoology.

Vanatta, E. G.
1903.   A list of shells collected in western Florida and Horn Island, Mississippi. *Proc. Acad. Natur. Sci. Phila.* 55: 756–759.

Verrill, A. E.
1897.   A study of the family Pectinidae, with a revision of the genera and subgenera. *Trans. Conn. Acad. Sci.* 10:41–95.

Vokes, Harold E.
1967.   Genera of the Bivalvia: A systematic and bib-
liographic catalogue. *Bull. Amer. Paleontol.* 51(232): 111–394.

Voss, Gilbert L.
1955.   The cephalopods obtained by the Harvard-Havana expedition off the coast of Cuba in 1938–39. *Bull. Marine Sci. Gulf & Caribbean* 5(2): 81–115.
1956.   Gulf of Mexico cephalopods. *Bull. Marine Sci. Gulf & Caribbean* 6(2): 359–363.
1960.   Biologist collects shells. *Sea Frontiers.* Bulletin of the International Oceanographic Foundation. 6(2): 66–77.

Wagner, Robert J. L., and Abbott, R. Tucker
1965.   *Van Nostrand's standard catalog of shells.* Princeton: D. Van Nostrand Co., Inc.

Warmke, Germaine L., and Abbott, R. Tucker
1961.   *Caribbean seashells.* Narberth, Penn.: Livingston Publishing Co.

Wasson, Jean Andrews
1963.   Notes on the finding of an addition to the marine mollusks of Texas. *Texas J. Sci.* 15(1): 119–120.

Weisbord, Norman E.
1962.   Venezuelan Cenozoic Gastropods. *Bull. Amer. Paleontol.* 42.
1965.   Venezuelan Cenozoic Pelecypods. *Bull. Amer. Paleontol.* 45.

Wells, Harry W.
1959.   Notes on *Odostomia impressa* (Say). *Nautilus* 72(4): 140–153.

Wells, Harry W., and Wells, Mary Jane
1961.   Three species of *Odostomia* from North Carolina, with description of new species. *Nautilus* 74(4): 149–157.
1962.   The distinction between *Acteocina candei* and *Retusa canaliculata. Nautilus* 75(3): 87–93.

Wenz, Wilhelm
1959–1961.   Gastropoda. *Handbuch Paläozool.* 6(1,2).

Whitehouse, Eula
1962.   *Common fall flowers of the coastal bend of Texas.* Dallas: Privately printed.

Whitten, H. L.; Rosene, Hilda F.; and Hedgpeth, J. W.
1960.   The invertebrate fauna of Texas coast jetties: A preliminary survey. *Pub. Inst. Marine Sci.* 1(2): 53–99.

Wilbur, Karl M., and Yonge, C. M.
1964.   *Physiology of Mollusca.* Vols. I and II. New York and London: Academic Press.

Wood, Amos L.
1967.   *Beachcombing for Japanese glass floats.* Portland, Ore.: Binfords and Mort, Publishers.

Writer's Round Table
1950.   *Padre Island.* San Antonio: Naylor Co.

Yonge, C. M.
  1963.   *The seashore.* New York: Atheneum.
  1969.   General characters of Mollusca. In *Treatise on
           invertebrate paleontology*, ed. R. C. Moore.
           Part I, Mollusca 1, pp. 13–136. Boulder, Colo.:
           Geological Society of America, Inc.
Yount, John T.
  1967.   *Bottle collector's handbook and pricing guide.*
           San Angelo, Tex.: Action Printery.
Zim, Herbert S., and Ingel, Lester
  1955.   *Seashores.* New York: Simon and Schuster.

MAPS AND TIDE TABLES

General highway maps, Texas counties. Texas State High-
  way Department, Planning Survey Division, 1962.
Maps, Gulf coast, United States. U.S. Department of Com-
  merce, Environmental Science Service Administration,
  Coast and Geodetic Survey, Washington, D.C., 1968.
Navigational maps of the Gulf Intracoastal Waterway,
  Port Arthur to Brownsville, Texas. U.S. Army Corps
  of Engineers, Galveston, Texas, 1955.
Tide tables, coast and geodetic survey. U.S. Department of
  Commerce, Washington, D.C., Annual.

INDEX

Abra: 15
 aequalis: 201
Acar: 151
acicula, Creseis: 137
acicularis, Neosimnia: 97–98
Acteocina
 candei: 136
Acteon: 133
 punctostriatus: 133
Acteonacea: 133–134
Acteonidae: 133–134
acuta, Nuculana: 147–148
acute mangelia: 122–123
acutidens, Odostomia: 129
acutus, Nassarius: 116
Adams' baby-bubble: 133
Adams' miniature cerith: 82
adamsi, Seila: 82
aequalis, Abra: 201
Aequipecten
 amplicostatus: 27, 164–165
 gibbus: 164
 irradians amplicostatus: 165
 mucosus: 165
Aequipecten s.s.: 165
affinis cruenta, Tricolia: 56
Aglossa: 92–93
Agriopoma: 205–206
Alaba
 incerta: 82–83
alatus, Isognomon: 162–163
alatus, Strombus: 93–94
alba, Anadontia: 173
albicanus, Ocypode: 265
albidum, Epitonium: 84–85
Aligena
 texasiana: 178–179
alternata, Luidia: 264
alternata, Tellina: 194, 196, 197
alternate tellin: 196
alterniflora, Spartina: 8
Amaea
 mitchelli: 84
Amaea s.s.: 84
americanus, Modiolus: 155–156
americanus, Spondylus: 166, 167
amethyst gem clam: 209
amiantus, Lucina: 174–175
Amphineura: 47–48, 53
amplicostatus, Aequipecten: 27, 164–165

Amygdalum
 papyria: 157–158
Anachis: 113
 avara: 112
 avara semiplicata: 111–112
 obesa: 112
 ostreicola: 112
 transliterata: 112
Anadara: 47
 baughmani: 154
 brasiliana: 152, 153, 155
 campechiensis: 153
 chemnitzi: 152–153
 incongrua: 152
 lienosa: 153–154
 lienosa floridana: 154
 ovalis: 153, 154
 secticostata: 154
 transversa: 152, 154
Anadara s.s.: 153–154
Anadarinae: 152–154
Anadontia
 alba: 173
 philippiana: 173–174
 schrammi: 174
Anaspidea: 139–142
Anatina
 anatina: 183–184
 canaliculata: 184
anatina, Anatina: 183–184
Andira
 inermis: 267
angel wing: 20, 25, 26, 220–221
angulate periwinkle: 59
angulate wentletrap: 85
angulatum, Epitonium: 85, 86
angulifera, Littorina: 59
Angulus: 193, 195
angustifolia, Baccharis: 14
Anomalocardia
 cuneimeris: 209–210
Anomia
 simplex: 167–168
Anomiacea: 167–168
Anomiidae: 167–168
Anticlimax
 pilsbryi: 64–65
antillarum, Haminoea: 135
Antillean lima: 169
Antillean nerite: 57

Antiquities Act: 13, 262
apicina, Marginella: 120
apicina, Prunum: 120
Aplacophora: 47
Aplysia
 dactylomela: 139–140
 floridensis: 140
 protea: 140–141
 willcoxi: 27, 139, 141
Aplysiacea: 139–142
Aplysiidae: 139–142
apple murex: 111
Aransas Bay: 8, 9, 20
Aransas National Wildlife Refuge: 8, 240
Aransas Pass: 7, 10, 11, 13
Aransas Pass, Texas: 241
Arca: 151
 imbricata: 149
 umbonata: 149
 zebra: 149–150
Arca s.s.: 149–150
Arcacea: 149–155
Archaeogastropoda: 54–59
Architectonica
 granulata: 75
 nobilis: 74–75
Architectonicacea: 74–75
Architectonicidae: 74–75
Arcidae: 149–154
Arcinae: 149–152
Arcoida: 149–155
arctic saxicave: 218–219
arctica, Hiatella: 218–219
arcuata, Melanella: 92–93
Arenaeus
 cribarius: 265
Argopecten: 164
argophyllus, Helianthus: 14
aristata, Lithophaga: 160
artist's mussel: 158–159
asphalt: 269
Astartedontina: 180–204
Astrangia
 astreiformis: 264
astreiformis, Astrangia: 264
Astropecten
 duplicatus: 264
Astyris: 112–113
Atakapan Indians: 5
Atlantic cyclinella: 208–209

Atlantic deer cowrie: 97
Atlantic distorsio: 107
Atlantic flat lepton: 177–178
Atlantic hairy triton: 105
Atlantic left-handed jewel box: 172
Atlantic modulus: 77–78
Atlantic pearl oyster: 161–162
Atlantic rocellaria: 218
Atlantic rupellaria: 214
Atlantic sanguin: 199
Atlantic surf clam: 187
Atlantic thorny oyster: 167
Atlantic wing oyster: 162
atlanticus, Glaucus: 142–143
Atrina: xvi, 260
    rigida: 160
    seminuda: 160–161
    serrata: 160, 161, 171
atrostyla, Mangelia: 121
Atyidae: 135
Aurelia
    aurita: 263
aurita, Aurelia: 263
aurora, Macoma: 191
Aury, Louis: 6
avara, Anachis: 112
avara semiplicata, Anachis: 111–112
awl miniature cerith: 80–81
aztecus, Penaeus: 17

Baccharis
    angustifolia: 14
    halimifolia: 14
Baffin Bay: 20
Bahia Azul Marina: 241
Balcis
    conoidea: 93
Bales' false dial: 69–70
balesi, Pleuromalaxis: 69–70
Balli, Juan José: 6
Balli, Padre Nicolás: 6, 7, 8
bamboo: 266
banded tulip shell: 116–117
Bankia
    gouldi: 224
Bankiella: 224
Barbatia
    cancellaria: 150
    candida: 150–151
    domingensis: 151
    tenera: 151–152
Barbatia s.s.: 150–151
Barnea: 20
    truncata: 219–220
barrattiana, Corbula: 214–215
Barratt's corbula: 214–215
bartschi, Teredo: 224–225
Bartsch's shipworm: 224–225
Basommatophora: 143–144
Batis
    maritima: 14
baughmani, Anadara: 154
bay scallop: 164–165
Baytown, Texas: 269
beach croton: 14
beach evening primrose: 14

beautiful little caecum: 76–77
beautiful truncatella: 62
bellastriata, Semele: 202–203
bicolor, Isognomon: 163
bidentata, Cylichna: 134
bidentatus, Melampus: 143–144
Big Shell: 11, 21, 44
bigelovii, Salicornia: 14
bilineata, Strombiformis: 92
biscaynense, Teinostoma: 71, 72
bisulcata, Heliacus: 74
bisulcata, Lithophaga: 159–160
bisuturalis, Odostomia: 130
bitruncata, Panopea: 219
Bittium: 81, 130
    varium: 79–80
Bittium s.s.: 79–80
Bivalvia: 44–47, 54, 147–227
black-lined triphora: 83
blakei, Solariorbus: 70
blood ark: 153
blue crab: 265
blue glaucus: 142–143
Bob Hall Pier: 10, 245, 246
Boca Chica Beach: 13, 247
Bolivar Pass: 8, 13, 233
Bolivar Peninsula: 233
Bolivar Roads: 233
bonariensis, Hydrocotyle: 14
bonasus, Rhinoptera: 264
borbonia, Persea: 14
Borrichia
    frutescens: 14
bottles: 268–269
Botula
    castanea: 159
Brachidontes: 47
    citrinus: 157
    exustus: 156–157
    recurvus: 29, 157
Brachidontes s.s.: 156–157
brasiliana, Anadara: 152, 153, 155
brasiliana, Mactra: 185
brasiliensis, Narcine: 264
Brazoria: 234
Brazos Island: 13, 247
Brazos River: 6, 9, 234
Brazos Santiago Pass: 5, 13
Brazosport Museum of
    Natural Science: 234–235, 259
brevifrons, Macoma: 190–191
brevis, Lolliguncula: 228
brief squid: 228
broad-ribbed cardita: 179–180
brown-banded wentletrap: 86–87
Brown Cedar Cut: 9, 235
brown moon shell: 100
brown rangia: 186
brown sargassum snail: 80
brown sea snail: 91
browniana, Rissoina: 64
Brownsville, Texas: 3, 14, 247
Bryan Beach: 234
Buccinacea: 111–118
Buccinidae: 113–114

Bulla
    striata: 134
Bullacea: 134–137
Bullidae: 134
Bursatella
    leachi plei: 141–142
Busycon: 27, 252, 265
    contrarium: xvi, 4, 36, 114, 115
    pulleyi: 114
    sinistrum: 114
    spiratum: 115
    spiratum plagosus: 36, 115
Busycotypus
    plagosus: 115
buttercup lucina: 173
by-the-wind-sailor: 263

cabbage bark: 267
cabbagehead: 28, 263
Cabeza de Vaca, Alvar Núñez: xvi, 3, 267
Cadulus
    carolinensis: 145–146
Caecidae: 75–77
Caecum
    cooperi: 76
    glabrum: 75–76
    imbricatum: 76
    pulchellum: 76–77
Caenogastropoda: 59–126
calico clam: 206–207
calico crab: 265
calico scallop: 164
Callinectes
    sapidus: 265
Calliostoma
    euglyptum: 55
Calliostomatinae: 55
Callocardia
    texasiana: 205–206
Calyptraeacea: 94–96
Calyptraeidae: 94–96
Camargo, Diego de: 3
Cameras: used in photographing shells,
    258–259
Campeche angel wing: 221
campechiensis, Anadara: 153
campechiensis, Mercenaria: 212–213
campechiensis, Pholas: 221
campechiensis texana, Mercenaria: 213
canaliculata, Anatina: 184
canaliculata, Retusa: 135–136
Cancellaria
    reticulata: 119–120
cancellaria, Barbatia: 150
Cancellariidae: 119–120
cancellarius, Cantharus: 113, 114
cancellata, Chione: 210
cancellate cantharus: 113
cancellate semele: 202–203
candida, Barbatia: 150–151
Caney Creek: 235
canrena, Natica: 100–101
Cantharus: 36
    cancellarius: 113, 114
    tinctus: 113–114
Cardiacea: 180–183

Cardiidae: 180–183
Cardiinae: 180–183
Cardita
    floridana: 179–180
Carditacea: 179–180
Carditamera: 179–180
Carditidae: 179–180
Carditinae: 179–180
Cardium: 25, 47
Carolina cadulus: 145–146
Carolina marsh clam: 204–205
carolinensis, Cadulus: 145–146
caroliniana, Polymesoda: 204–205
carrier shell: 37
Cassididae: 102–103
castanea, Botula: 159
castaneus, Lioberis: 159
cat tail: 14
Cate's risso: 63–64
catesbyana, Rissoina: 63–64
Cavolina
    longirostris: 138–139
    uncinata: 139
Cavolinidae: 137–139
cayenensis, Diodora: 54
Cayenne keyhole limpet: 54
Cedar Bayou: 9, 10, 13, 240
Cedar Lane, Texas: 235
Cephalaspidea: 133–137
Cephalopoda: 47, 48–49, 228–230
cerina, Mangelia: 121–122, 123
cerinella, Kurtziella: 122
cerinella, Mangelia: 122
Cerithiacea: 75–83
Cerithidea
    pliculosa: 78
Cerithideopsis: 78
Cerithiidae: 78–80
Cerithiinae: 78–80
Cerithiopsidae: 80–83
Cerithiopsis
    emersonii: 80–81
    greeni: 81, 83
    iota: 81–82
    subulata: 81
    vinca: 81
Cerithiopsis s.s.: 81–82
Cerithium
    floridanum: 78–79
    variabile: 79–80
cervus, Cypraea: 97
chalky buttercup: 173–174
Chama: 47, 167
    congregata: 171
    macerophylla: 171–172
Chamacea: 171–172
Chamidae: 171–172
channeled barrel-bubble: 135–136
channeled duck clam: 184
checkered pheasant: 56
chemnitzi, Anadara: 152–153
Chemnitzia: 131
Chemnitz's ark: 152–153
chesneli, Rissoina: 64
Chione: 47
    cancellata: 210

clenchi: 211
    grus: 210–211
    intapurpurea: 212
    latilirata: 211
Chioninae: 209–213
chitons: 47–48, 53
Chlamydinae: 164–166
chlorostomum, Cymatium: 105
Chrysallida: 129–130
Cibedezebina: 64
Cinctura: 116–117
cinerea, Hastula: 125
citrinus, Brachidontes: 157
clams: general discussion of, 44–47;
    recipes for, 260–261
clathrata, Distorsio: 107
clathrata, Luidia: 125–264
Clathrodilla
    ostrearum: 121
Clathrus: 87
Clavinae: 120–121
cleaning methods: 254–255
clenchi, Chione: 211
Clench's chione: 211
Clibinarius
    vittatus: 265
Clidophora: 226
climate and salinity: 17–20
Cline's Point: 24, 68, 241
cloudy periwinkle: 61
Coahuiltecan Indians: 5
coastal waters: 17–30
Cochliolepis
    nautiliformis: 65
    parasitica: 65
    striata: 65–66
coconuts: 266
Codakia
    orbicularis: 174
Codakia s.s.: 174
coffee melampus: 143–144
coffeus, Melampus: 143–144
Coleoidea: 48, 228–230
Colorado River: 9, 51, 235
colorful Atlantic natica: 100–101
Columbellidae: 111–113
colymbus, Pteriacea: 161–163
common Atlantic abra: 201
common Atlantic auger: 125
common Atlantic marginella: 120
common Atlantic octopus: 229–230
common Atlantic slipper shell: 94–95
common baby's ear: 102
common eastern nassa: 115–116
common egg cockle: 181–182
common jingle shell: 167–168
common nutmeg: 119
common purple sea snail: 90
common rangia: 38, 186
common rice olive: 118–119
common spirula: 228–229
common sundial: 74–75
common West Indian simnia: 97–98
Conacea: 120–126
concentric ervilia: 187–188
concentric nut clam: 148

concentrica, Ervilia: 187–188
concentrica, Nuculana: 148
Congeria
    leucophaeta: 204
congregata, Chama: 171
conical melanella: 93
conoidea, Balcis: 93
Conrad's false mussel: 204
Conrad's paper-bubble: 135
constricta, Macoma: 190
constricted macoma: 190
contracta, Corbula: 215
contracted corbula: 215
contrarium, Busycon: xvi, 4, 36, 114, 115
convexa, Crepidula: 95
cooperi, Caecum: 76
Cooper's caecum: 76
Copano Bay: xvi, 240
coquina shell: xvi, 189–190
Corbiculacea: 204–205
Corbiculidae: 204–205
Corbula
    barrattiana: 214–215
    contracta: 215
    dietziana: 215–216
    krebsiana: 216–217
    swiftiana: 217
Corbulidae: 214–218
cornuta, Echinochama: 172
Corpus Christi, Texas: 3, 5, 7, 10, 13, 17,
    246, 247, 269
Corpus Christi Museum: 259
corrugated razor clam: 199–200
costata, Cyrtopleura: 26, 220–221
costatus, Cymatium: 107
Costoanachis: 112
cownose ray: 264
crabs: types of, listed, 265
Crassatellacea: 180
Crassatellidae: 180
Crassatellinae: 180
Crassinella
    lunulata: 180
Crassispira
    tampaensis: 120–121
Crassispirella: 120–121
Crassostrea: 176
    virginica: xvi, 29, 130, 157, 169–170
crenulata, Pyramidella: 130
Crepidula: 25, 129, 130
    convexa: 95
    fornicata: 94–95
    plana: 95–96
Crepidula s.s.: 94–95
Creseis
    acicula: 137
crested oyster: 170
cribarius, Arenaeus: 265
cristata, Tellidora: 194, 198–199
cross-barred venus: 210
Croton
    punctatus: 14
cruenta, Sanguinolaria: 199
cucumber melanella: 92
Cumingia
    tellinoides: 201–202

cumingianus, Solecurtus: 199–200
Cunearca: 152–153
cuneata, Rangia: 62, 186, 187
cuneiformis, Martesia: 222–223
cuneimeris, Anomalocardia: 209–210
cuprea, Diopatra: 264–265
currents: 20–22
curved melanella: 92–93
cut-ribbed ark: 153–154
cuttlefish: 228
Cyamiacea: 178–179
Cyclinella
    tenuis: 208–209
Cyclininae: 208–209
Cyclostremella
    humilis: 66
Cyclostremiscus
    jeannae: 66–67
    pentagonus: 67
    suppressus: 67–68
    trilix: 67
Cylichna
    bidentata: 134
Cylichnella: 134
Cymatiidae: 103–107
Cymatium
    chlorostomum: 105
    costatus: 107
    muricinum: 103–104
    nicobaricum: 104–105
    parthenopeum: 106–107
    pileare martinianum: 105
    poulsenii: 105–106
Cymatriton: 104–105
Cyphoma: 258
    intermedium: 98, 99
    mcgintyi: 98–99
Cypraea: 254
    cervus: 97
Cypraeacea: 97–99
Cypraecassis
    testiculus: 103
Cypraeidae: 97
Cypraeinae: 97
Cyrtopleura
    costata: 26, 220–221

Dactylmetra
    quinquecirrha: 263
dactylomela, Aplysia: 139–140
Dallocardia: 183
dark pyram: 127–128
dealbata, Olivella: 118–119
Decapoda: 48, 228–229, 265
Decros Point: 240
DeKay's dwarf tellin: 195
demissus, Geukensia: 156
demissus granosissimus, Modiolus: 156
Dendronotoidea: 142
Dentaliidae: 145
Dentalium
    eboreum: 145
    texasianum: 145
Dentalium s.s.: 145
Depressiscala
    nautlae: 89

Devil's Elbow: 5
Diacria
    quadridentata: 138
    trispinosa: 137–138
Diberus: 159–160
dietziana, Corbula: 215–216
Dietz's corbula: 215–216
Dinocardium: 260
    robustum: 180–181
Diodora
    cayenensis: 54
Diodorinae: 54–55
diomedea, Rubellatoma: 124
Diopatra
    cuprea: 264–265
Diplanthera
    wrightii: 16, 30
Diplodonta
    punctatus: 177
    semiaspera: 176–177
    soror: 177
Diplothyra
    smythi: 223
discus, Dosinia: 207–208
disk dosinia: 207–208
dislocata, Terebra: 125–126
Distorsio
    clathrata: 107
divisus, Tagelus: 200, 201
Doc Bales' ark: 151–152
domingensis, Barbatia: 151
domingensis, Typha: 14
Donacacae: 189–190
Donacidae: 189–190
Donax: xvi, 125, 261
    tumidus: 189
    variabilis roemeri: 189, 190
    variabilis texasiana: 189–190
Dosinia: 200, 260
    discus: 207–208
    elegans: 208
Dosinidia: 207–208
Dosiniinae: 207–208
Dosyatis
    sabina: 264
double-sutured (two-seamed)
    odostome: 130
Dreissenacea: 204
Dreissenidae: 204
driftwood: 267
drummondii, Malvaviscus: 16
drummondii, Oenothera: 14
dune sedge: 14
dune sunflower: 14
duplicatus, Astropecten: 264
duplicatus, Polinices: 99–100, 101, 251
dux, Odostomia: 129
dwarf cerith: 79
dwarf purple sea snail: 89
dwarf surf clam: 185–186

eastern oyster: 169–170
eastern white slipper shell: 95–96
eboreum, Dentalium: 145
Echinochama
    cornuta: 172

edulis, Ostrea: 35
eelgrass: 16, 30
egg cases: 36–37, 265–266
electric ray: 264
elegans, Dosinia: 208
elegans, Turbonilla: 131
elegant dosinia: 208
elegant paper-bubble: 135
elegant turbonilla: 131
elegantula, Turbonilla: 131
Ellobiacea: 143–144
Ellobiidae: 143–144
emaiginata, Libinia: 265
emerald nerite: 58–59
emersonii, Cerithiopsis: 80–81
Ensis
    minor: 188, 189
Entada
    gigas: 266–267
    scardens: 266–267
Eolidoidea: 142–143
Eontia: 155
epheliticus, Hepatus: 265
Episcynia
    inornata: 68
Epitoniacea: 83–91
Epitoniidae: 83–89
Epitonium: 264
    albidum: 84–85
    angulatum: 85, 86
    humphreysi: 85–86
    lamellosum: 87
    multistriatum: 87–88
    novangliae: 88
    rupicola: 86–87
    sericifilum: 88–89
    tollini: 86
Epitonium s.s.: 84–86
equestris, Ostrea: 157, 170
Eratoidae: 96–97
Ervilia
    concentrica: 187–188
euglyptum, Calliostoma: 55
Eugorgia
    setacea: 264
    virgulata: 98, 264
Eulimastoma: 127
Eulimidae: 92–93
Eurytellina: 196–197
Euthecosomata: 137–139
Euthyneura: 126–143
Euvola: 163–164
euzonus, Solariorbis: 71
exustus, Brachidontes: 156–157

faded slipper shell: 95
fallen angel wing: 219–220
false angel wing: 213–214
false limpet: 24, 25
false willow: 14
fargoi, Vermicularia: 75
Fargo's worm shell: 75
fasciata, Tegula: 56
Fasciolaria: 36, 265
    hunteria: 116–117, 258

lilium: 117
tulipa: 117
Fasciolariidae: 116–118
Fasciolariinae: 116–118
fat dove shell: 112
fat little donax: 189
fat odostome: 128
fighting conch: 93–94
file fleshy limpet: 54–55
filiforme, Syringodium: 16
Fimbristylis
　spadicea: 14
fine-ribbed auger: 126
Fissurellacea: 54–55
Fissurellidae: 54–55
flame auger: 126
flares, naval: 267
flat tree oyster: 162–163
flexuoso, Rangia: 186–187
floats: 268
flora and fauna: 14–17
floralia, Olivella: 119
Florida cerith: 78–79
Florida horse conch: 117–118
Florida lucina: 175–176
Florida lyonsia: 227
Florida marsh clam: 205
floridana, Cardita: 179–180
floridana, Lucina: 175–176
floridana, Pseudocyrena: 205
floridana, Vitrinella: 72–73
floridanum, Cerithium: 78–79
floridanus, Pagurus: 265
Florida rock shell: 108–109
floridensis, Aplysia: 140
fornicata, Crepidula: 94–95
four-toothed cavoline: 138
fragile Atlantic mactra: 185
fragile martesia: 221–222
fragilis, Mactra: 185
fragilis, Martesia: 221–222
Freeport, Texas: 8–9, 234, 269
frutescens, Borrichia: 14
Fugleria: 151–152
fulgurans, Nerita: 57
fulvescens, Murex: 110–111
Fulton, Texas: 241

galea, Tonna: 107–108
Galeommatacea: 177–178
Gallardia
　pulchella: 14
Galveston, Texas: 3, 6, 7, 8, 17, 233–234
Galveston Island: 3, 5, 7
Garcitas Creek: 6
Gastrochaenacea: 218
Gastrochaenidae: 218
Gastropoda: 41–44, 54–144
Garay, Francisco de: 3
Gemma
　purpurea: 209
Gemminae: 209
geoduck: 219
Geukensia
　demissus: 156
ghost crab: 265

giant Atlantic cockle: 180–181
giant eastern murex: 110–111
giant squid: 33
giant tun shell: 107–108
gibbosa, Odostomia: 128
gibbosa, Plicatula: 166–167
gibbus, Aequipecten: 164
gigantea, Pleuroploca: 117–118
gigas, Entada: 266–267
Gilbert, W. S.: 17
glabrum, Caecum: 75–76
Glaucidae: 142–143
Glaucus
　atlanticus: 142–143
globe purple sea-snail: 91
globosa, Janthina: 89, 91
goatfoot morning-glory: 14
gold-mouthed triton: 104–105
gooseneck barnacle: 265
gouldi, Bankia: 224
Gould's shipworm: 224
Grant, Ulysses S.: 7
granulata, Architectonica: 75
granulatum, Phalium: 102–103
Graptacme: 145
gray pygmy venus: 210–211
green jackknife clam: 188–189
Green Lake, Texas: 240
greeni, Cerithiopsis: 81, 83
Green's miniature cerith: 81
Gregariella: 158–159
Gregory, Texas: 241
groundsel: 14
grus, Chione: 210–211
Guadalupe River: 6
Gutturnium: 103–104
Gyroscala: 86–89

haemastoma floridana, Thais: 108–109, 110
haemastoma haysae, Thais: 109–110
hairy vitrinella: 68
half-folded dove shell: 111–112
half-smooth odostome: 129–130
halimifolia, Baccharis: 14
Halter, R. E.: 7
Haminoea
　antillarum: 135
　succinea: 135
Harbor City: 241
Harbor Island: 13
Hastula
　cinerea: 125
　maryleeae: 124–125
　salleana: 125
Hay's rock shell: 109–110
Heliacus
　bisulcata: 74
Helianthus
　argophyllus: 14
helicoidea, Vitrinella: 73
helix vitrinella: 73
hemphilli, Turbonilla: 132
Hemphill's turbonilla: 132
Henry, W. S.: 6
hepaticus, Polinices: 100

Hepatus
　epheliticus: 265
hermit crab: 104, 251, 265
Heterodonta: 171–227
hians, Rocellaria: 218
Hiatella
　arctica: 218–219
Hiatellacea: 218–219
Hiatellidae: 218–219
Hippuritoida: 171–172
hooked mussel: 157
Horace Caldwell Pier: 10
horse conch: 33
horse oyster: 170
Houston, Texas: 3, 269
Houston Museum of Natural History: 259
Houston Ship Channel: 233
humilis, Cyclostremella: 66
humphreysi, Epitonium: 85–86
Humphrey's wentletrap: 85–86
hunteria, Fasciolaria: 116–117, 258
hyalina floridana, Lyonsia: 227
Hydrobiidae: 62–63
Hydrocotyle
　bonariensis: 14

imbricata, Arca: 149
imbricatum, Caecum: 76
impressa, Odostomia: 130
impressed odostome: 130
incerta, Alaba: 82–83
Incirrata: 229–230
incongrua, Anadara: 152
incongruous ark: 152
Independence Trail: 233
Indian blanket: 14
Indianola, Texas: 7, 13, 240
inequale, Periploma: 226–227
inermis, Andira: 267
infracarinata, Solariorbis: 70–71
Ingleside, Texas: 241
inornata, Episcynia: 68
intapurpurea, Chione: 212
intermedia, Melanella: 93
intermedium, Cyphoma: 98, 99
interrupta, Niso: 92
interrupta, Turbonilla: 131
interrupted turbonilla: 131
interruptus, Parviturboides: 69
Intracoastal Waterway: xvi, 3, 9, 14, 16, 235, 240, 241, 247
iota, Cerithiopsis: 81–82
Ipomoea
　pes-caprae: 14
　stolonifera: 14
iris tellin: 198
irradians amplicostatus, Aequipecten: 165
irrorata, Littorina: 60
Ischadium: 157
Ischnochiton
　papillosus: 53
Ischnochitonidae: 53
Ischnochitonina: 53
Isla Blanca Park: 13
isocardia, Trachycardium: 182–183

Isognomon
  alatus: 162–163
  bicolor: 163
Isognomonidae: 162–163
Ispidula: 118
ivory tusk: 145

jackknife clam: 188
jacobeus, Pecten: xvi
jamaicensis, Melanella: 93
Janacus: 95–96
Janthina: 25, 143
  globosa: 89, 91
  janthina: 90, 91
  megastoma: 89
  pallida: 90
  prolongata: 90, 91
  umbilicata: 89
janthina, Janthina: 90, 91
Janthina s.s.: 90
Janthinidae: 89–91
jeannae, Cyclostremiscus: 66–67
Jeanne's vitrinella: 66–67
jellyfish: 263
Jodina: 89
Jouannetia
  quillingi: 223–224

Karankawa Indians: xvi, 3, 5, 6, 16,
  267, 269
Kelliidae: 177–178
kitten's paw: 166–167
knobbed triton: 103–104
Kombologion: 55
Krebs' corbula: 216–217
krebsiana, Corbula: 216–217
Kurtziella
  cerinella: 122
  limonitella: 121, 123–124

Labiosa
  lineata: 184
lady-in-waiting venus: 212
Laevicardium
  laevigatum: 181–182
  mortoni: 182
laevigata, Odostomia: 128–129
laevigata, Rissoina: 64
laevigatum, Laevicardium: 181–182
Lafitte, Jean: 6
Laguna Madre: 9, 10, 11, 13, 20, 51, 247
Lake Jackson, Texas: 234, 235, 259
Lamellaria
  rangi: 96
Lamellariacea: 96–97
Lamellariidae: 96
Lamellariinae: 96
lamellosum, Epitonium: 87
La Quinta: 241
LaSalle, René Robert Cavelier, Sieur de:
  6, 7, 13, 240
Laskeya: 80–81
lateral musculus: 158
lateralis, Mulinia: 185–186, 187
lateralis, Musculus: 158
latilirata, Chione: 211

Lavaca Bay: 6, 7, 240
leachi plei, Bursatella: 141–142
leafy jewel box: 171–172
León, Alonso de: 6
Lepas: 265
lepidum, Lepton: 178
Leptegouana: 120
Leptogorgia
  setacea: 98, 250
Lepton
  lepidum: 178
Leptonidae: 178
leremum, Teinostoma: 72
lettered olive: 118
leucophaeta, Congeria: 204
Libinia
  emaiginata: 265
lienosa, Anadara: 153–154
lienosa floridana, Anadara: 154
lightning whelk: 112
lilium, Fasciolaria: 117
Lima
  lima: 168–169
  pellucida: 169
lima, Lima: 168–169
Lima s.s.: 168–169
Limacea: 168–169
Limaria: 169
limatula, Lucapinella: 54–55
Limidae: 168–169
limonitella, Kurtziella: 121, 123–124
Linatella: 105–106
lindheimeri, Opunta: 16
lineata, Labiosa: 184
lineata, Tellina: 194, 196–197
lineolata, Littorina: 24, 60–61
Linné, Karl von: 38, 39
linneana, Porpita: 263
lintea, Quadrans: 198
Lioberis
  castaneus: 159
lion's paw: 166
Lithophaga: 47
  aristata: 160
  bisulcata: 159–160
Litiopa
  melanostoma: 80
Litiopinae: 80
little corrugated jewel box: 171
little horn caecum: 77
Little Shell: 11, 45, 247
Littoraria: 59–60
Littoridina
  sphinctostoma: 62–63
Littoridinae: 62–63
Littorina: 15, 25
  angulifera: 59
  irrorata: 60
  lineolata: 24, 60–61
  meleagris: 61–62
  nebulosa: 61
  zic zac: 61
Littorinacea: 59–62
Littorinidae: 59–62
live oak: 14
livida, Sayella: 127–128

Lolliginidae: 228
Lolliguncula
  brevis: 228
Lonchaeus: 127
longirostris, Cavolina: 138–139
long-snout cavoline: 138–139
lovely miniature lucina: 174–175
Lucapinella
  limatula: 54–55
Lucina
  amiantus: 174–175
  floridana: 175–176
  multilineata: 175
Lucinacea: 173–177
Lucinidae: 173–176
Lucinina: 173–180
Lucininae: 173–176
Luidia
  alternata: 264
  clathrata: 125, 264
luminescence: 49, 265
lunar dove shell: 112
Lunarca: 153
lunata, Mitrella: 112–113
lunate crassinella: 180
lunulata, Crassinella: 180
Lydia Ann Channel: 10
Lyonsia
  hyalina floridana: 227
Lyonsiidae: 227
Lyrodus
  pedicellata: 225
Lyropecten
  nodosus: 166
  subnodosus: 166

macerophylla, Chama: 171–172
McGinty's cyphoma: 98–99
Machaeranthera
  phyllocephala: 14
Macoma
  aurora: 191
  brevifrons: 190–191
  constricta: 190
  mitchelli: 191
  tageliformis: 191–192
  tenta: 192
Macrocallista
  maculata: 206–207
  nimbosa: xvi, 207
Macromphalina
  palmalitoris: 68–69
Mactra: 25, 47
  brasiliana: 185
  fragilis: 185
Mactracea: 183–188
Mactridae: 183–187
Mactrinae: 183–187
maculata, Macrocallista: 206–207
maculated baby's ear: 101–102
maculatum, Sinum: 101–102
mahogany date mussel: 159–160
Malaquite Beach: ii, 246
Malvaviscus
  drummondii: 16
manatee grass: 16

Mangelia
   atrostyla: 121
   cerina: 121–122, 123
   cerinella: 122
   oxyata: 122–123
   plicosa: 123
Mangelliinae: 121–124
Mansfield jetties: 11, 12, 13, 246, 247
Mansfield Ship Channel: 247
many-lined lucina: 175
margaritaceum, Periploma: 227
Marginella
   apicina: 120
Marginellidae: 120
Marine Biological Laboratory: 241
marine grass flats: 27, 30
maritima, Batis: 14
maritima, Ruppia: 16, 27
marsh periwinkle: 60
Martesia: 28
   cuneiformis: 222–223
   fragilis: 221–222
   striata: 222
Martesia s.s.: 221–222
maryleeae, Hastula: 124–125
Matagorda Bay: 6, 9, 13
Matagorda Island: 7–8, 9, 16, 240, 246
Matagorda Peninsula: 9, 235, 240
Matamoros, Mexico: 247
mcgintyi, Cyphoma: 98–99
megastoma, Janthina: 89
Meioceras
   nitidum: 77
Melampus
   bidentatus: 143–144
   coffeus: 143–144
Melanella
   arcuata: 92–93
   intermedia: 93
   jamaicensis: 93
Melanellidae: 92–93
melanostoma, Litiopa: 80
Melarhaphe: 60–61
meleagris, Littorina: 61–62
meleagris, Stomolophus: 263
Mellita
   quinquiesperforata: 264
Melongenidae: 114–115
Menestho: 130
Menippe
   mercenaria: 265
Mercenaria: 15, 28, 176, 260
   campechiensis: 212–213
   campechiensis texana: 213
   mercenaria: xv, xvi, 213
mercenaria, Menippe: 265
mercenaria, Mercenaria: xv, xvi, 213
mesh-pitted chiton: 53
Mesodesmatidae: 187–188
Mesogastropoda: 59–108
Mesopleura: 200–201
Mesquite Bay: 240
Micranellum: 76–77
miniature natica: 101
minor, Ensis: 188, 189
minuta, Olivella: 119

minute dwarf olive: 119
mirabilis, Pedipes: 144
mirabilis, Strigilla: 192–193
mitchelli, Amaea: 84
mitchelli, Macoma: 191
Mitchell's macoma: 191
Mitchell's wentletrap: 84
Mitrella
   lunata: 112–113
Modiolus: 47
   americanus: 155–156
   demissus granosissimus: 156
   tulipa: 156
Modulidae: 77–78
Modulus
   modulus: 77–78
modulus, Modulus: 77–78
mollusks: basic pattern of, 31–37; identification of shell of, 37–38; classification and nomenclature of, 38–41; types of, 48–49
Monodontinae: 56
Monoplex: 106–107
moon jelly: 263
mooreana, Solariorbis: 71
mortoni, Laevicardium: 182
Morton's egg cockle: 182
mossy ark: 149
mucosus, Aequipecten: 165
mud balls: 268
Mulinia
   lateralis: 185–186, 187
mulleri, Renilla: 263–264
multilineata, Lucina: 175
multiribbed wentletrap: 87–88
multistriatum, Epitonium: 87–88
Murex: 252, 256
   fulvescens: 110–111
   pomum: 36, 111
Muricacea: 108–111
muricatum, Trachycardium: 183
Muricidae: 108–111
Muricinae: 110–111
muricinum, Cymatium: 103–104
Musculus
   lateralis: 158
   opifex: 158–159
Mustang Island: 10, 241, 244
mutica, Olivella: 119
mutica, Velella: 90, 263
Myacea: 214–218
Myina: 214–219
Myodia: 214–225
Myoforceps: 160
Mysella
   planulata: 177–178
Mytilacea: 155–160
Mytilidae: 155–160
Mytiloida: 155–161
Mytilopsis: 204

Narcine
   brasiliensis: 264
Narváez, Pánfilo de: 3
Nassariidae: 115–116

Nassarius
   acutus: 116
   vibex: 115–116
Nassarius s.s.: 116
Natica
   canrena: 100–101
Naticacea: 99–102
Naticarius: 100–101
Naticidae: 99–102
Naticinae: 100–101
National Park Service: 7
nautiliformis, Cochliolepis: 65
Nautiloidea: 48
nautlae, Depressiscala: 89
navalis, Teredo: 225
navy shipworm: 225
nebulosa, Littorina: 61
Neogastropoda: 108–126
Neoluricata: 53
Neosimnia: 251
   acicularis: 97–98
   uniplicata: 33, 98, 250
Nerita
   fulgurans: 57
Neritacea: 57–59
Neritidae: 57–59
Neritina
   reclivata: 57–58
   virginea: 58
Neritrema: 61–62
Neverita: 99–100
New England wentletrap: 88
nicobaricum, Cymatium: 104–105
nimbosa, Macrocallista: xvi, 207
Niso
   interrupta: 92
Niteoliva: 119
nitidum, Meioceras: 77
nobilis, Architectonica: 74–75
Nodipecten: 116
nodosus, Lyropecten: 166
Noetia
   ponderosa: 155
Noetiidae: 155
Noteiinae: 155
nopal: 16
North Pass: 10, 11
notched pyram: 127
novangliae, Epitonium: 88
Nucula
   promixa: 147
Nuculacea: 147
Nuculana: 47
   acuta: 147–148
   concentrica: 148
Nuculanacea: 147–148
Nuculanidae: 147–148
Nuculidae: 147
Nuculoidea: 147–148
Nudibranchia: 37, 142–143
Nueces County Park: 10, 244, 245
Nueces River: 6
nutans, Sargassum: 267
nut clam: 147
nuts: 267

obesa, Anachis: 112
Octopoda: 48, 229–230
Octopodidae: 229–230
octopus: 37, 48, 229–230
Octopus
   vulgaris: 229–230
Ocypode
   albicans: 265
Odostomia: 127
   acutidens: 129
   bisuturalis: 130
   dux: 129
   gibbosa: 128
   impressa: 130
   laevigata: 128–129
   seminuda: 129–130
   trifida: 130
Odostomia s.s.: 128–129
Oenothera
   drummondii: 14
Offatts Bayou: 234
Oliva: 4, 25, 254
   sayana: xvi, 118
olive nerite: 57–58
Olivella
   dealbata: 118–119
   floralia: 119
   minuta: 119
   mutica: 119
Olividae: 118–119
Olivinae: 118–119
operculata, Varicorbula: 217–218
opifex, Musculus: 158–159
Opisthobranchia: 126–143
Opuntia
   lindheimeri: 16
Orange, Texas: 269
orbicularis, Codakia: 174
Orbigny's baby-bubble: 134
Orbigny's sundial: 74
Ostrea
   edulis: 3–5
   equestris: 157, 170
Ostreacea: 169–170
ostrearum, Clathrodilla: 121
ostreicola, Anachis: 112
Ostreidae: 169–170
Ostreina: 169–170
oval corbula: 217–218
ovalis, Anadara: 153, 154
Ovulidae: 97–99
oxytata, Mangelia: 122–123
oxytata, Volvulella: 136
oyster grass: 8
oyster piddock: 223
oyster turret: 120–121
oysters: 29, 260

Packery Channel: 244
Padre Island: 5–13 *passim*, 46, 244,
   246, 247
Padre Island National Seashore: 8, 10,
   233, 244, 246, 262
Pagurus
   floridanus: 265

Palacios, Texas: 235, 240
Palaeotaxodonta: 147–148
pale purple sea snail: 90
pallida, Janthina: 90
palmalitoris, Macromphalina: 68–69
Palmito Hill: 7, 13
Pandora
   trilineata: 226
Pandoracea: 226–227
Pandoridae: 226
paniculata, Uniola: 9, 13
Panopea
   bitruncata: 219
Pánuco River: 5
paper mussel: 157–158
papillosus, Ischnochiton: 53
papyria, Amygdalum: 157–158
parasitica, Cochliolepis: 65
Paravolvulella: 136–137
parthenopeum, Cymatium: 106–107
Particoma: 222–223
parvicallum, Teinostoma: 72
Parvilucina: 175
Parviturboides
   interruptus: 69
Pass Cavallo: 9, 13, 15, 240
patens, Spartina: 9
pear whelk: 115
Pecten: 25, 47, 129
   jacobeus: xvi
   raveneli: 163–164
Pectinacea: 163–167
pectinata, Siphonaria: 24, 143
pectinatus, Phacoides: 176
Pectinidae: 163–166
Pectininae: 163–164
pedicellata, Lyrodus: 225
Pedipes
   mirabilis: 144
Pegurus
   floridanus: 265
pelagica, Scyllaea: 142
Pelecypoda: 147–227
pellucida, Lima: 169
pen shell: xvi
Penaeus
   aztecus: 17
   setiferus: 17
pennywort: 14
pentagonus, Cyclostremiscus: 67
Periploma
   inequale: 226–227
   margaritaceum: 227
Periplomatidae: 226–227
Persea
   borbonia: 14
Persephona
   punctata aquilonaris: 265
persimilis, Volvulella: 136
perspectivum, Sinum: 102
perversa nigrocinta, Triphora: 83
pes-caprae, Ipomoea: 14
Petricola
   pholadiformis: 213–214
Petricolidae: 213–214
petroleum tars: 269

Phacoides
   pectinatus: 176
Phacoides s.s.: 176
Phalium
   granulatum: 102–103
Phasianellidae: 56
philippiana, Anadontia: 173–174
Phlyctiderma: 176–177
Pholadacea: 219–225
Pholadidae: 219–224
pholadiformis, Petricola: 213–214
Pholadina: 219–225
Pholadomyina: 226–227
Pholadomyoida: 226–227
Pholas: 213
   campechiensis: 221
Phrontis: 115–116
phyllocephala, Machaeranthera: 14
Phyllonotus: 110–111
Physalia
   physalia: 263
physalia, Physalia: 263
pileare martinianum, Cymatium: 105
pilsbryi, Anticlimax: 64–65
pimpled diplodon: 176–177
Pinctada
   radiata: 161–162
Piñeda, Alonso de: 3
Pinnacea: 160–161
Pinnidae: 160–161
Pisostrigilla: 192–193
Pitar
   texasiana: 206
Pitarinae: 205–207
Pitoscala: 87
Plagioctenium: 164–165
plagosus, Busycotypus: 115
plana, Crepidula: 95–96
planulata, Mysella: 177–178
plastic pellets: 269
plebeius, Tagelus: 26, 200–201
Pleuromalaxis
   balesi: 69–70
Pleuroploca
   gigantea: 117–118
Pleurotomarioidea: 54–59
plicate horn shell: 78
plicatella, Raeta: 184
Plicatula
   gibbosa: 166–167
Plicatulidae: 166–167
plicosa, Mangelia: 123
pliculosa, Cerithidea: 78
Pododesmus
   rudis: 168
pointed nut clam: 147–148
pointed venus: 209–210
Polinices: 37, 266
   duplicatus: 99–100, 101, 251
   hepaticus: 100
Polinices s.s.: 100
Polinicinae: 99–100
Pollia: 113–114
Polygireulima: 93
Polymesoda
   caroliniana: 204–205

Polyplacophora: 47
Polyschides: 145–146
Pomacea sp.: 144
pomum, Murex: 36, 111
ponderosa, Noetia: 155
ponderous ark: 155
Ponocyclus: 66–68
porpita: 263
Porpita
  linneana: 263
Port Aransas, Texas: 7, 10, 24, 240, 244,
  247
Port Arthur, Texas: 259
Port Bolivar, Texas: 233
Port Isabel, Texas: 6, 13, 19, 20, 247
Port Lavaca, Texas: 17, 240
Port Mansfield, Texas: 10, 11, 13, 246, 247
Port O'Connor, Texas: 52, 240
portoricana, Turbonilla: 132
Portuguese man-of-war: 90, 263
portulacastrum, Sesuvium: 14
Potamididae: 78
Potamidinae: 78
poulsenii, Cymatium: 105–106
Poulsen's triton: 105–106
prickly cockle: 182–183
prickly pear: 16
proficua, Semele: 203
prolongata, Janthina: 90, 91
Prosobranchia: 54–126
protea, Aplysia: 140–141
protexta, Terebra: 126
proxima, Nucula: 147
Prunum
  apicina: 120
Psammobiidae: 199
Pseudochama: 47
  radians: 172
Pseudocyrena
  floridana: 205
Pseudoirus
  typica: 214
Pseudomiltha: 175–176
Pteria
  columbus: 162
Pteriacea: 161–163
Pteriidae: xvi, 161–162
Pteriina: 161–169
Pteriinae: 161–162
Peterioida: 161–170
Pteriomorphia: 149–170
Pteropoda: 137–139
Puerto Rican turbonilla: 132
pulchella, Gallardia: 14
pulchella, Truncatella: 62
pulchellum, Caecum: 76–77
pulleyi, Busycon: 114
Pulmonata: 143–144
pumice: 267
punctata aquilonaris, Persephona: 265
punctate mangelia: 123–124
punctatus, Croton: 14
punctatus, Diplodonta: 117
punctostriatus, Acteon: 133
purplish semele: 203–204
purplish tagelus: 200

purpuracens, Semele: 203–204
purpurea, Gemma: 209
purse crab: 265
pusilla, Tectonatica: 101
Pyramidella: 130
  crenulata: 127
Pyramidellidae: 34, 126–133
Pyrenidae: 111–113
Pyrgiscus: 131
Pyrgocythara: 123

quadridentata, Diacria: 138
quahog: 25
Quadrans
  lintea: 198
Quercus
  virginiana: 14
quillingi, Jouannetia: 223–224
quinquecirrha, Dactylmetra: 263
quinquiesperforata, Mellita: 264

radians, Pseudochama: 172
radiata, Pinctada: 161–162
ragged sea hare: 141–142
Raeta
  plicatella: 184
railroad vine: 14
Raja
  texana: 264
rangi, Lamellaria: 96
Rangia
  cuneata: 62, 186, 187
  flexuosa: 186–187
Rangianella: 186
Rang's lamellaria: 96
Ransom Island: 241
raveneli, Pecten: 163–164
Ravenel's scallop: 163–164
razor clam: 25
recipes for cooking mollusks: 260–261
reclivata, Neritina: 57–58
Recluzia
  rollaniana: 91
recurvus, Brachidontes: 29, 157
red brown ark: 150
reddish mangelia: 124
Redfish Bay: 30, 241
Renilla
  mulleri: 263–264
reticulata, Cancellaria: 119–120
reticulated cowrie helmet: 103
Retusa
  canaliculata: 135–136
Retusidae: 135–137
Rhinoptera
  bonasus: 264
Rhizorus: 136
Rhysena: 107
ribbed mussel: 156
rigida, Atrina: 160
Rio Grande: 3, 5, 6, 7, 8, 9, 13, 20, 247
Rissoacea: 62–74
Rissoidae: 63–64
Rissoina
  browniana: 64
  catesbyana: 63–64

chesneli: 64
laevigata: 64
Rissoinae: 63–64
robustum, Dinocardium: 180–181
Rocellaria
  hians: 218
rock shell: 25
Rockport, Texas: 16, 241
rocky shores: 24
Roemer, Ferdinand von: 233
rollaniana, Recluzia: 91
rose petal tellin: 196–197
rough jingle shell: 168
rough scallop: 165
Rubellatoma
  diomedea: 124
rudis, Pododesmus: 168
Rupellaria
  typica: 214
rupicola, Epitonium: 86–87
Ruppia
  maritima: 16, 27

sabina, Dosyatis: 264
Sabine, Texas: 17
Sabine Lake: 51
Sabine Pass: 2, 13
Sabine Pass, Texas: 8, 233
Sabine River: 3, 5, 8
sacahuiste: 14
St. Joseph Island: 6, 7, 10, 11, 16, 240,
  241
Salicornia
  bigelovii: 14
salleana, Hastula: 125
Salle's auger: 125
salt marsh flats: 15–16
saltwort: 14
Saluria Bayou: 240
San Antonio, Texas: 5
San Antonio Bay: 17
San Bernard River: 9, 235
San Luis Pass: 8–9, 13, 234
sand dollar: 103, 264
sand flats: 26–27
Sanguinolaria
  cruenta: 199
sapidus, Callinectes: 265
Sargassum
  nutans: 267
sargassum: 267
sargassum nudibranch: 142
Sargent Beach: 9, 235
sawtooth pen shell: 161
sayana, Oliva: xvi, 118
Sayella
  livida, 127–128
Say's chestnut mussel: 159
Say's pandora: 226
Say's tellin: 193–195
Scaphopoda: 47, 48, 145–146
scardens, Entada: 266–267
schrammi, Anadontia: 174
Schwartziella: 63–64
scissor date mussel: 160
Scissula: 198

Scobinopholas: 220–221
scorched mussel: 156–157
Scotch bonnet: 102–103
Scrobiculariidae: 201
sculptured top-shell: 55
Scyllaea
    pelagica: 142
Scyllaeidae: 142
Scyphozoa: 263
sea beans: 266–267
Sea Gun Lodge: 240, 241
sea hare: 35, 37
sea nettle: 263
sea oats: 9, 11, 14
sea pansy: 263–264
sea pork: 158
sea purslane: 11, 14
sea slug: 37
sea urchins: 103
sea whip coral: 96, 98, 250, 264
Seadrift, Texas: 269
secticostata, Anadara: 154
seeds: 267
Seila
    adamsi: 82
Semele
    bellastriata: 202–203
    proficua: 203
    purpuracens: 203–204
Semelidae: 201–204
semiaspera, Diplodonta: 176–177
Semicassis: 102–103
seminuda, Atrina: 160–161
seminuda, Odostomia: 129–130
Sepioidea: 228–229
Septa: 105
sericifilum, Epitonium: 88–89
serrata, Atrina: 160, 161, 171
Sesuvium
    portulacastrum: 14
setacea, Eugorgia: 264
setacea, Leptogorgia: 98, 250
setiferus, Penaeus: 17
7½ fathom reef: 67
shark's eye: 99–100
sharp-knobbed nassa: 116
shell clubs: listed, 259
shell collections: listed, 260
shipworm: xvi-xvii, 28, 30
short macoma: 190–191
simplex, Anomia: 167–168
Singer, John V.: 7, 8
single-toothed simnia: 98
Sininae: 101–102
sinistrum, Busycon: 114
Sinum: 96
    maculatum: 101–102
    perspectivum: 102
Siphonaria
    pectinata: 24, 143
Siphonariacea: 143
Siphonariidae: 143
Siphonodentaliidae: 145–146
skate: 264, 266
Smaragdia
    viridis viridemaris: 58–59

smooth Atlantic tegula: 56
smooth caecum: 75–76
smooth duck clam: 183–184
smooth odostome: 128–129
smooth risso: 64
smythi, Diplothyra: 223
snails: general discussion of, 41–44;
    land and fresh-water, 266
Solariorbis
    blakei: 70
    euzonus: 71
    infracarinata: 70–71
    mooreana: 71
Solariorbis s.s.: 70–71
Solecurtidae: 199–201
Solecurtus
    cumingianus: 199–200
Solen
    viridis: 188–189
Solenacea: 188–189
Solenidae: 188–189
solidissima raveneli, Spissula: 187
solidissima similis, Spissula: 187
sooty sea hare: 140
soror, Diplodonta: 177
southern quahog: 212–213
Southwest Foundation for Research
    and Education: xvii
spadicea, Fimbristylis: 14
Spanish dagger: 14
Spartina
    alterniflora: 8
    patens: 9
    spartinae: 14
Spartina s.s.: 16
spartinae, Spartina: 14
speckled crab: 265
sphinctostoma, Littoridina: 62–63
spider crab: 265
spiny lima: 168–169
spiny pen shell: 160–161
spiral-staircase shells: 83
Spiratellacea: 137–139
spiratum, Busycon: 115
spiratum plagosus, Busycon: 36, 115
Spirula
    spirula: 48, 228–229
spirula, Spirula: 48, 228–229
Spirulidae: 228–229
Spissula
    solidissima raveneli: 187
    solidissima similis: 187
Spondylidae: 167
Spondylinae: 167
Spondylus: 47
    americanus: 166, 167
Sportellidae: 178–179
spotted periwinkle: 61–62
spotted sea hare: 139–140
squid: 33, 48, 228–229
starfish: 264
stepping shell: 144
Stenoglossa: 108–120
Stewart Beach: 234
sting ray: 264
stingaree: 264

stolonifera, Ipomoea: 14
Stomolophus
    meleagris: 263
stone crab: 265
stony coral: 264
storing shells: 255–260
stout tagelus: 200–201
straight-needle pteropod: 137
Stramonita: 108–111
Streptoneura: 54–126
striata, Bulla: 134
striata, Cochliolepis: 65–66
striata, Martesia: 222
striate bubble: 134
Strigilla
    mirabilis: 192–193
Strioterebaum: 125–126
Strioturbonilla: 131, 132
striped false limpet: 143
Strombacea: 93–94
Strombidae: 93–94
Strombiformis
    bilineata: 92
Strombus: 37, 254
    alatus: 93–94
subnodosus, Lyropecten: 166
substratum: 24–30
subulata, Cerithiopsis: 81
succinea, Haminoea: 135
suffusa, Trivia: 96–97
suffuse trivia: 96–97
sunray venus: 207
suppressus, Cyclostremiscus: 67–68
Surfside, Texas: 6, 234
sweet bay: 14
swiftiana, Corbula: 217
Swift's corbula: 217
Syringodium
    filiforme: 16

tageliformis, Macoma: 191–192
Tagelus
    divisus: 200, 201
    plebeius: 26, 200–201
tagelus-like macoma: 191–192
Tampa tellin: 193, 194
tampaensis, Crassispira: 120–121
tampaensis, Tellina: 190, 193, 194
tars: 269
taurinus, Terebra: 126
Taylor, Zachary: 6, 7
tayloriana, Tellina: 194, 196, 197
Tectibranchia: 139–142
Tectonatica
    pusilla: 101
Tegula
    fasciata: 56
Teinostoma
    biscaynense: 71, 72
    leremum: 72
    parvicallum: 72
Tellidora
    cristata: 194, 198–199
Tellina: 25, 47, 260
    aequistriata: 194, 196, 197–198
    alternata: 194, 196, 197

iris: 194, 195, 198
lineata: 194, 196–197
sayi: 195
tampaensis: 190, 193, 194
tayloriana: 194, 196, 197
texana: 193, 195
versicolor: 194, 195
Tellinacea: 190–204
Tellinidae: 190–199
tellin-like cumingia: 201–202
tellinoides, Cumingia: 201–202
temperature of coastal waters: 17
tenera, Barbatia: 151–152
tenta, Macoma: 192
tenta macoma: 192
tenuis, Cyclinella: 208–209
Terebra
    dislocata: 125, 126
    protexta: 126
    taurinus: 126
Terebridae: 124–126
Teredinidae: 224–225
Teredo
    bartschi: 224–225
    navalis: 225
testiculus, Cypraecassis: 103
testudinum, Thalassia: 16, 30
Teuthoidea: 228
Texadina: 62–63
texana, Raja: 264
texana, Tellina: 193–195
texana, Vitrinella: 73–74
Texas A&M University: 234
Texas City, Texas: 269
Texas Game and Fish Commission: xvi
Texas Highway Department: 233
Texas lepton: 178–179
Texas Parks and Wildlife Department:
    xvi, 240; Marine Laboratory of, 259
Texas quahog: 213
Texas Trails: 233
Texas tusk: 145
Texas venus: 205–206
texasiana, Aligena: 178–179
texasiana, Callocardia: 205–206
texasiana, Pitar: 206
texasiana, Volvulella: 136–137
texasianum, Dentalium: 145
Thaididae: 108–110
Thais: 24, 36, 37, 251
    haemastoma floridana: 108–109, 110
    haemastoma haysae: 109–110
Thalassia
    testudinum: 16, 30
Thecosomata: 137–139
Thericium: 78–79
thick lucina: 176
Thovana: 221
three-spined cavoline: 137–138
tides: 22–23
tiger lucina: 174
tinctus, Cantharus: 113–114
tinted cantharus: 113–114
tiny-calloused teinostoma: 72
tollini, Epitonium: 86
Tollin's wentletrap: 86

Tonna
    galea: 107–108
Tonnacea: 102–108
Tonnidae: 107–108
Toxoglossa: 120–126
Trachycardium
    isocardia: 182–183
    muricatum: 183
Trachycardium s.s.: 182–183
transliterata, Anachis: 112
transversa, Anadara: 152, 154
transverse ark: 154
treculeana, Yucca: 6
Tricolia
    affinis cruenta: 56
trifida, Odostomia: 130
trilineata, Pandora: 226
trilix, Cyclostremiscus: 67
trilix vitrinella: 67
Triphora
    perversa nigrocinta: 83
Triphoridae: 83
trispinosa, Diacria: 137–138
Trivia
    suffusa: 96–97
Triviinae: 96–97
Trochacea: 55–56
Trochidae: 55–56
Trona: 97
Tropical Trail: 233, 247
true spiny jewel box: 172
true tulip shell: 117
truncata, Barnea: 219–220
Truncatella
    pulchella: 62
Truncatellidae: 62
Truncatellinae: 62
tulipa, Fasciolaria: 117
tulipa, Modiolus: 156
tulipa, Volsella: 156
tulip mussel: 155–156
tumidus, Donax: 189
turbidity: 23–24
Turbonilla: 27, 131, 132
    elegans: 131
    elegantula: 131
    hemphilli: 132
    interrupta: 131
    portoricana: 132
    typica: 130
turkey wing: 149–150
Turk's-cap: 16
Turridae: 120–124
Turritellidae: 75
turtle grass: 16, 30
two-lined melanella: 92
two-toned tree oyster: 163
type terms: 39
Typha
    domingensis: 14
typica, Pseudoirus: 214
typica, Rupellaria: 214
typica, Turbonilla: 130

umbilicata, Janthina: 89
umbonata, Arca: 149

uncertain miniature cerith: 82–83
uncinata, Cavolina: 139
unequal spoon clam: 226–227
Ungulinidae: 176–177
unicate cavoline: 139
Uniola
    paniculata: 9, 14
uniplicata, Neosimnia: 33, 98, 250
U.S. Air Force: 8, 9
U.S. Coast and Geodetic Survey: 23, 252
U.S. Navy: 234, 267
University of Texas Marine Science
    Institute: 10, 241, 259

Vampyromorpha: 48
variabile, Cerithium: 79–80
variabilis roemeri, Donax: 189, 190
variabilis texasiana, Donax: 189–190
variable bittium: 79–80
Varicorbula
    operculata: 217–218
varium, Bittium: 79–80
Velasco, Texas: 9
Velella
    mutica: 90, 263
Vellucina: 174–175
Veneracea: 205–214
Veneridae: 205–213
Venerina: 204–214
Veneroida: 173–214
Vermicularia
    fargoi: 75
Vermiculariinae: 75
versicolor, Tellina: 194, 195
vibex, Nassarius: 115–116
vinca, Cerithiopsis: 81
Violetta: 91
virgin nerite: 58
virginea, Neritina: 58
virginiana, Quercus: 14
virginica, Crassostrea: xvi, 29, 130,
    157, 169–170
virgulata, Eugorgia: 98, 264
viridis, Solen: 188–189
viridis viridemaris, Smaragdia: 58–59
Vitrinella: 265
    floridana: 72–73
    helicoidea: 73
    texana: 73–74
Vitrinellidae: 64–74
Vitta: 57–58
vittatus, Clibinarius: 265
Volsella
    tulipa: 156
Volutacea: 118–120
Volvulella
    oxytata: 136
    persimilis: 136
    texasiana: 136–137
von Salis' Triton: 106–107
vulgaris, Octopus: 229–230

Wadsworth, Texas: 235
Washington Beach, Mexico: 247
waves and surf: 23
wax-colored mangelia: 122

weak-ridge cyphoma: 99
wedge-shaped martesia: 222–223
wentletraps: 83–89
West Beach (Galveston): 234
whelk: 25
White Atlantic semele: 203
white bearded ark: 150–151
white-crested tellin: 198–199
white miniature ark: 151
white strigilla: 192–193

white wentletrap: 84–85
widgeon grass: 16
willcoxi, Aplysia: 27, 139, 141
Willcox's sea hare: 141
wiregrass: 9
worm tube: 264–265
wrightii, Diplanthera: 16, 30

Xenophora: 37

Yarborough Pass: 11
yellow aster: 14
yellow cockle: 183
Yucca
    treculeana: 16

zebra, Arca: 149–150
zebra periwinkle: 60–61
zic zac, Littorina: 61

The Incredible
A. J. Foyt

The Incredible
A. J. Foyt

Produced by Lyle Kenyon Engel
and the Editors of Auto Racing magazine

ARCO PUBLISHING COMPANY, INC.
New York

Published by ARCO PUBLISHING COMPANY, INC.
219 Park Avenue South, New York, N.Y. 10003

Second Printing, 1971

Copyright © Lyle Kenyon Engel, 1970
All Rights Reserved
Library of Congress Catalog Number 76-103078
ISBN 0-668-02195-0
Printed in the United States of America
Typography by Elroy

Editorial Staff

Ross R. Olney
Donald C. Davidson
George Engel
Marla Ray

Photography

Kenneth H. Coles
Gene Crucean
Arnie de Brier
Eric della Faille
George S. Engel
Lyle Kenyon Engel
Joe Farkas
Ford Motor Company
Goodyear Tire & Rubber Company
C. V. Haschel
Indianapolis Motor Speedway

Bud Jones
Pete Lyons
John R. Mahoney
John P. May
Leroy Patton
Photo Graphics
John A. Pietras
John Plow
Dennis Torres
Dave Underwood
Jerry Vass

Contents

A Victory 9

A. J., Personally 20

On to Indy 38

The Developing Champion 52

The Good Years 66

The Bad Years 77

The Return of Tough Tony 91

The Professional Race Car Driver 106

Foyt is Championship Auto Racing 130

The Statistical Side of A. J. Foyt 140

Major Events Won 141

Summary of Indianapolis Record 145

Summary of Winnings at Indianapolis 145

USAC Record 146

Summary of USAC Starts and Top Three Finishes . . . 147

USAC Money Winnings 1956-69 147

Foyt's 12 Years at Indy 148

The incredible A. J. Foyt, perhaps the greatest racing driver in the history of the sport of automobile racing.

A Victory

It was a fine day for a race that Memorial Day morning in 1961, especially for a race on the grand old 2½-mile track at Indianapolis, Indiana. Everyone there was celebrating the Golden Anniversary of the Indianapolis "500," the grinding, long auto race which was first held in 1911. The sun was bright, the temperature warm, and the fans in a festive mood. Track officials had done everything possible to make this running of the Indy classic a motor race to be remembered.

But nothing the officials could think of doing to make this a memorable event came close to what two men did out on the track. Although 33 drivers were in the race and six or eight of them had a chance to win, it was two champions who made the show worth remembering.

One of these, Eddie Sachs, was a man with a personality that probably would have thrust him into view no matter what his occupation. Wherever he went, he created a stir. He was voluble, loquacious, a well-loved braggart and, more to the point, a superb handler of powerful racing cars.

He had never won the famous 500-mile race at Indianapolis, although he had tried a number of times.

The other leading driver on that warm afternoon was A. J. Foyt. He was driving a sleek Offenhauser roadster, white with red and black. The car carried a huge number "1" on its nose, showing that A. J. was the current National Driving Champion. He had won the title and the number by being the best driver along the Championship Trail the year before. He was quiet, serious, businesslike and cool—a professional who allowed his driving to do his talking. He, too, had never won the great 500-mile race.

9

Foyt leads Eddie Sachs during their intense driving duel at the 1961 "500" (above) and takes the checkered flag (below) for his first Indy win!

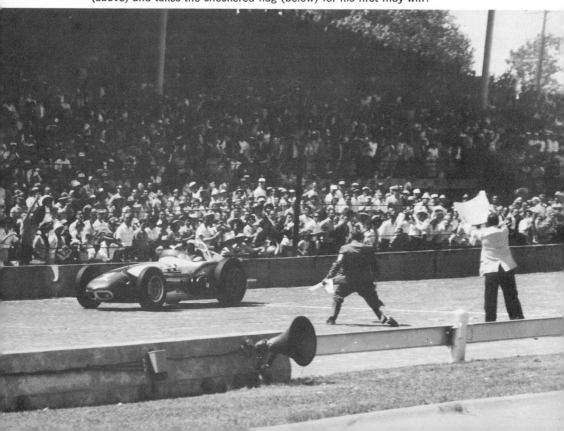

Foyt got an early start. This is his first car, built by his father.
The grim three-year-old looks ready to **race.**

There was something else riding on this race. The great Eddie had promised to retire upon winning a "500" at Indy. This, according to many fans, was the day the promise would be kept. For Eddie was the fastest qualifier in his Dean Van Lines Special, an Offy roadster much like Foyt's. He had captured the pole position, and so was the first man in the field. He had served his apprenticeship well in previous races. Many fans felt he was certain to win this one.

Most of the fans basking in the morning sun with their fried chicken and beer acknowledged that other great drivers like Jim Hurtubise, Rodger Ward, Don Branson, Jim Rathmann and Parnelli Jones must be reckoned with, at least for a close finish. But surely Eddie the Great would win the golden classic.

Other fans weren't so sure that Eddie would be able to beat the charging young champion who had made child's play of the Championship Trail the previous year. To win this race Eddie would have to beat Foyt, who had won

11

The youngster learned quickly. In this three-horsepower midget
he beat Doc Cossey, shown here beside the five-year-old.

four of those tough races, placed second twice and third in two more. After
all Foyt was the kind of driver who would either place at the top, or break.
There was no middle ground for him.

Foyt wanted desperately to add Indy to his growing list of victories. This
was his fourth year at the "500." He was, by any man's account, an experienced Indianapolis driver. He had served his grueling apprenticeship. His
Bowes Seal Fast Special was ready, poised on the inside of the third row as
the traditional ceremonies of Indy rolled on toward the eleven o'clock start.

Much of Foyt's mechanical skill came from his father, A. J. Foyt, Sr., here showing his son a tuning tip.

During this long time of mounting excitement, Foyt appeared unperturbed. He stood waiting calmly through the long renditions of "Taps," "Back Home Again in Indiana" and the nerve-shattering bombs which have come to be identified with the "500." Casually he adjusted his helmet, and slowly and carefully seated himself in his racer. Unconcerned, he watched as his pit crew buckled and tightened his seat belt and shoulder harnesses.

The volatile Sachs, on the other hand, was a bundle of nerves. His hands trembled, as they always did just before a race. His famous grin flashed

13

on and off like a hotel sign outside a rainy window at night. As always, he wept into his sleeve during the playing of the traditional music.

But at the command "Gentlemen, start your engines," both men, along with the other 31 on the tense grid, became professional racing car drivers, ready, waiting, nervous, skilled and eager to go. Every driver of racing machines—from cow-pasture jockeys to the gentlemen of the Grand Prix Circuit—wants to win this race.

At the beginning, soon after the thunderous start, it was a contest between Hurtubise, Sachs, Ward and Rathmann. Branson was in the middle until valve trouble sidelined him almost immediately. Foyt remained in the shadow of these racers, but Parnelli Jones joined the elite group up front. Gradually, with the advent of mechanical ills, crashes and miscues in the pits, the picture began to change.

For a while it was a fight between Foyt, Jones, Sachs and Ward, with the drivers in that order. Then Foyt, Sachs and Ward. Finally it became a Foyt/ Sachs duel. At first Foyt would lead, then Sachs would lead. Race record speeds were set, then broken as Sachs in his white and blue number 12 car and Foyt in his white, red and black number 1, pushed each other faster and faster.

On lap 141 the two cars screamed down the long straightaway side by side, within touching distance. The fans shouted at first. Then they grew silent in the face of the sheer ferocity of the contest. The race became almost unendurable in its intensity as the cars switched positions, never more than a car length apart. So close were they that one slip, one error by either driver, and both would go spinning into disaster. Sachs led, then Foyt, then Sachs, then Foyt again. It was magnificent racing.

On lap 152 Prince Eddie roared into his pit for his final fuel and service

15

stop. It was a smooth, precise operation, and in 24 seconds he was screaming back onto the track to chase a disappearing A. J. Then Foyt pitted for what he hoped was his final stop. Back on the track 22 seconds later, he was once again within a car length of Eddie, the Clown Prince.

Rodger Ward, still to make his final stop, was in the lead. Then his stop was ordered, and the Foyt/Sachs duel was out front again.

As lap 170 thundered into history, it was Sachs and Foyt, then Foyt and Sachs. Still the cars seemed glued together in this 500-mile sprint. Finally, gradually, calling upon every bit of his great skill, Foyt stretched his advantage over Eddie to three full seconds by lap 180 . . . to four seconds by lap 182.

But at that moment a Foyt pit man rushed out with a sign as A. J. streaked down the straightaway battling to hold his slim lead.

"FUEL LOW" it announced. Foyt nodded.

There was a hasty conference in the Foyt pit. Then they hung out a second sign on lap 185. This one ordered him in. Foyt skidded into his pit on command. His previous stop had not been a smooth one after all. A malfunction in the fueling equipment had prevented a full flow of fuel, so his tanks had not been completely filled. As repaired equipment forced just enough fuel into his nearly empty tanks, Foyt watched Prince Eddie thunder around turn one and out of view. With him, thought Foyt, went the greatest victory in motor racing.

But Foyt had never been a man to give up, even in the face of almost certain defeat. With Eddie 25 full seconds ahead and the race in its closing minutes, Foyt pushed his car to new record speeds. On the 195th lap the main straightaway was strangely silent. Only 12 cars remained in the race. Then from far down around turn four came Foyt. Larger he grew, then larger, shattering the silence with the thunder of his engine as his car streaked by straining to cut down Eddie's lead. But, with only five laps to go, the gallant effort of Foyt seemed wasted. For Sachs was on the backstretch maintaining his own record speeds.

Then with only three laps to go, Sachs started to slow down on the main straightaway. The strain of keeping up to Foyt's high speed challenge had apparently ruined Eddie's right rear tire. White fabric was showing through the thin tread. Eddie studied the tire from the cockpit, and made his decision. It would not last through another three laps. He needed a new tire to finish.

Into his pit he roared, where his frantic crew was ready with tire changing

18

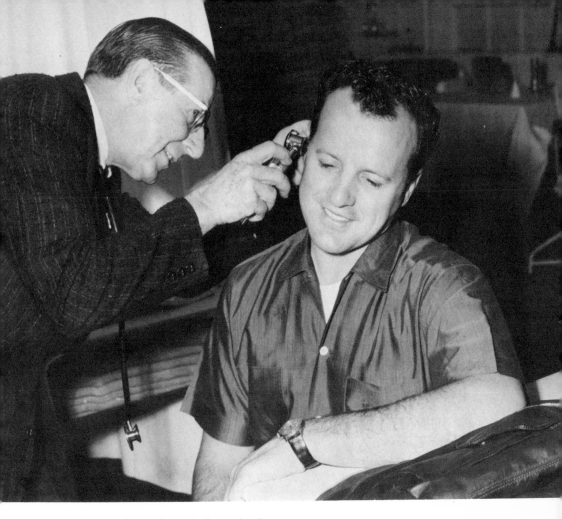

Every Indy driver gets a thorough physical examination.

equipment. In seconds he was back on the track, but he was too late. Foyt had thundered by into first place, a casual glance at the Sachs pit his only indication that he was aware of the reversal of positions once again. Meanwhile, thousands of fans screamed themselves hoarse, first for A. J., then for Eddie.

With one lap to go a jubilant Bowes Seal Fast crew, leaping and jumping and dancing, hurried to Victory Lane. A frustrated Dean Van Lines crew watched them go. Then, with a burst of sound and a jaunty wave of his hand, A. J. Foyt took the checkered flag for his first Indianapolis victory. Moments later he was being mobbed in Victory Lane, while Eddie Sachs, a grin still on his face, was acknowledging the cheers of the fans for his well-earned second place as he guided his oil-stained car into the garage area.

19

A. J., Personally

An A. J. Foyt is a one-of-a-kind item. He is physically impressive. He looks the part he plays, and he has the temperament to match. He is an absolutely professional racing driver

Any man who lives and works in one of the world's most dangerous professions and who has become one of its top men might be temperamental. And Foyt is. One moment he will snub you cold, the next he will welcome you with a Texas-sized bear hug and a yelp of a greeting. At times he is moody and completely unwilling to talk. At other times he is the center of the group, regaling everyone around with a wild racing story . . . or some other type of story.

Foyt has grown with his increasing wealth. He is a far more polished man than he was at the beginning of his career, although his hot temper still gets out of hand on occasion. Now he can meet famous people on their own grounds without stammering, and he can stand before an audience without blushing.

Still "Tough Tony" (as some fans call him) can back up any of his outbursts with hard fists and honed muscles if the unpleasantness must go that far. He is in fine physical shape—the only tougher sport than driving a racing car is supposed to be heavy-weight boxing—his only problem being a hint of plumpness. He doesn't have to squeeze himself to get into the narrow confines of a racing car cockpit, but he does wiggle a bit more as the good life increases his girth. Why not, his fans ask. He could tow a trailer to carry any flab he cares to add on (and he is by no means flabby) and still beat most of the other hot shots.

During a break in practice at Mosport, Canada, prior to the running of a stock car race, A. J. tells former USAC Stock Car Champion Don White a story. Don listens attentively and then starts to break up. . . .

But Foyt tells him there's more to come. White quiets down and when
A. J. comes to the punch line, they both crack up completely. . . .

Since Don was such a good audience, A. J. calls over NASCAR ace Richard Petty to try it out on him . . . and it worked.

That could almost be true. In any case, a Foyt fan is a rabid fan. He can see no other man on the track but A. J. He cannot imagine Foyt's losing a race. To him the battle on any track where Foyt runs is for second place. Recently, a bedsheet sign count at California's Riverside International Raceway during the Rex Mays Memorial 300 road race for Indy cars, turned up as many for A. J. as for hometown-boy Dan Gurney. Dan practically owns Riverside, and he's so popular that he could win an election for the presidency

23

Foyt can also take a joke. Challenged to allow this pet lion to knock him down, Foyt runs as instructed . . .

and is quickly brought down. . . .

However, the 385-pound kitten wouldn't get off A. J.'s back and began mouthing at his uniform, as the trainer runs to his rescue. . . .

of the United States if it were held there. Wouldn't take more than a little gerrymandering either.

Foyt stands five feet eleven inches tall and weighs a solid 185. He lives in Houston, Texas with his striking blond wife, Lucy, a native Texan. They have three children: A. J., III, age 14; Terry Lynn, age 11; and Jerry, age 7. They live in a lavish ranch house with all the comforts.

As a gentleman ranch owner, Foyt is smooth, polished and gracious. As a racing driver though, he's been called names a lot worse than "no gentleman."

Freed from the lion's clutches, Foyt hastily scrambles to his feet. . . .

Although he laughed about the entire incident—sort of —Foyt did admit to having been a little concerned and **was quite** relieved when it was all over.

The fact that A. J., the most recognized driver in auto racing, must answer a roll call seems to amuse other ace drivers.

He always has been, and remains, difficult to work either for or with, according to the men who've tried. A. J. is a skilled mechanic in the very best sense of the word, and he knows it. He has a delicate, accurate and absolute feel for engines and racing cars. Along with this rare ability, he has an equally absolute inability to tolerate errors or lack of judgment or even failing enthusiasm in others. Foyt is always up for a race, and everyone working with him must share this enthusiasm.

He may hire the best chief mechanic in the business (he can certainly afford to), only to plunge into the engine himself shoving aside the man hired for the job. The better the man, the less apt is he to like this kind of treatment. To make matters worse, Foyt does this with an almost total lack of diplomacy because he is so certain that only A. J. himself can find and fix the problem. Adding injury to insult, he's often right.

Only Foyt's own father, A. J. Senior, has stuck with the hot-tempered champion· on a long-term basis. A. J. Senior has crew-chiefed for his now famous son off and on from the beginning. At this writing he appears to have the job firmly in hand for good.

Foyt will drive anything. A hefty A. J. leads a sulky race (above) and shows a trophy (right) for a turbine-powered truck speed record.

Racing mechanic George Bignotti endured the champion for six years, and they were good years, before leaving the Foyt team permanently. The two would argue and make up. They broke up. Then they discussed their differences, the good each was doing for the other, and got back together. When Foyt was on a winning streak, they worked together as smoothly as any two such temperamental men could. When Foyt lost due to car failures, they fought constantly. Bignotti finally quit for good after a violent argument at Langhorne in 1965.

Then Foyt hired the great Johnny Pouelson, an ex-Parnelli Jones mechanic, and this arrangement lasted for about a year. Then, at Indy, the Foyt crew worked late into the night preparing two cars for the upcoming races. Still, one of the cars wasn't just right . . . and A. J. was unhappy about it—heartily and vocally unhappy about it. He wanted the racer fixed, and he wanted it fixed NOW.

Although he was beat and dirty after two sleepless nights spent working on

As a "USAC All Star" he competes in a basketball game to benefit the USAC Benevolent Fund. Action is at the Indianapolis Coliseum in 1969.

Clowning with photographers, he attempts to reflect sunlight at them with a silver platter award received at Indy before the '69 race.

Foyt particularly enjoys younger fans, often lifting them over the fence to pose for a quick snapshot.

the cars, Pouelson agreed to keep on working. Since it was then early morning and he had gone straight through without even a break for meals, he wanted to slip away for a quick bite and a shave first.

Angrily Foyt said, "No!"

Pouelson had his meal anyhow, and then returned to pick up his personal gear. He sold his tools on the spot and went home for good.

Foyt has lost other crewmen in much the same way. Yet, almost without exception, they later say that they think he is the greatest driver in the history of the sport. And they say it with no apparent rancor. They just admit that—like any other great artist in any field—Foyt is very, very difficult to work with.

It's hard to imagine how a man with such a hot, demanding temper, such an unshakable faith that he is right and everyone else is wrong, can be as cool as Foyt is on a race course. His temper hasn't cost him any fans or many hangers-on from all appearances. He still attracts droves of people wherever

uring a race Foyt is a serious, "take charge" man. Out himself at the 1969 'over Downs 200, he becomes a pit man for teammate Roger McCluskey.

le can get angry, as many fans know. Here he grumbles when e learns that youngsters have broken his motorcycle.

he goes, and generally—unless a tough race is just moments away—he is cordial and friendly.

Most fans agree that you can ask little more from a real champion than this. They have been around other drivers on race-day morning, and have seen them grump and grouse—even cold-shoulder personal friends. These men are prima donnas. Foyt is not.

To A. J. a job of racing is a job. Certainly he enjoys his work, but he's not one to get all worked up over it. He races for a living, period. In the end he knows he'll win more races than most of the prima donnas put together.

Before the 1969 Hoosier 100, he steps between his father (left) and car builder Wally Meskowski (far right) to adjust tuning. Foyt won the race.

And along the way he will probably have shattered the nerves and egos of not only the other drivers but of his pit crew and sponsors as well. He will calmly change sponsors as well as crew for a better deal in an instant, and they all know it. At the 1964 Firecracker 400, Foyt switched from Ford to Dodge in a last minute move that drove the Ford people wild . . . and wilder still when he brought the Dodge home in first place. Yet Ford welcomed him back, because they knew they could get none better.

Foyt will play one tire company against another to get a better deal for himself, and they must accept it. In 1964 he won the "500" on Firestones, then popped out of his car wearing a Goodyear driving suit. A few weeks later he was riding on Goodyears and wearing a Firestone cap.

Maybe the final answer for Foyt would be to do what the Unser family did. Since one Unser son was riding on Goodyears and the other on Firestones, the parents wore logos which said "Goodstone" and "Fireyear."

In spite of all his changes, Foyt is quite honest in his dealings. He will not go behind the back of a sponsor to deal. He will merely tell them straight out that another man has offered a better deal to use this or that tire or other piece of equipment. The first man can either beat the offer or get out.

Foyt is probably the original fish-or-cut-bait-man. Play or don't play, it doesn't matter to him.

He is just as brutal in his dealings with people who want his endorsement

34

of a product. Many of the top drivers in auto racing earn a fat extra income by endorsements, as do many champions in all forms of sports. Foyt must actually use and like a product before he will endorse it. Then he may do so completely free of charge, to the great delight of the manufacturer. One time he turned down several thousand dollars of such easy money. He didn't use the product, so he refused to say he did.

It was that simple to A. J. How many baseball stars do you know who actually eat the crunchy stuff for breakfast, or use the hair grooming cream they praise on television? If Foyt says he does, then you can believe him.

To add to the Foyt paradox, he will go out and drive hard for low purses, even though he is by now a millionaire sportsman. He will show up at an obscure dirt track, having spent more to get there than he could possibly win. Then he will drive flat out against young drivers who have everything to gain by beating him. They seldom do. They would like to win, but Foyt *must* win. Winning is his life, the air he breathes and the food he eats. Nothing else matters, no matter where he is racing, or in what.

He is content if he has won. He is happy and congenial. When he has lost, he is ill-mannered and ill-tempered and unhappy, regardless of the importance of the race. They are all important to A. J. His inner drive to win keeps him from ranking them.

He is wealthy, he has won more open cockpit races than any other driver, he needs nothing. Still he entered the risky Ted Horn Memorial Race at DuQuoin, Illinois in 1967, and he won it. His winnings were a little more than $7000. He had made over 25 times that much at the Indy "500" a short time before. He stood to double that amount if he stayed alive and healthy for endorsements. But he still raced flat out to win. In 1964, one of his finest years, he made over a quarter of a million dollars in purses alone, yet he went to Ascot Speedway in Los Angeles, a rugged little dirt track, to race with local boys for a few dollars. He does that sort of thing to this day.

The day before the "500" at Indianapolis most of the qualified drivers have but one thought in mind—how to win tomorrow. In 1961 Foyt must have been having those same thoughts, but he reported to a muddy track nearby to race for a total prize of only $800. He won it, of course. But what if he had broken a finger, or sprained a wrist? He is sure he won't. He has been extremely fortunate in this respect throughout his career. Still the hungry young drivers would have a lot to gain by pushing Foyt through a fence and beating him.

Later that same year, after Foyt had won at Indy, he pulled into the pits of a dirt track at Terre Haute, Indiana for a relatively minor race. He was one of two or three drivers who would brave the slick track in qualifying attempts. Later the sun came out and dried the mud. The other drivers qualified on a

much better surface and were quicker. Super Tex was bumped from the field. Angry, he paid the last man in the field $100 for his ride. The man, who had no real chance of winning, quickly and happily accepted Foyt's offer, and Foyt jumped into the strange car. In an amazing display of driving skill he passed everybody else in the race and won it. He lost money, but he won the race—and that was what mattered to Foyt.

Warming up for a stock car race at Milwaukee in 1964, he blew his engine when it was too late for repairs. Again he took over the last machine in the field rather than watch from the sidelines. In the first two laps he passed 13 of 29 other cars. By the end of the race he was far ahead, and in an inferior machine.

With A. J. Foyt at the wheel, no machine can be considered inferior. He seems to have the capacity to add whatever the car needs just by being in the driver's seat.

If you see A. J. Foyt lose a race, you know that he is going to suffer until the next one. So will everybody else around him. He is not a diplomat or a good loser.

You have met people who give and people who take. The takers seem to know exactly what they want and, within their own limits, they grasp for these things. They lay everything on the line. They take chances time after time. They go all out, and if they survive they take the fruits of victory. This is true in life, and especially true in auto racing.

Foyt is one of the takers. He knows what he wants—to be the best, to be the champion, to become wealthy from racing and then to walk away from it when he chooses to leave—and he has taken it so far. He is a major threat in any race, anywhere, anytime and in any car.

Yet Foyt is not a "get the hell out of my way or I'll put you through the fence" kind of driver. People who know him and who have raced against him say that he has one of the strictest personal codes of honesty in the sport. He'll beat you, but he'll beat you fair—or at least what he considers to be fair.

37

On to Indy

A. J. Foyt was born on January 16, 1935. His father, Anthony Joseph, Senior, was a racing driver and a skilled mechanic. Young A. J. could not stay away from his father's garage, his father's cars and his father's racing friends. When he was only three he was driving his own miniature racing car.

At five he had his own midget racer which he drove between feature events at local tracks—at speeds up to 50 miles an hour. Once he took his three-horsepower racer to the Houston Speed Bowl and, with his father as his crew chief, challenged the local Doc Cossey.

Much to Cossey's chagrin, the youngster won the race. Cossey might have felt a little better about it if he had been able to see into the future. For his record now shows that Foyt has an inborn skill, impossible to learn and possessed by very few men. This is the skill to drive a racer faster than anybody else.

Foyt couldn't keep away from racing cars. Once his father and mother took one of their midget cars to an out-of-town track for a racing event. They left their backup car (and young A. J.) at home. When they returned they found the back yard torn into great ruts, with grass thrown against the fence and the house. The racer, scorched and dirty and still smoking, was in the garage. In his bedroom "sound asleep" was an innocent-looking A. J. He had raced the car around the yard until it caught fire.

Young Foyt quit school in the 11th grade to race cars for a living. Though this was a decision which would be a disastrous one for nearly any other teen-ager, it was not for Foyt. He has been racing cars ever since. By now he has become rich beyond his own wildest expectations. He worked his

Ace mechanic George Bignotti stayed with A. J. for six years through break-ups and make-ups

From unknown rookie to great champion, sometimes called "Super Tex."
Bobby Allison jokes, "There really **is** an "S" on your chest!"

hree great drivers. Mario Andretti, Jim Clark and A. J.

way up through the dangerous dirt tracks of Texas in midgets, sprints and stock cars until he became the area's leading driver.

Like every other young driver, he dreamed of racing at the track in Indianapolis, and he got there in 1958.

This was not Foyt's first visit to the old Brickyard. The three years before he had saved his hard-earned winnings to attend the race as a spectator. Each of these years he watched and studied from the stands. He drifted around the track, examining the curves and the surface flaws. He talked to the drivers, and they advised the youngster who wanted to be a great champion.

Then, thanks to his showing in other races, he was offered an Indy driver's test in the Dean Van Lines Special. This was the car formerly driven by National Champion Jimmy Bryan.

Foyt was an unknown commodity at Indy in 1958. His previous racing career was shorter than most of the other rookies. Yet it was impressive. His smooth, easy driving style had pushed him into the headlines during the 1957

41

At the 1966 12 Hours of Sebring Foyt discusses strategy with John Holman.

driving season. That year he had campaigned midgets, but soon he was noticed by big car owners and offered rides both in Midwest sprints and in Championship cars. He did well in both divisions. He had not finished at the top in either division, but he was showing fine promise. By the end of 1958 he had won enough to finish 10th in the Championship division and second in Midwest sprints.

In May though, it's Indianapolis time. This Midwestern city draws speed lovers from all over. Many drivers consider it the most important race in the world. They can lose everywhere else, but if they win here the season is a smashing success. However, this does not happen often. This race is one where experience counts.

The track must be learned. To the casual fan it seems to be a simple matter of four straightaways—two long ones and two short ones—and four identical curves. It may also look that way to the rookie. But the experienced driver will quickly explain that Indy is a tough track to learn. It is a very demanding

Ex-champion and two-time Indy winner Rodger Ward interviews him for television program "Wide World of Sports."

track. Each curve is slightly different, just different enough so that a driver can never relax, and each straightaway has its own unique personality. In almost every case, with only one exception in recent years, a veteran will win at Indy.

That exception was rookie Graham Hill in 1966, if Hill could honestly be considered a rookie. He was new to Indy, but he was a former World Driving Champion. And unlike many Indy rookies, he took the track slowly and gently. Rather than battling the track, he wooed it. He didn't try to push. He didn't try to run with the leaders. He sped along biding his time as the men before him dropped out with mechanical ills (after a first lap crash eliminated several favorites). In the final 10 laps of the 200-lap race, the last man in front of him dropped out on the backstretch with an oil pressure problem. And a surprised Hill took the checkered flag.

This seldom happens at traditional old Indy. Foyt's first Indy race was eight years before Hill's, so he probably didn't really expect to win. Generally a first-timer at Indy races for experience, hoping that in the not-too-distant future he can bring his car home in front. Foyt was a shrewd driver from the start, so he must have been aware of his very slim chances for victory.

Indy rookies must take a driver's test before they are allowed to even practice at high speeds. This test consists of many supervised laps at pre-determined speeds as a way of showing a new man's driving ability. Foyt passed the tests with ease, and went on to practice and qualify.

Thirty-three drivers and cars race at Indy on Memorial Day, no matter how many show up to practice during the month. It is not unusual for 80 or 90 drivers to enter, but only the 33 fastest are permitted to race for the big money. The officials feel that more drivers than that would overcrowd the track during the race.

The qualification runs are to cut down the huge entry list. As a general rule (barring rain or other unforseen circumstances), four days are set aside for qualifications. Usually these are the two Saturdays and the two Sundays before Memorial Day. On these four days everyone is given a chance to run alone on the track for four laps. The fastest average speed for those four laps on the first day of qualifications will win the driver the pole position in the starting field. That is the inside spot on the front row. The second fastest driver wins the middle of the front row for the start, and the third fastest the outside position in the front row. And so it goes. The qualifiers on the second

Being presented with one of his strangest trophies, a 100-pound watermelon, for winning the 1969 Hoosier 100

day fill in behind the first day men. Then the third day, and the fourth day. Obviously the field for the start is not entirely arranged by average speed. Some of the faster men are farther back in the starting field, men who for some reason or other were not ready to make their qualification run on the first day.

After the field has been filled, the "bumping" starts. The 33rd fastest man, the slowest man in the original field, is in the bump position. Other drivers still waiting may bump this man by averaging a faster speed than he did. In the end, the 33 fastest are in the starting field.

A. J. Foyt's first year at Indy is a fine example of the bitterness of these qualifying speed duels . . . and of what that bitterness can cause.

It had been a battle for speed honors between Dick Rathmann and Ed Elisian, both Indy veterans, since the beginning of May. First one would raise the average speed for the year, then the other. They were the speed kings of the month. No other driver came close to their speeds. Foyt was on the track, of course, but neither his speeds nor those of anyone else were of any great note. So everyone watched Rathmann and Elisian to see who would break the speed record next.

This duel carried over into qualifications, with Rathmann barely winning the pole position from Elisian, who took the middle front row spot. They were the two fastest men in the field. Beside them was a relatively inexperienced young driver, Jimmy Reece, who had captured the third fastest spot.

Foyt, in a fine qualifying run, won for himself the outside position in the fourth row ahead of 21 other drivers including some noted veterans. It was a marvelous performance for a rookie. But few people were watching Foyt in those days. He was unproven and almost unknown. Fans were watching the great drivers of the day—Rathmann (both Dick and Jim), Elisian, Pat O'Connor, Johnnie Parsons, Jimmy Bryan, Tony Bettenhausen and Johnny Thomson.

The spectators were waiting eagerly for the start of the race. This is one of the most exciting moments in all of sports, when 33 finely tuned machines thunder into the first turn of the first lap in an ear-shattering display of noise, power and speed.

Some people were worried. Never before had two drivers fought such a bitter speed duel during practice and qualifications. What would Rathmann and Elisian do at the start of the race? They were side by side in the front row

A portrait of A. J. Foyt, taken at the 1968 Riverside 500 for stock cars.

and next to them was young Reece, who had never been known as a driver who would back off from a challenge.

The normally perfect start of the "500" became a near comedy. In an attempt to make the start more interesting for the spectators, the officials tried out a complicated new system. The result was that the front row, perfectly aligned, was far ahead of the rest of the field, also perfectly aligned, with the start of the race only seconds away. An extra lap for alignment was quickly allowed, and the front row was barely in position when the green flag dropped.

Some drivers were confused. Others were forging ahead. Had the race started, or not? It must have started, for the leaders were screaming through the first turn at a record pace, fighting for an inch or so of advantage. The 33 threaded their way through the first turn in a lengthening line, then the second turn.

On the backstretch, the duel that had been fought all month flared up again. Rathmann and Elisian were wheel to wheel, only this time Jimmy Reece was close behind them.

The third turn rushed forward to meet them, but neither man would back off. Farther behind, the rest of the field thundered along in a close bunch. It was a harrowing moment, a moment in which veteran fans along the backstretch felt chills under their scalps. It was almost as if they could see what was going to happen.

And *still* neither man slowed.

Finally, at the last possible instant, Rathmann took his foot off the accelerator. He had been challenging Elisian on the outside, and realized that Ed would not let him pass. Elisian was in the "groove," the best part of the track for the upcoming turn. At that instant Elisian also backed off a bit . . . but his move came a split second too late. His car wobbled slightly, a sign that it was almost completely out of control. Skittering up out of the groove, it seemed to brush the speeding car of Rathmann. Both cars screamed into the outside wall. Elisian's car was completely destroyed. Rathmann's was torn in two. The real carnage was to follow, for the rest of the drivers were rushing onto the scene, Jimmy Reece in the lead.

50

Reece, seeing the wreckage before him, slammed on his brakes to avoid the cars and car parts spinning about. He was nudged by Bob Veith, who was also fighting for control. Pat O'Connor, directly behind Reece, slammed into Reece's skidding machine at full speed. O'Connor's car leaped over Reece's, struck the track upside down, flipped back over, skidded and burst into flames. Pat didn't know any of this. He was killed the moment his car struck the track.

The other drivers behind fought to avoid the spreading accident. Johnnie Parsons crashed into Bob Veith. A. J. Foyt spun hard to miss the cars ahead. Jerry Unser, elder brother of current champions Bobby and Al, climbed over Paul Goldsmith's car, leaving a clear tire mark on Goldsmith's helmet and sailed on over the outside wall. Other cars bounced off each other and spun into the infield. Safety men rushed back and forth, dodging the spinning cars, trying to see which drivers needed help the most.

Fourteen cars were directly involved in the terrible crash. A hush fell over the 200,000 spectators as they waited for news. The third turn at Indy looked more like an auto wrecking yard than a race course. There were smashed race cars and car parts strewn over the track and the infield.

Foyt, who had managed to skid out of the worst of the wreck, would be able to continue in the race. He was one of the lucky ones.

Eight other cars had been destroyed, and nine more required varying amounts of work in the pits before they were able to go on. Pat O'Connor, a fine young driver and a favorite of the fans, was dead. Others were injured, shaken or stunned. For 21 long, slow laps the remaining racers ran while the track was being cleared. Then finally the green flag came out once more, and the race resumed at full speed.

While Ed Elisian, uninjured, sat on the pit wall with his head in his hands, a picture of dejection, Jimmy Bryan won the 1958 500-mile race at the Indianapolis Motor Speedway.

A. J. Foyt lost control and spun out once again on the 148th lap. He ended up 16th, a finish well back in the field. It was the last time A. J. Foyt ever left any race unnoticed by the fans. But he had survived his baptism under Indy's fire, and would return.

The
Developing Champion

A. J. Foyt was 24 years old and a fast rising star. He impressed many European experts after the 1958 "500" when he joined a group of American drivers who invaded the race at Monza in Italy. The Americans, in this race of "Indy-type" cars, overwhelmed the European racing drivers, breaking every track record. Jim Rathmann was the winner, with Jimmy Bryan second and Foyt sixth.

The Championship Trail was Foyt's real challenge during the remainder of 1958, and he finished the year with an excellent record for a yearling driver— or any driver, for that matter. He drove in 10 Championship races, and in three of these he finished in the top 10. He finished the season in 10th place, far ahead of many more experienced drivers.

Foyt also drove on the Midwest sprint car circuit, and ended up in second place behind the great Eddie Sachs.

The Foyt team competed relentlessly throughout 1959, and with increasing success. In May A. J. reported in at the Indianapolis Motor Speedway determined to improve on his previous year's showing.

One of the handicappers of this race, W. F. Fox, the sports editor of the Indianapolis News, said of the Foyt team, "Driver popular. Good sprint car man. Second start here [at Indy]. Ran 148 laps, spun out [in 1958]. New Kuzma car now has engine out of Ansted car which is reported to have 60 extra horsepower over other Meyer-Drake power plants. Clint Brawner, top ranking mechanic. Owner Al Dean spent $8,000 for Ansted engine. Not hungry, but determined."

Foyt was still relatively new, and reporters hadn't learned yet that he is always hungry for every victory in sight. Fox went on to rate Foyt at 25 to 1

52

Foyt won his first National Driving Championship in 1960. Here he
congratulates Al Unser for winning his first Championship race.

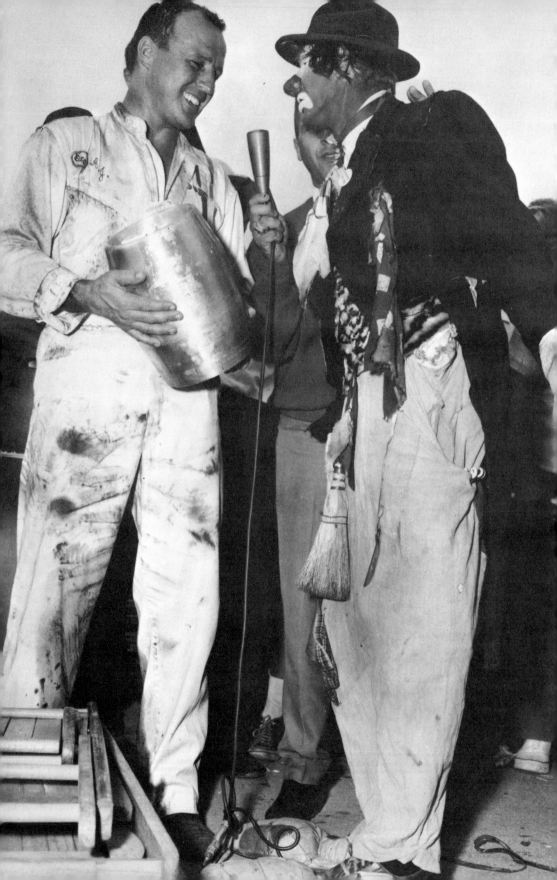

yt with Jackie Stewart at Indianapolis Raceway Park
ior to the 1967 Yankee 300.

yt accepts a keg of beer from Yano the Clown
ter winning the 1965 Hoosier 100.

on his chances for a victory. Rodger Ward was rated at 9 to 1, the third best (according to Fox) in the field. Johnny Thomson was rated at 8 to 1 and Jim Rathmann at 7 to 1.

As things worked out, Fox wasn't far wrong. Foyt's first qualification run was too slow, and he was flagged off after three laps so that the run would not count. Later in the day, the third of qualifications, he pulled onto the track again and this time posted a safe 142.648 miles per hour. This put him in the middle of the sixth row for the start.

Foyt ran a good race in 1959. By lap 105 he had worked his way into fifth place. His pit stops were as smooth as glass. Each one of the three he made was for the same things . . . two rear tires, the right front tire and fuel. He finished in 10th place and took home, for the first time, a good amount of the attractive Indy gold. Rodger Ward won the race after a blistering duel near the end with Jim Rathmann, who placed second, setting a new speed record along the way.

A typical sight at race tracks—Foyt personally checking
on every detail of his car before a race.

56

motorcycle is the fastest means of transportation around the tracks.
Here he stops to chat with a fan.

Enjoying a rare moment
of relaxation with Joe Leonard
before a race at Langhorne.

Rodger Ward, who was carrying the number "1" of the National Champion on his car the following year, probably realized that his position was somewhat shaky. He had been paying close attention to Foyt's climb. Toward the end of 1959, Foyt began to win with greater and greater regularity. He won the USAC sprint car race at Salem, Indiana in September—100 laps on a ½-mile track. In October he won the USAC sprint car race held on the ½-mile paved track at Houston, his hometown. In November he won the USAC midget car race at Corpus Christi, Texas. Foyt was learning.

Although he dropped out of the 1960 "500" with a broken clutch after only 90 laps, and was placed 15th at the finish, he stormed the other circuits. A. J. had come into his own, and his superb natural skill was finally meshing with the experience he had gained. In 12 Championship races, he placed in the top three eight times, winning four of them. In the Midwest sprints, the Eastern sprints and the midget division, he was almost as consistent.

The final race of the 1960 season was at Phoenix, Arizona on November

Discussing race strategy with long-time rival Parnelli Jones. Though still an active driver, Parnelli no longer races at Indianapolis.

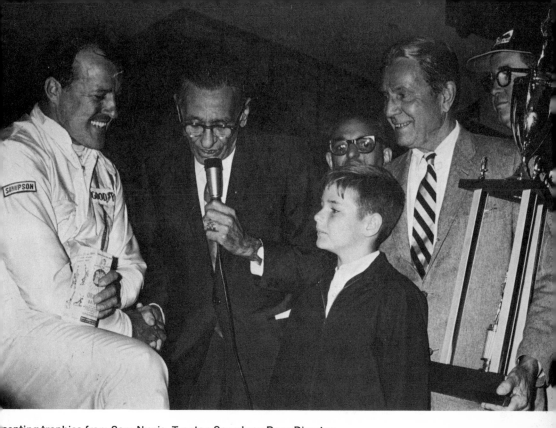

cepting trophies from Sam Nunis, Trenton Speedway Race Director,
Tony Hulman, Indianapolis Motor Speedway owner, looks on
bove) and from Linda Vaughn, Miss Hurst Golden Shifter (left).

20th. This Championship course is a one-mile oval track. The race was
scheduled for 100 miles, or 100 laps. All the top drivers were there.

The two top ones were Foyt and Rodger Ward, the only ones with enough
points to win the National Championship. Foyt was leading in the standings,
but Ward was a close second. Ward needed a clear-cut victory at Phoenix to
keep his championship.

On the other hand, Foyt could finish in fifth place if Ward did not win the
race, and still take the championship. If Foyt was fifth or better, Ward just
would not get enough points from a second place finish.

Ward, of course, arrived at Phoenix ready to drive as hard as he could to
win. He could only hope that in the meantime Foyt would either drop out,
or at least place poorly.

Foyt's strategy was typical of Foyt. If he won the race, then Rodger
couldn't. That would settle the matter cleanly and simply.

By the 22nd lap Foyt was in the lead, and Ward, in a gallant effort to stay

61

with the charging A. J., had blown his engine. Ward was out of the race and out of the championship. With his chief competitor no longer a threat, Foyt could afford to slow down and nurse his car along. The championship was his no matter what happened.

Instead, Foyt charged on. A champion drives to win, and so Foyt did. He increased his speed and his lead, leaving no doubt in the minds of the spectators that he was indeed the new National Champion. At the end of the contest Foyt was leading his nearest rival, Jim "Herk" Hurtubise, by two full laps.

In only five years A. J. Foyt had thundered to the top of the sport. Most drivers have never reached it at all. Foyt did it with marvelous consistency. In his first year of competition he was ranked 109th (in midgets). The next year he climbed to 26th in Championship cars, then 10th, then fifth and then first. On his cars and on his helmet, he could proudly carry the number "1."

Foyt enjoyed being number one. He needed winning to live, to be happy and content. He has that spark inside which will drive him to the ragged edge of disaster before it will allow him to quit. This spark can be seen at every race, important or not so important. In 1960 he won nearly $60,000, yet he drove like a poor man anxious for a $50.00 purse. He hadn't won at Indianapolis, but he was one of the most recognized drivers in the sport. His picture began to appear in ads and on television. Reporters flocked to interview Foyt at a race track, ignoring past Indy winners. If ever a driver was destined to win the "500," it was A. J. Foyt. It was only a question of time.

And not much time at that, for in 1961 he fought his dramatic duel with Eddie Sachs and won the race at Indianapolis. Nobody was surprised to see A. J. at the pinnacle of the sport, especially not A. J. himself. He had known all along that he would be a great winner.

Winning didn't change A. J. one bit. He was more difficult to get to, to interview and to chat with. This is true of any great winner. But he was just as difficult, or easy, to deal with as ever before. If his mood was right, he would give you the shirt off his back. He would recognize fans he hadn't seen in a year, calling them by name and stopping for a friendly chat. He would play on the track, and cavort in the pits at any race. Nothing, it seemed, was sacred to A. J. Foyt—other drivers, USAC officials, anyone or anything else. A practical joke usually had A. J. somewhere in the middle of it.

At other times, he would sit morosely and stare through anybody who approached him. Silent, brooding and uncommunicative.

trolling through the pits with Mario Andretti, his chief competitor on the hampionship Trail. Only once in a seven-year span (1963-69) did one f these two drivers fail to become the National Driving Champion.

In 1961 Foyt won four Championship races, including Indianapolis, six sprint car races and nine midget races. In actual purses, he took home more than $130,000. He was rapidly becoming one of the top money winners in the sport. His other, related business interests were making him a wealthy man.

His success didn't slow him a bit on the track. He continued to charge against all competition. Other top drivers have been known to take appearance money for driving a quick lap or two of a cowtown track before pulling in with "mechanical trouble." If Foyt appeared (and he would take the money just as fast as any man), he appeared to win. More often than not, he did. If, as happened occasionally, he really did break down, he would brood and sulk and wonder if he had cheated the fans.

His personality has remained much the same throughout his career. He may grin and wave at you from the starting line, but he may not even be seeing you. He may kid around in the pits, but his preoccupation is with winning the coming race. That is what marks A. J. Foyt, a total preoccupation with winning by any means possible. He has been fined for unsportsmanlike conduct, but most of his competitors say that he is totally fair and honest on the race course. He just drives to the sound of his own drummer.

He is known to be explosive. One time he might laugh off a remark or an incident, the next time he might take a swing at the offender. Except for his ability with a racing car, he is unpredictable.

Many motorsports writers dread the job of interviewing A. J. He is not a pal with whom you can talk. He has no ready answers at hand, which makes him difficult to interview. He may answer questions with one word, or even look away as though he had more important things on his mind. The easy drivers to interview are the ones who elaborate, who give quotable quotes, who add sparkle to an answer. As a general rule, Foyt is not one of these.

An interview with A. J. usually runs something like this.

"Well, A. J. how does it feel to be the winner of this great race?"

"Fine."

"Tell us about your car, A. J. How did it perform?"

"Just fine."

"Any problems on the track? I mean what about that near miss on the 60th lap, or that battle with so-and-so. Tell us about those, A. J. Did they worry you?"

"No."

"Yes . . . well, I see . . ." At about that point the interviewer begins to feel as if he's talking to himself. Desperately he goes on with the interview. Meanwhile, A. J. is looking at a pretty girl in a tight dress a few feet away. He is smiling coolly, even if the interviewer is not.

"Well, A. J. where do you go from here? What are your plans?"

"Langhorne," answers Foyt, still not looking at the harried reporter or his microphone.

"Ahem . . . yes . . . and what about your crew, A. J.?"

"They're just fine . . . the best."

"Well . . . er . . . thanks very much, A. J. Foyt."

"Thank you," will be A. J.'s response. His mind as he turns away will already be either 100 miles away on the next track and the next race, or on that pretty girl, a few feet away.

osing for photographers after winning the 1964 Hoosier Hundred.

The Good Years

After his Indy victory in 1961, A. J. Foyt could not be stopped. He entered a period in his career where it was Foyt against everyone else in the business of auto racing. No other driver was such a threat in every race, in any type of equipment. Wherever he went he was mobbed, but he continued his cool, calm battle to remain exactly where he was—on top.

He even expanded his field. He had driven stock cars before, of course, but never as a champion against champions. So he took up this branch of racing in 1962.

Stock cars are sedans and so look like standard passenger cars. This is not to say that they *are* like standard sedans, but only that they *look* like them. The stockers are reshaped for better aerodynamics, completely rebuilt inside and mounted on huge racing tires. The engine is super-powered and designed specifically for racing. The championship stock car is a racing car, and nothing else.

This is the type of machine which should be in the Foyt domain, and soon it was. After his fantastic 1961 record of wins and championships, Foyt started racing the stockers too. In February 1962 he won the USAC stock car event on the ½-mile dirt track at Los Angeles. He entered 16 stock car races in 1962. Of these he won two, and placed in the top three in four more. He finished the year in fifth place in the stock division of USAC, and added some more money to his growing bank account.

As always though, it was in the open cockpit division that Foyt really shone. He was entering the greatest period of his career, and he was enjoying every moment of it.

In the Championship division Foyt won the races at Trenton, Milwaukee,

66

Foyt was becoming a cool, calm, masterful race car driver in any type of
equipment. He co-drove this Ford GT prototype with Lloyd Ruby
to second place at the 12 Hours of Sebring in 1967.

Langhorne and Sacramento. He was second in three more. Between the Championship events, he was racing sprints and midgets. He was in the top three in seven sprint car races, and though he only entered seven midget races that year, he won one of them.

It was only at Indy that he ran into trouble which cost him his National Championship in the end.

The Europeans, who had been watching this race for years and who had participated in it occasionally, were now studying it carefully. Foyt's victory at Indy in 1961 had overshadowed an event which became very important in retrospect. Jack Brabham entered a rear-engined Cooper Climax, an English car. Most of the mechanics and spectators laughed at it, until the end of the race. Then when the final results were tabulated, they were surprised to see that Brabham had placed ninth. Behind him, still running or sitting sick around the course and in the garages, were a large number of the traditional roadsters with their Offenhauser engines in the front.

After a few sincere congratulations, the lightweight little car, which was thought to be underpowered and of the wrong design for the brute "500," was forgotten. A few men didn't forget, but the car passed from the minds of the fans as a freak which just happened to do well in the race.

Then the next year, 1962, another rear-engined car was entered. Dan Gurney, until then a competitor on the European tracks, drove the Buick-powered racer, and did reasonably well until he finally retired at 93 laps with mechanical problems. Gurney eventually was awarded 20th place with his then-strange machine designed by Mickey Thompson.

Parnelli Jones was the man to beat that year. He had urged his Watson-built roadster to a new track record and shattered the so-called 150-mile-per-hour barrier with a qualifying speed of 150.370 mph.

Parnelli finished the race in seventh place after problems with his brakes caused him to make a very slow pit stop, but he had led the race for the first 59 laps, and again from the 65th to the 125th lap, when his problems began.

A. J. qualified at a very creditable 149.074 mph, and started the race from the middle of the second row. He led laps 60 and 61, after finally getting by Jones. It was like old times, for two laps anyway, but then a bolt dropped from his brake mounting. His pit stop was long and involved. His engine died and was re-started only as the crew, believing the car out of the race, was pushing it back to the garage.

A. J. poses with other Indy winners Troy Ruttman, Jim Rathmann and Rodger Ward. Only Foyt is still racing today.

Foyt finally re-entered the race far behind the leaders. Then as he was on his 70th lap, the left rear wheel fell off his car. Only his great skill enabled him to bring the three-wheeled racer down onto the safety apron without incident, but his car was out of the race for good. One of the crewmen had neglected to tighten the wheel during the long frantic pit stop.

Cursing his luck, and the negligent crewman, Foyt stalked back to his pit, ignoring reporters and fans along the way. He took over the car that Elmer George was driving, and went back out on the track again. This car went out 20 laps later. So, from the pits, Foyt watched Rodger Ward edge out Len Sutton by only 11 seconds for Ward's second "500" victory. He may not have realized it at the time, but he was also watching a large part of his championship go to Ward.

Foyt finished the year behind Ward in the point standings. He could carry, if he wanted to, the number 2 on his car to show he was in second place. To many drivers this would be an honor. To Foyt, who had been number 1 for two years in a row, it was an insult. He became obsessed with winning number 1 back in 1963. So he charged into the year at full tilt.

By 1963 the first real shots in the great rear-engine revolution were being heard. Into Gasoline Alley at the Speedway came Scotsman Jimmy Clark, and when his car was unloaded everyone gasped. First of all, though certainly the least important, it was painted British racing green. No respectable (or sane) American would ever paint his car that color. Everybody knew that the color green just invites bad luck. Green was almost as bad as peanuts around a racing car.

The car was a long, low, sleek-looking Lotus, a familiar brand name on the European circuits but never before seen at Indy. Finally, it was powered from the rear by a big Ford racing engine. It looked fast just sitting there in the garage area, and Clark had proved his ability on courses around the world.

This was probably the first time a rear-engined car was considered a real challenger. Certainly nothing had been spared to make it a winner.

Foyt studied the Clark car carefully. So did Jones, Ward, Branson, Marshmann, Hurtubise and all the other great American drivers. What if Clark should actually win the race? How would that look? The Indianapolis Speedway Race is even more American than apple pie.

To make matters worse, Jimmy Clark didn't even *look* like a racing car driver. Jones looked tough, Ward looked smooth and polished and Foyt looked

utolite's Chick Hirashima checks a Foyt plug. Chickie is a amiliar figure at the Speedway, having debuted there in 935 as a riding mechanic for Rex Mays.

like a rugged athlete, handsome but hard. Clark. however, looked as if he would have been more at home on a tennis court, or playing golf on a Scottish moor. He was good looking with a fine white set of teeth, but his hands seemed to be gentle. He was thinner than most of the American drivers, and more slightly built. He didn't look like a man who had been fighting big, oily, bucking, snorting Offy roadsters for a living.

Of course he was a winning driver, but that was in Europe where the gentlemen race car drivers play. This was tough, hard Indy, where you race in a moat and where the hard walls are always waiting. The Americans chuckled. After all in Europe they had bales of hay to catch you if you made a mistake.

The Indy drivers would stay with their proven roadsters that were heavier, front-engined and able to go the distance in most cases.

Jones captured the pole position for the second year in a row. He was driving the Agajanian Willard Battery Special, number 98, old faithful "Calhoun" as the Jones crew fondly called the car. Hurtubise took the second fastest spot on the starting grid with his powerful Novi, the Hotel Tropicana Special. Branson was third in the starting field. All three qualified at over 150 miles per hour. Next was Rodger Ward in his Kaiser Aluminum Special at 149.800 mph.

Jim Clark took to the track on the first day and qualified his Lotus at 149.750 which gave him the middle second row spot for the start. His low bomb screamed around the course with a sound the old-timers likened to the original Novis. Then Clark left to tend to other "more important" matters in Europe, such as Grand Prix racing. That was absolutely unheard of. You don't just qualify at Indy and then go home. You stay for the entire month and work on your car.

On the second day of qualifications, A. J. qualified his Sheraton Thompson Special at 150.615 mph to take the middle of the third row for the start on Memorial Day. Well, at least he had outgunned the Lotus. He grinned. He would outgun it on race day too.

And so the race began. On the very first lap the Lotus dropped back. It looked very small and frail in the field of thundering roadsters. Hurtubise grabbed the lead and set a blistering pace, followed by Jones and then Foyt. By the 10th lap the picture had started to change. Jones had passed Herk and was in the lead. Herk was second, and Foyt had dropped back to seventh

One of the reasons for the Foyt success—he is a fine mechanic and the sight of him working on his cars has become familiar at race tracks.

place. Clark was in eighth place and running as smooth as glass. He looked unworried. He didn't seem to be wrestling with his car. He merely seemed to be motoring along, enjoying the fine sunny afternoon.

By lap 20, with Jones still leading, A. J. had moved up to fourth place. Clark was down to 10th behind Gurney, who was also in a Ford-powered Lotus. Neither Clark nor Gurney looked at all concerned.

They weren't. They knew exactly what they were doing. They wouldn't need many of the services required by the roadsters. The standings of the first 10 drivers at the end of 80 laps (after the first pit stops) were flashed out over the wires from the Indy press box. And the shock came home to America. It was Clark, Gurney, Foyt, Sachs, Lloyd Ruby, Hurtubise, Ward, Roger McCluskey and Jim McElreath. The two European cars were showing the Americans the quickest way around the most American track in the country.

And so the 1963 Indy race went for many, many laps with American spectators on the edges of their seats. Clark screamed on, cool and calm and still not fighting his car or the track. Close behind, but still behind, came Jones and Foyt and the rest trying hard to catch up.

Halfway through the race, Clark pitted and Jones took the lead. Clark was second when he went back out, thanks to an amazingly smooth and efficient pit stop. Foyt was fourth at this point.

At 150 laps it was Jones, Clark, Foyt, and then Gurney in fourth place. At 160 it was Jones, then Clark 12 seconds behind, then McCluskey who had streaked past during a Foyt pit stop, then Foyt. Gurney had dropped back.

Then came the great Indy oil spill controversy to muddy the situation. Unmistakable puffs of smoke came out of Jones' exhaust, and streaks of black oil appeared on the tail of old faithful Calhoun. The black flag came out at the starting line to call Jones in. Oil makes the course so slippery that not even the leader of the race is allowed to spill oil on the track at Indy. Calhoun's owner, flamboyant J. C. Agajanian, rushed to the starting line to plead with the officials to take another look, and they did. The flag was withdrawn, and just as McCluskey was spinning out of the race in Jones' oil, Parnelli took the checkered flag.

Clark streaked by seconds later to take second place. Foyt was a few seconds later still and third. Ward was fourth.

Later USAC officials measured the oil remaining in Jones' car, and ruled that it had not lost enough "to be a hazard" to the other cars on the track. The finish was official, and controversial. Many people thought that Parnelli would have been disqualified by the officials if an American driver had been behind him instead of Clark.

In any case, the Americans had been badly frightened. Already the frail-looking little lightweight had taken a huge sum of Indy gold back to Europe. Second place at Indy pays a handsome fee. And the car had almost won the big race. The handwriting was on the wall at Indy. The day of the rear-engined cars had dawned.

Foyt raced on along the Championship Trail during the rest of 1963. In spite of Ward's winning as many as A. J., Foyt took the National Championship back. He did this in his usual manner, by consistently finishing in the top three throughout the year. Once again he could carry the number 1 on his car.

In the Championship division, he won the Trenton 100-mile race, the Langhorne 100, the Trenton 150, the DuQuoin 100 and the Trenton 200. He also won five sprint races and placed second in that division. Then in December, as always ready to experiment with new kinds of equipment, he entered two sports car races in the Bahamas. He won both, including the Nassau Trophy race.

A. J. carried his number 1 proudly into the 1964 "500" at Indy. He qualified his Sheraton Thompson roadster early, and fast. With a speed of 154.672 mph, he won the middle spot in the second row, a good place to be for this year's start.

Foyt had a rear-engined car available and ready, but he chose to use his old faithful Watson-built racer after all.

Eddie Sachs, on the other hand, had decided to drive a new rear-engined car. He encountered some problems in handling and brushed the wall with it. Finally he was ready on the second weekend of qualifications. He easily qualified fastest of the day, but this still put him back in the field for the start. He ended up in the middle of the sixth row. Directly in front of him was a promising young driver named Dave MacDonald.

At the start of the race, Jim Clark, who had captured the pole position, screamed away at new record speeds. The rest of the field strung out slightly, battling to keep up with the Flying Scot. The first lap was run at a new record speed.

Then the second lap began. The field thundered in an ever-lengthening line through the first turn, with Clark increasing his lead in the little Lotus Ford. Near the middle of the pack, Sachs increased his own speed preparing to pass the cars ahead. He wasn't accustomed to running so far back, and the fans were screaming for him to move up. Through the second turn flew the long line of cars. Through the third turn they went, with Clark's Lotus leading the way.

Then coming out of the fourth turn and onto the main straightaway MacDonald wavered and seemed to lose control of his hurtling car. The odd rear-engined machine skidded sideways, bounced off the inside wall, burst into flames and spun toward the outside wall right across the track. Prince Eddie, unable to find room on the track, thundered right into MacDonald's spinning car. Both men perished in the explosion and flames. Several other

Foyt does it again at Indy in 1964, though it was otherwise
a dark day at the Speedway.

drivers were burned in the resulting tangle of cars.

A quiet, subdued field started the race once again more than an hour later
before hushed fans. The contest soon became an intense duel between Foyt
and Jones. Clark had only lasted 48 laps. A chunking tire caused his
suspension to fail. Jones' car caught fire after a routine pit stop, just as he was
pulling away to reenter the race. That put Jones out. Foyt's only other real
challenger was Rodger Ward in a Watson-built rear-engined racer, but Ward
had to make five pit stops during the race, which kept him from being a
serious threat.

So A. J. Foyt won his second 500-mile Indy race, a feat that few men had
accomplished before, or since. He was the National Driving Champion again,
and a two-time winner at Indianapolis as well. He was at the peak of the sport.

76

The Bad Years

From the very top there is only one way to go. No matter how good you are, you can only stay on the peak for so long. Even A. J. Foyt was due for a slide. He went from being unable to lose to being unable to win.

There were two reasons for Foyt's fall. The first was the race at Riverside, California in January of 1965. The second was the complete takeover of the sport by the lightweight little rear-powered racers.

Super Tex, as he was then being called, entered the gaudy Riverside 500-mile race for NASCAR Grand National stock cars. This is a thundering race of huge stockers over a twisting road course which attracts 100,000 fans to the high desert country of Southern California. Everybody is there. The top Indy drivers are there, and so are the legendary Southern stock car aces. And in 1965 Foyt was there to win, and to preserve the near-legend that had been built around him.

As the colorful cars bellowed through the sweeping last turn of the course, Foyt was closing in on the leaders. Just ahead were the cars of Junior Johnson and Marvin Panch, two stock car racing stars from the South. A. J. reached for the brake pedal as he approached the corner. He shoved it down, and it plunged uselessly to the floor. He was hurtling toward the rear ends of the slowing cars ahead at 140 miles per hour. He had an instant to make a decision, to ram or swerve away. Ramming them would possibly kill three drivers, including Foyt, while swerving away would at least confine the carnage to Foyt's car.

He swerved, plunged over a 35-foot embankment and rolled end over end before slamming to an abrupt stop. It was one of those frightening racing

Greer

Foyt battles it out with National Champion Rodger Ward at Indy in 1963. Foyt finished third, Ward fourth.

You can't win them all. An unhappy A. J. discusses a losing situation with J. C. Agajanian.

accidents with car parts flying all about, from which you don't expect a man to walk away.

A. J. didn't, although he was alive. He was carried gently to an ambulance and rushed to the nearest hospital. Long weeks of recuperation forced him to reassess his position.

He hated the new lightweights which were sweeping the sport, but he knew he would have to switch to them. They were the certain winners of the future. They had survived their development period and proven themselves. They were faster and demanded less from the drivers. The Offy-powered dinosaur roadster (though certainly not the Offy engine) had died. Foyt would prefer to race on forever in the fine old roadsters, but he knew that he would have to change. He would have to go to a rear-engined car.

He was ill at ease in those he had tried. They broke up too easily in a

◀ Eddie Sachs and A. J. battling for position in the 1963 Indy "500" race.

A. J. just before his serious crash during the Riverside 500 in 1965.

Jim Clark leads A. J. (1) and Parnelli Jones (98) during the 19
500-mile race. Clark and Jones finished one-two while Fc
dropped out with mechanical difficultie

In 1964 A. J. switched from Ford to Dodge at the last minute and won the Firecracker 400. In 1965 he switched back to Ford and won it again.

Sometimes he won, but didn't get credit for it. He relieved Marvin Panch at Atlanta 500 in 1965, won the race, but the record books list it as Panch's victo

collision, and they tended to catch fire too quickly. Foyt had never been uneasy in any other racing car, but he was in the rear-engined machines. They wouldn't even perform properly for him. Problems he had never had before cropped up in the lightweights.

Before he was completely well, Foyt went back to racing. His insides were still sorting themselves out. His smashed heel and the broken vertebra in his back were still knitting and hurting.

For the first time in his career, he really began to know the meaning of DNF (did not finish). Although he was racing with the leaders more often than not, he kept dropping out with mechanical problems. Foyt was trying hard, perhaps too hard, in 1965. To add to his problems, there was a new young rising star in USAC, Mario Andretti. Andretti, they said, was the new Foyt. Long live the new king, for the old one had fallen.

Still Andretti was not yet winning that many races. In fact he was winning precious few on the Trail. Meanwhile Foyt, suffering through DNFs as he

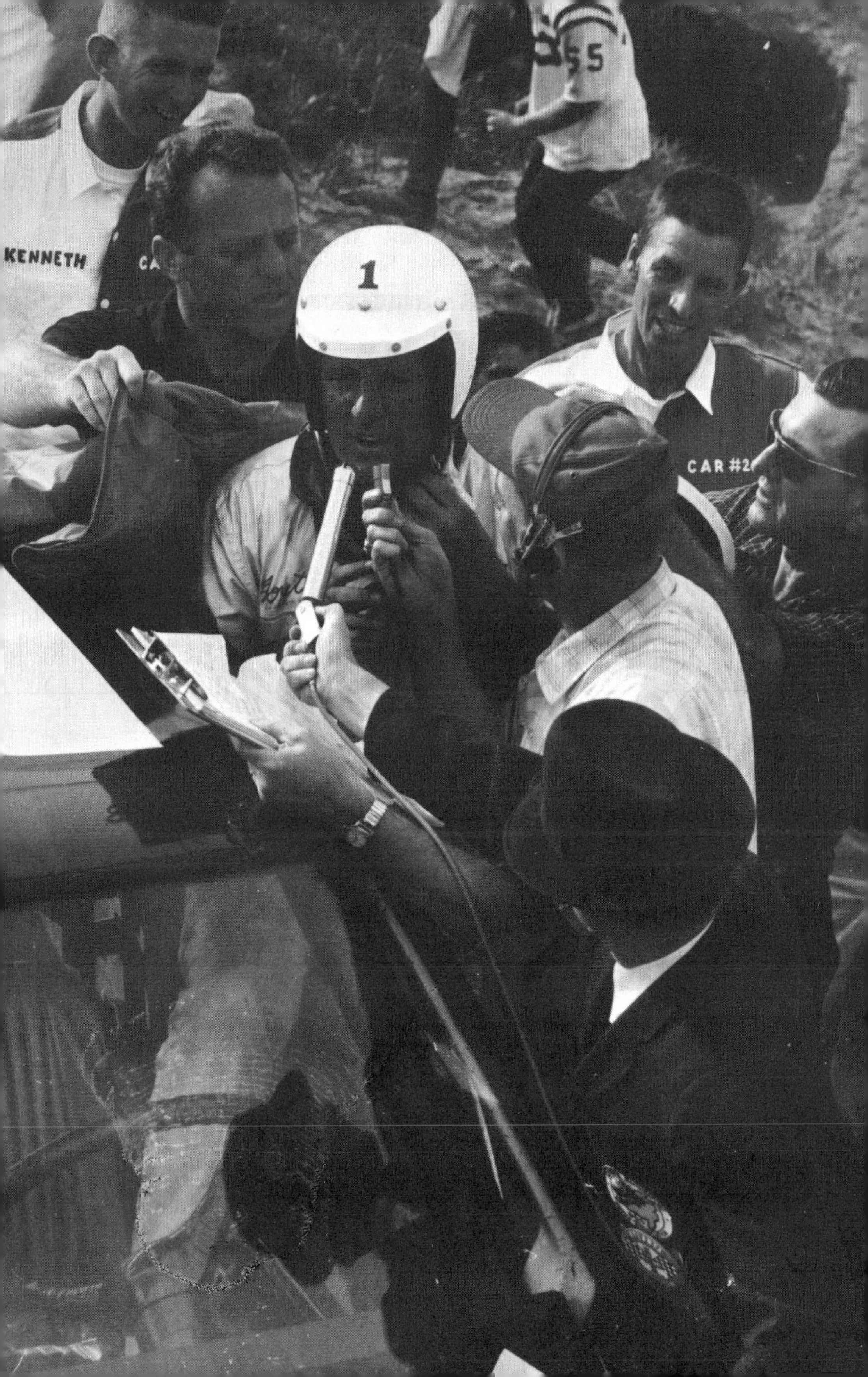
KENNETH
CAR #2
1
55

never had before, managed to compile an astounding record in his untrusted cars. For Foyt it was a disappointing record, for any other driver it would have been phenomenal. He won five Championship races, and placed in the top three in nine that season. He was the pole position winner at Indy in the new Sheraton Thompson Special, a lightweight rear-engined car. He was runner-up for the National Championship at the end of the year. But if Foyt is not a winner, he thinks of himself as a loser.

In 1965 Jim Clark finally won at Indy. Parnelli Jones placed second in a fuel-starved car and Mario Andretti, the brilliant new rookie, was third. Foyt finished in 15th place after the gears in the rear end of his car broke. Andretti became the new National Champion thanks to his many top finishes along the Trail, and he was a popular champion. It had been all Foyt, and now he had been soundly beaten. The fans cheered the new champion, but the wise, old railbirds nudged each other and said, "Foyt may be down, but he isn't out. Not by a damn sight!"

If 1965 was a bad year for A. J., 1966 was even worse. Nothing seemed to go right, no car would hold together long enough to finish. He won a USAC midget race at Ascot Park in Los Angeles and a sprint car race at Altamont, California. That was it.

He participated in 12 Championship races, and won not one of them. He was a DNF with almost perfect consistency. It was the worst year of his career.

In an effort to counter the trend, he designed and built his own car for the Indianapolis race. It was called a Coyote and was built to run fast. It had a Ford engine, in the rear.

He crashed the car into the wall at Indy before the first day of qualifications. Foyt turned to his backup car, a Lotus Ford, and qualified in it. But he was well back in the field, in the sixth row on the outside. Up front were the shining new stars of USAC, Mario Andretti on the pole and Jim Clark in the middle of the front row. Mario had A. J.'s number 1 on his car now. That rankled A. J.

Foyt could have quit. Perhaps he considered this possibility. He had more money than he would ever need, and he had his health. He had recovered from the serious crash at Riverside, but he had been through several others since then. They must have shaken him and given him thoughts of retiring. If he had quit, he would still have been at the top of nearly any all-time list of great racing drivers. He didn't though.

"It's how I make my living," he would drawl, and then crawl into another racing car.

Almost every fan is afraid of the first lap at Indy, and of what might happen if one driver makes one small error in judgment. The 33 cars hurtle down the straightaway in 11 rows of three, picking up speed all the way. Then as the green flag drops, they all attempt to funnel down into the groove of the

A. J. discusses 24 Hours of Daytona strategy with Bruce McLaren (far left), Dan Gurney (next), Ronnie Bucknum (right center) and other Ford team members.

first turn, while still increasing their speeds. It is a wildly exciting moment for the fans, and a hazardous one for the drivers.

In 1966 one driver in the middle of the thundering field lagged badly just as the green flag came out. This created a hole in the perfect alignment, which two other men tried to fill to better their position. Cars touched, and the grinding wreck was on. There was no room for the drivers behind to avoid the other cars on the straightaway between the hard walls. Shortly after the green flag had fallen and before the front runners had hurtled into the first turn, the main straight was a jumble of wrecked cars. Bits of cars were strewn all over. Wheels were sailing through the air. Several flew into the crowd causing minor injuries there.

Sixteen of the 33 cars were involved in the gigantic crash, with 11 of them out of the race for good. It looked as if several drivers had been badly injured, but only one was hurt at all. Foyt cut a finger scrambling out of his wrecked car and over the fence beside the track.

87

Foyt's car was a torn pile of junk along the outside wall by the first turn. He hadn't even completed the first turn of the first lap. Angry and frustrated, he was at the bottom. He complained bitterly and harshly to the men he blamed for causing the accident.

Foyt felt things couldn't get any worse, but they did.

He quickly bought another racer, a rear-engined model, in time to join the other USAC drivers at Milwaukee for the race which traditionally comes the week after the "500." There he smashed into the wall and his latest car burned. He had second and third degree burns on both hands, his face and neck and 15 per cent of his body.

He tried again in a stock car race, but his hands were still tender. They blistered so painfully, they put him out of the race. Back again to Indiana he came, to the Hoosier 100, a track he knew very well. He had won this classic race in 1964 and again in 1965. Near the end of the 1966 race it seemed like old times, with A. J. leading going away. But this was 1966. On the next-to-last lap his brake pedal came off, simply fell off, and he went into the wall.

Mario Andretti won the National Championship for the second year in a row. Foyt was in 13th place for the year, his lowest standing since 1957.

What drove Foyt on? He alone knew why he was punishing himself so severely and needlessly. He didn't need the victories or the money. He had already proved that he was one of the greatest in the history of the sport. Everybody gets older and suffers somewhat in slower reflexes. In auto racing this aging often comes much earlier than in other careers. Still the greater experience counts for something. And A. J. was only 31 in a business where the greatest champions were often in their middle and late 30s.

Meanwhile he had a beautiful wife, three children and a fine house with spacious lawns. Several automobiles are in the garages, and an airplane is kept in a nearby hangar. Foyt, a private pilot, used the plane to go from race to race. By anyone's yardstick, Foyt had it made.

So why go on? Because he "made his living" that way. Because he had to drive to live and win to be happy.

Perhaps this is what makes a champion. Perhaps this is what carries some men to the top of whatever field they choose. Perhaps it is this ability to take a beating, and come back for more, and more, and to finally emerge victorious. Certainly this is a trait of A. J. Foyt. He will not stay beaten. He will come back until he wins, or until he can no longer come back.

89

A familiar sight at race tracks during Foyt's bad years, as he and
Wally Meskowski analyze his search for an elusive victory.

90

The Return of Tough Tony

It must be said of the fans during this low period in Foyt's career that although they cheered on the young Andretti and some of the foreign drivers invading the tracks of America, they didn't forget A. J. Wherever he went, as a winner or a loser, he was noticed by the people who love the sport.

Foyt, with a prematurely receding hairline and a hard glint in his eye, felt he was back in shape for the 1967 season. He was healthy, happy and getting plumpish. And he had conquered the rear-engined cars at last. He had done it the hard way, first by learning about them on the track and then by designing and building his own. His Ford-powered Coyotes were a recognized threat, even before Foyt won a race in one.

The 1967 Indy "500" was supposed to be a Parnelli Jones rout. For Parnelli was driving the famous Andy Granatelli Turbocar, a car which convinced every other driver that the race would be for second place that year.

Jones was sure to win the race, even though he only qualified sixth fastest in the field. His car, powered by the controversial turbine engine, was built to go fast almost forever, with little service or assistance needed. It was driven by all four wheels, which gave it precise control and extra speed through the turns. It had quick acceleration and almost no limit in top speed. Some drivers felt it was more an airplane than a race car. Indy, they complained, was becoming an airport runway.

Still, with his choice of cars, Foyt stayed with his Ford-powered Coyote, and he qualified it at 166.289 mph for the inside spot on the second row. Mario Andretti was on the pole that year, directly in front of Foyt. Mario was carrying number 1 on his car. Foyt wanted it.

J. accepts an award for his 1967 Indy victory from 1966
ener Graham Hill as Foyt, Sr. looks on happily.

y 1967, A. J.'s **third** victory in the "500."

...yt has many fans. They know that he is the greatest and these signs prove it.

...rd Motor Company named Foyt their 1967 "Man of the Year" in auto racing.
...th A. J. is Jacque Passino, head of the company's racing program.

Jones wanted his second win at the track. He went after it with a vengeance. At the drop of the green flag he shot into the lead, passing all five of the cars in front of him in less than half a lap. By the end of the first lap he had drawn to a considerable lead. His car seemed glued to the track in the turns. Shortly after the beginning of the race, a heavy rain began to fall and the race was postponed to the following day.

This made little difference to Jones. He merely picked up where he had left off, continuing to extend his lead. Soon, as many drivers had predicted, the race was for second place. The flat tail of the Turbocar, with the two "tail-lights," became a familiar sight to the rest of the field.

For the first 50 laps the position of the first three cars didn't change. It was Jones, Gurney and Foyt. Andretti had dropped out almost immediately with a clutch problem. Then it became Jones, Foyt and somebody else.

So it remained, with Jones flying and Foyt in dogged pursuit. Any shifting

After winning at Le Mans, Foyt jokingly asked if they picked a Rookie of the Year, so one of the mechanics complied to the best of his ability under the circumstances and turned one of his Goodyear wipe cloths into an award for A. J. It's hanging above co-driver Dan Gurney's head. Manager of the team, Carroll Shelby, looks rather doubtfully at his glass of champagne.

of the field came behind these two. Except for one brief period when Jones pitted and Foyt took the lead, the story of almost the entire race was Jones followed by Foyt.

From the start Foyt had not felt that the turbine would finish the race, and he was running his own race with this thought in mind. If he attempted to challenge the quickly running Jones, he would merely have forced Jones to even greater speeds. This might overstress both the hissing turbine and his own car. So he waited. His strategy, though almost cut too fine, was correct.

On lap 197 of the 200-lap race, Jones coasted slowly into his pit, out of the race with a broken gearbox bearing. The way was clear for Foyt to become only the fourth man in the history of the "500" to win three times. The other three-time winners are Lou Meyer, Wilbur Shaw and Mauri Rose. No man has won the race four times, and only Foyt has a chance to do so, since the other three are either dead or long since retired.

As Foyt was cruising happily toward his great victory, he had a sudden

Out of the 1968 "500," Foyt leaps to help on Jim McElreath's car after it had collided with Johnny Rutherford's. While the front of the Coyote is being repaired, A. J. removes the cowling to check the engine.

Despite an early spin to avoid a crash involving Andretti and Ruby,
A. J. went on to win the Trenton 200 in 1967.

oyt dices with Mario Andretti before a frozen wheel bearing
orced him out of the 1968 Milwaukee 150.

The last lap accident at the 1967 "500." Foyt, arm in air,
threads his way through the spinning cars to victory.

100

premonition which told him to slow down. Around the corner on the main straightaway, where he could not yet see them, cars were crashing. On his very last lap, within sight of the checkered flag, Foyt had to slow down to a crawl and thread his way through the wrecked cars.

But he made it, and waving to the fans he took the coveted Indy checkered flag. He was the only man in the entire field to complete the full 500 miles. As soon as he crossed the line, the red flag came out and the few remaining cars were stopped because of the wreckage on the main straight.

Foyt won a total of $171,527 for this victory. He was heading for the top again.

Probably the best known circuit in all of Europe is the one at Le Mans. Everybody has heard of this race course and the 24-hour race, even non-enthusiasts. Foyt decided to tackle it in 1967 with Dan Gurney as his teammate. Their race car was the super-fast, reliable Ford Mark IV. Also in the race as part of the Ford contingent were such drivers as Mario Andretti, Lucien Bianchi, Roger McCluskey, Jo Schlesser, Ronnie Bucknum, Lloyd Ruby and Denis Hulme. No less than 13 Fords were entered in the 1967 Le Mans, including two Mirages and a Shelby Mustang.

Le Mans is like no other race in the world. In the infield is "The Village," a potpourri of exhibits, hot dog stands, rock and roll music, the Musée de l'Automobile, accessory shops and souvenir stands. Along the outside of the esses is the amusement area with thrill rides, sideshow exhibits, motorcycle stunt shows and a dance hall, but nothing is ever out of earshot of the powerful race cars that roar on through the night in their 24-hour quest for victory.

In spite of the extra added attractions, the track is where the real action is. It was on the track that rookie Foyt hoped to do well enough to keep the Europeans from sneering at "that Indianapolis driver."

It was a thrilling race, a showdown between Ford and Ferrari (although the Commendatore did withhold his new P4s for some reason). In one accident alone, three of the Fords were knocked out of the race. Andretti, in a car with brand new brake pads, thundered into the esses. He hit his brakes, they grabbed, and Mario spun and slammed backwards into the earthen embankment. No caution flag or light warned the oncoming cars, and Roger McCluskey roared into the turn in his Mark IIB Ford. He saw Andretti at the last moment. Trying to miss him, he too spun and hit the bank. Still there was no caution warning for the others. Jo Schlesser in another Mark IIB streaked onto the scene. He too wrecked his car trying to avoid the others. In a few seconds three of the Fords were out of the race.

One Ford was not. Driving 14 of the 24 hours of the race, A. J. Foyt brought the Mark IV home a winner. Gurney and Foyt had beaten the finest European drivers on their own home ground to win going away.

Back home the battle for National Champion of 1967 was as close as any in the history of auto racing. Mario Andretti and A. J. Foyt were the leaders, with Mario trying to keep his title and Foyt trying to get it back again. Foyt had held the title four times before, and Andretti had had it two years running.

The last race along the Trail is at Riverside, California each year. This is the Rex Mays Memorial 300, named after one of the great racing drivers. It is a wild affair with Indy cars on a road circuit. It is one of the most thrilling races of the year for spectators and drivers alike.

The 1967 championship would be decided at Riverside. Although Foyt was leading in points going into the race, Mario could overtake him by beating him in the race. They were that close in the standings.

Everything looked rosy for Foyt, and to Foyt, who optimistically assumed that he would win and so settle the matter. His car was in excellent shape, fast and ready to go, and so was he. He was relaxed and laughing in the pits prior to the race. Before the race was over, he would be dour, exhausted and covered with mud. With a championship at stake, Foyt would probably fight his way through hot coals to the flag. The only trouble was that Mario would too.

Mario was dicing for the lead with McCluskey. If Mario won, that was that. So Foyt had to increase his own speed to a dangerous level to keep up his third-place position. Around the turns, right and left, and down the straight-aways they roared.

Foyt and Al Miller brushed together in the middle of some traffic which Andretti and McCluskey had already managed to get through. Both cars spun into the mud alongside the esses. Axle-deep in the mire, they came to a mud-showering stop, and to all appearances both cars and drivers were out for the day. They were mired over half a mile from the starting line.

Foyt's impossible drive to catch up had ended in the mud, but then no driver had ever been National Champion five times.

Foyt didn't know when he was finished. He threw off his seat belts, slid out of his cockpit and leaped out of the car, sinking ankle deep in the sticky mud. He eyed the starting line in the distance and then began to run. Splashing mud, he ran all the way back to the pits.

Foyt had entered a backup car in the race for just such an emergency and had arranged to have capable Jim Hurtubise drive it so that it would be among the front runners. Unfortunately this car was already out. It had sputtered into the pits just one lap before. Foyt did not know this.

When he came panting into the pits, he saw Hurtubise's car and his heart leaped. They had anticipated his wish and pulled the car in to be ready for him. No, he was given the disheartening news. It seemed that Foyt was out of it once and for all, and that he had lost the championship since Mario was solidly in with the front runners. Of course Mario might spin out or suffer mechanical problems, and then Foyt would be champion. Foyt could just sit

102

In Roger McCluskey's car, A. J. is on his way to a fifth place finish
at Riverside in 1967 and his fifth USAC driving championship.

in the pits and watch and hope. But that is not the active kind of gamble that Foyt is used to taking. He does his gambling on the course, not in the pits.

Without a second glance at his own backup car, he hurried to Roger McCluskey's pit. A quick conversation with the crew, and the pit board went out. On the board was the call-in sign. McCluskey was being pulled out of the race, although he was leading at the moment.

For friendship, for a past favor in a sprint car race and for other considerations unknown, McCluskey came into his pit to turn his car over to Foyt. Out leaped Roger, in leaped A. J., and with a great roar Foyt was back in the race once again.

The car was no longer in the lead due to the stop. This did not matter, nor did the broken half shaft that hit the Foyt/McCluskey car later. Foyt drove masterfully in the unfamiliar car and managed to keep it going despite the mechanical trouble. He salvaged fifth place in the race and earned enough points to win his fifth National Championship.

A. J. Foyt had returned to the top of the sport once more.

The Professional
Race Car Driver

With number 1 on his car again, for the fifth time, Foyt kept on racing. He had accomplished nearly everything he could hope to accomplish in racing. He had become an all-time money winner and had won more Championship races than any other driver in history—42 as of this writing. No other driver is close to this figure. Even if Foyt should retire today, the records he has set might stand for some time. Few men have been able to stay alive and healthy in this hazardous business long enough to have time to compile such a long list of victories.

His first win in 1968 was not a Championship event, but a stock car race. He took time off from Indy practice and entered the 250-mile race at Indianapolis Raceway Park in Indianapolis. Then it was back to the Speedway on 16th Street.

1968 was Bobby Unser's year. No one could stop him, not Foyt, not Andretti, not anyone else. Bobby had already won Championship races going into the "500," and he confidently expected to add the "500" to his list.

The turbines were the talk of Gasoline Alley. Several were entered, including the sleek wedge-shaped racers of Andy Granatelli and Colin Chapman. Only

Foyt is interviewed by "Wide World of Sports" after his Hoosier 100 victory in 1969. Indy owner Tony Hulman stands by.

(Left) Foyt and crew change an engine in two hours and qualify for the Trenton
300 . . . but the car (above) is finally out with a broken valve after leading the race.

Indy 1968. Foyt's teammate Jim McElreath pulls in at the end of the race.
Foyt had dropped out with broken gears, McElreath finally placed 14th.

Foyt gets under Andretti for second position, lapping Sam Sessions (14) in the
process, during the Milwaukee 150 of 1968. The race was won by Lloyd Ruby.

the year before, the turbine had dominated the race until its last minute
breakdown. It was certain to be a threat in 1968.

Super Tex once again decided to stay with his proven piston-driven Coyote
racers that he had personally designed and built. Just as he had once remained
with the Offy roadster in the face of the Lotus-powered-by-Ford invasion.
Foyt's Indy machine for 1968 had a non-turbocharged double overhead cam
Ford engine, a "standard" power plant. As insurance, Foyt also entered a
backup car. This was another Ford-powered Coyote and was to be driven by
Carl Williams. The Foyt stable was ready with proven machines.

The rumor flew around the stands that Foyt was quitting. "If Foyt wins,
he'll retire."

It was impossible to imagine auto racing without Foyt. Yet he finally
admitted that if he won the fourth victory he was seeking, he would at least
retire from all but Indy racing.

The turbines, however, streaked to new track records during qualifications.

◀ Three-wheeling it and leaving in his wake roostertails of dirt, Foyt charges on to win the 1969 Hoosier 100—his sixth victory in this event.

The powerful Foyt-driven, Watson-built Ford sprint car,
shown here at Terre Haute in 1968.

Foyt's Coyote is pushed from the track after failing in the 1968 "500."
Chris Economaki of "Wide World of Sports" questions the crew
as to the reason for the retirement.

They captured the first two places in the field of 33. The third turbine and
the only other one left in the field (Carroll Shelby had withdrawn his turbines
as not yet ready for competition, and only three of the Granatelli Turbocars
remained after accidents during practice) was qualified by Art Pollard as 11th
fastest in the field.

A. J. Foyt took to the track on the first day and qualified eighth fastest for
the middle spot in the third row. His teammate, Carl Williams, finally qualified
in the 10th row of starters. Memorial Day approached and though the odds
were strongly in favor of the turbines, the Foyt team was ready.

The Coyotes were proven racers, and A. J., Sr. was a skilled and enthusias-
tic pit chief. A. J. himself had won more races here than any other driver on
the track. The turbines were fast, but they had a record of unreliability. Foyt
was confident.

The fans were confident of his ability too. Foyt has a certain magic with
the fans. All drivers, of course, are respected and many are even idolized by

115

Foyt blows an engine just as he is trying for the lead at Terre Haute
while in a three-way battle for the USAC stock car title
with McCluskey and White. McCluskey took the 1969 title.

arting in 23rd place, Foyt charges through for a second place finish in the first
)0-lap race at the Houston Astro Grand Prix for Midgets in 1969.

J.'s first Championship win of 1968 was at the 2.66-mile road course
Continental Divide Raceways in Colorado.

...ading Al Unser (24) and Dan Gurney during
...e early stages of the 1968 Indy "500."

119

ith stopwatch in hand car owner/builder Foyt checks teammate Carl
illiams' lap times during practice for the 1968 Indy "500."...

Pleased with the speeds Carl is getting and the performance of the Coyote, he crosses his fingers that it continues (above). As he awaits the call for Williams to attempt his qualification run (right), he nervously swings his stopwatch back and forth. . . .

He buckles Carl in . . .

nd gives him some last-minute information. . . .

The engine has been started, Williams is ready and the car is being pushed off....

He qualified at 162.323 mph and was awarded 15th finishing posit
after a crash on his 164th lap eliminated h

the fans, but Foyt has an added charisma. It is one thing to be one of the top men in any field, but it is something else to be *the* top man with a record number of victories to prove it. All Indy drivers, even the tail-enders, are mobbed by the fans for statements about the race, autographs, endorsements or just a quick handshake. Foyt gets more than this. The fans often watch him instead of approaching. Just as a baseball fan might not approach Joe DiMaggio with familiarity even though he knows Joe's record perfectly, or a football fan might hesitate to rush up to Joe Namath, so racing fans hesitate around A. J. He is something special even in the field of courageous racing drivers.

His "500" strategy was the same as that of several other drivers of piston-engined cars in the 1968 race. To drive hard, to stay in contention and to be ready when the obviously faster and better handling turbines failed—if they did. Many people felt this would be the year of the turbine when they would finally win the great "500."

Just as it was expected to do, the Turbocar of pole-winner Joe Leonard rocketed quietly into the lead at the start of the race. It remained there, or close on the heels of whoever else was leading (usually Bobby Unser) through-out most of the race. The 1968 race was almost a replay of the 1967 race, with a Turbocar apparently in complete control of the situation.

Leonard seemed to be able to pick his time and take the lead whenever he wanted. He was content to run along in second or third place while the piston-engined car in front of him thundered frantically lap after lap trying to stay ahead.

Meanwhile Foyt and Dan Gurney were battling for position. Co-drivers at Le Mans the year before, they were now bitter competitors. At first one would be ahead, and then the other. For lap after lap they fought.

The race between the two popular veterans was much more closely fought than the race for first place, but the strain finally told on Foyt's car. Foyt pulled slowly into his pit and out of the 1968 race. The rear end of the car had failed. The fans groaned, hoping against hope that the fault could be corrected and that Foyt would roar once again onto the track and after Gurney, but he was out for good.

The race continued, even if it wasn't the same to disappointed fans who had hoped to see a four-time winner. Toward the end of the race, Carl Williams smashed into the wall. He was uninjured, but the car was destroyed.

At the very end of the race, Joe Leonard's Turbocar failed while he was leading. Once again Foyt's strategy had been correct, but it was Bobby Unser who swept into the lead to win his first "500." With this victory (worth 1000 points on the Trail, Unser went on to win the National Driving Championship.

The only good thing about this race for Foyt fans was that he would not retire since he had not won.

128

He went on to finish up the year in sixth place in the Championship standings.

Bobby Unser carried A. J.'s number 1 into the 500-mile race at Indy in 1969, but it was Mario Andretti who took the victory and Foyt who provided some of the great drama.

The first weekend of qualifying was entirely rained out—something that isn't supposed to happen at all. The second weekend Foyt easily took the pole position with a speed of 170.568 mph, in spite of problems with the waste gate of his turbocharger. Andretti was next to him at the start and Bobby Unser had the third spot.

During the race Foyt was running in front or with the leaders and seemed to have a good chance to win that unprecedented fourth victory. Just behind Foyt was his teammate, Roger McCluskey. Then mechanical difficulties in the turbocharger cost Foyt a long, long pit stop.

When he finally returned to the track he was more than 20 laps behind the leaders. He didn't have a chance of winning. Yet, driving on the ragged edge of control, A. J. battled to catch up. He was driving much faster than Andretti who was comfortably out in front of everybody else. Lap after lap Foyt pushed his car on his impossible quest. He was much too far behind to catch up by the end of the race, but he did manage to work his way back up to eighth place before Andretti took the checkered flag.

A. J.'s only Championship victory in 1969 was "his" Hoosier 100 at the State Fairgrounds at Indianapolis in September. They call it "his" race, because he has won the same annual event six times as well as taking five other victories on the track in stock cars.

"When are you going to give me the title to this place?" Foyt asked as he accepted his prize money from the racing director.

Foyt ended up seventh on the Championship Trail for 1969 after a disappointing series of mechanical failures. Mario Andretti had won enough points to secure the title halfway through the season.

Foyt is Championship Auto Racing

In spite of the mechanical failures and the loss of the Indy race, Super Tex kept battling for the Championship which he honestly believes belongs to him as long as he races. It is as though he might lend the title to another driver from time to time. Then he is driven from the inside to retake it. It is a fact that he seems much more content when his car carries the number 1, and that he drives ever more dangerously and desperately when it does not. It has not since 1968, and it will not through 1970. If Foyt continues though, it probably will once again.

Meanwhile Foyt drives everything. On one weekend he will be competing against the greatest drivers in the world at a famous track, and on the next he will show up at some grubby little track nobody ever heard of to race against the local boys. One day he will drive a sports car to victory, the next a stock car and the next a midget or a sprint car. He is master of them all.

If he were forced to pick one specific type of racing (and nobody forces Foyt to do anything) as his personal favorite, he would probably choose a wild and dangerous sprint car race on a half-mile or one-mile dirt track. In this type of race, certainly one of the most hazardous for the drivers, Foyt seems completely at home. You can almost see the grin behind the face mask as he broadslides his open cockpit racer around a turn, throwing out great roostertails of dirt from his spinning rear wheels.

He seems to get more of a real kick out of a hotly contested midget race on dirt than he does out of Indy. Of course Indy pays far more, but money

A. J. and Andy Granatelli at Indy in 1969.

131

The 1969 Indy first row, with Super Tex on the pole,
Andretti next and Bobby Unser on the outside.

Foyt's 1969 Indy car at speed during the "500."

Foyt rounds "Foyt Corner" during the 1968 Rocky Mountain 150 at the Continental Divide Raceway on his way to victory in the event.

Winners on NASCAR's superspeedways, Dave Pearson, Donnie Allis
Cale Yarborough, A. J. Foyt and LeeRoy Yarbrou

135

has long since become a secondary consideration to millionaire Foyt. Racing, and winning are all that seem to matter.

He will leave his lovely home and family without hesitation to race in an out-of-the-way event halfway across the country for a victory that will do him no good at all as far as money and prestige are concerned. Perhaps that is the secret of A. J. Foyt. He loves racing more than anything else. Racing is his food and drink. Without it, he will die. With it, he will perhaps do the same, but Foyt is sure he will survive until he is ready to quit. He has been hurt, seriously hurt, but these accidents have not slowed him a bit. He is sure that he will not be hurt again, for he is the total master of his fate on a race course.

Foyt does not feel that any odds are against his winning the "500" for the fourth time. In fact he feels that the odds are now strongly in his favor, since he has won three times and learned so much about coming in first at Indy. He feels that any winner has a better chance of winning again than any non-winner has of winning the first time.

Foyt is a winner. He may lose from time to time, but over the long haul he has won, and doubtless will continue to win, more races than any other driver. This will be true as long as he chooses to race.

It looks as if he will choose to race at least until he has become a four-time winner at Indy, and perhaps until he has set records which no driver could ever hope to equal.

Throughout, he will remain as difficult (or honest) as ever to deal with. After winning the 1967 "500" he was presented with his prizes. On the almost endless list of money and merchandise was the pace car—this year a Chevrolet Camaro—traditionally awarded to the winner of the race. Each year the sponsoring automobile company lends several hundred identical cars for use as official and celebrity cars and for use as the pace car which leads off the big race. Well, not exactly identical after all.

Foyt did not want to accept his car until air conditioning and an automatic top had been installed. He felt that if the many, many other cars used by the officials and the celebrities were equipped with such niceties (giving the manufacturer the publicity), then the winner's award certainly should be too.

So Foyt, in typical Foyt fashion, merely refused to accept the automobile. A settlement was quickly made and the items were installed. While everyone sighed with relief, Foyt took the car home—where it still sits in his garage with very low mileage.

It was just the principle of the thing as far as Foyt was concerned. After all, by now A. J. could afford to buy a noticeable part of the entire company if he so desired.

With the help of members of his pit crew A. J. gets into the cockpit of his Coyote, which is in position on the starting grid for the 1968 Indianapolis "500," as his father looks on. . . .

Securely buckled in, his car's being fired up and he's offered a last-minute
drink (left) which he refuses. Just before moving off for the
start (above) he glances up at his father.

The Statistical Side of
A. J. Foyt

On the south side of West 16th Street, across from the Indianapolis Motor Speedway, stands the "500" shopping center. This plot of land was once occupied by the West 16th Street Speedway, a ¼-mile paved midget track which was probably the finest of its type at the time it was pulled down, a few years ago.

Although there was racing at the track throughout each summer, the big occasion came each May 29th—when it played host to the classic "Night Before the 500" midget races. Three separate events would be held, each consisting of time trials, heats and the main event. The first would be held in the afternoon; and as soon as the track had been cleared after the feature was over, preparations for the evening program would commence. This in turn would be followed by the late night event, which would end in the small hours of the next morning. In 1956 the late Shorty Templeman made May 29th a memorable occasion by winning all three main events. That same night, almost unnoticed, a young driver was making his debut in USAC.

Too slow to qualify for either the afternoon or the evening races, this young driver came back to set sixth fastest time for the late night show, finished fourth in his heat, 13th in the main event and won $68.00 for his night's work. This was the first prize money ever paid at a USAC-sanctioned event to 21-year-old A. J. Foyt, Jr. Since that time Foyt's record has been fantastic.

He has won 42 national Championship events, more than any other driver. He is the all-time Championship point leader, and is the only man to have won the Indianapolis "500" three times since World War II. He has won the National Driving Championship five times and has been runner-up twice. He holds the record for the most consecutive Championship starts . . . 82 from the Trenton 100 in 1960 to the "500" in 1966. Burns received in practice at Milwaukee the following week ended the fantastic streak.

He holds the record for the most Championship victories in a single year . . . 10 in 1964. His streak of seven straight wins that same year is also a record. When he finished fourth in the Hanford 200 in March 1968, it marked the 100th time that he had finished in the top 12. He also has placed in the top 12 on four other occasions as a relief driver.

In 1963 he placed in the first three in 10 out of the 12 races toward the championship that year, placing fourth and eighth in the other two. From the Milwaukee 200 of 1960 through the Sacramento 100 of 1964, he placed in the top three in 40 of the 56 races in that span, and from Sacramento in 1962 until the same event in 1964 the figures are unbelievable. In the 26 events covering the two-year period, he placed first on 16 occasions, was second four times and third twice, thus failing to place in the top three in only four of those 26 events.

In addition, Foyt has the second highest number of sprint car victories in the 12-year history of USAC and has the third highest accumulated points total in the division.

He is third in accumulated stock car points, despite the fact that he has only been active in this division since 1962. He won the stock car championship in 1968 and was second in both 1963 and 1969.

He also won 16 midget races in 39 starts between 1960 and 1967.

The Indy "500," Championship races, sprints, stocks, midgets, NASCAR stocks, sports cars and even the Le Mans 24 hours race . . . A. J. Foyt has won them all.

Major Events Won

1957

May 12	USAC	Midget	Kansas City, Mo.	¼D	100 laps
Sept 11	USAC	Midget	Xenia, Ohio	¼P	50 laps

1958
None

1959

Sept. 13	USAC	Sprint	Salem, Ind.	½P	100 laps
			Pat O'Connor-Joe James Memorial		
Oct. 11	USAC	Sprint	Houston, Texas	½P	50 laps
Nov. 21	USAC	Midget	Corpus Christi, Texas	¼P	50 laps

1960

Feb. 7	USAC	Midget	Los Angeles, Calif.	½D	100 laps
Apr. 17	USAC	Sprint	Reading, Pa.	½D	30 laps
Apr. 24	USAC	Sprint	Langhorne, Pa.	1D	50 laps
June 10	USAC	Midget	Anderson, Ind.	¼P	50 laps
Sept. 5	USAC	Champ	DuQuoin, Ill.	1D	100 miles
Sept 17	USAC	Champ	Indpls. Fairgrounds, Ind.	1D	100 miles
			Hoosier Hundred		
Oct. 9	USAC	Sprint	Williams Grove, Pa.	½D	50 laps
Oct. 30	USAC	Champ	Sacramento, Calif.	1D	100 miles
Nov. 20	USAC	Champ	Phoenix, Ariz.	1D	100 miles
Nov. 24	USAC	Midget	Los Angeles, Calif.	½D	122 laps*
			Pacific Coast Grand Prix		

* Called because of fog.

1961

Mar. 19	USAC	Midget	San Bernardino, Calif.	¼D	50 laps
Mar. 26	USAC	Sprint	Reading, Pa.	½D	30 laps
Apr. 15	USAC	Midget	Los Angeles, Calif.	½D	50 laps
Apr. 30	USAC	Sprint	Salem, Ind.	½P	30 laps
May 28	USAC	Sprint	Indpls. Raceway Park, Ind.	⅝D	30 laps
May 30	USAC	Champ	Indianapolis, Ind.	2½P	500 miles
June 18	USAC	Champ	Langhorne, Pa.	1D	100 miles
June 30	USAC	Midget	Anderson, Ind.	¼P	50 laps
July 29	USAC	Midget	San Bernardino, Calif.	¼D	50 laps
Aug. 4	USAC	Midget	Lawrenceberg, Ind.	¼D	40 laps
Aug. 6	USAC	Sprint	Salem, Ind.	½P	30 laps
Sept. 4	USAC	Champ	DuQuoin, Ill.	1D	100 miles
Sept. 8	USAC	Sprint	Lancaster, N.Y.	½D	30 laps
Sept. 16	USAC	Champ	Indpls. Fairgrounds, Ind.	1D	100 miles
			Hoosier Hundred		
Sept. 17	USAC	Sprint	Reading, Pa.	½D	30 laps
Oct. 1	USAC	Midget	Terre Haute, Ind.	½D	75 laps
Oct. 21	USAC	Midget	Los Angeles, Calif.	½D	40 laps
Nov. 4	USAC	Midget	San Bernardino, Calif.	¼D	50 laps
Nov. 23	USAC	Midget	Los Angeles, Calif.	½D	150 laps
			Pacific Coast Grand Prix		

1962

Feb. 25	USAC	Stock	Los Angeles, Calif.	½D	100 laps
Apr. 8	USAC	Champ	Trenton, N.J.	1P	100 miles
May 27	USAC	Sprint	Indpls. Raceway Park, Ind.	⅝P	50 laps
June 10	USAC	Champ	Milwaukee, Wisc.	1P	100 miles
July 1	USAC	Champ	Langhorne, Pa.	1D	100 miles
Sept. 30	USAC	Sprint	Salem, Ind.	½P	100 laps
			Pat O'Connor-Joe James Memorial		
Oct. 7	USAC	Stock	Detroit, Mich.	1D	150 laps
Oct. 20	USAC	Midget	Los Angeles, Calif.	½D	40 laps
Oct. 28	USAC	Champ	Sacramento, Calif.	1D	100 miles

1963

Mar. 24	USAC	Sprint	Reading, Pa.	½D	30 laps
Apr. 7	USAC	Sprint	Williams Grove, Pa.	½D	100 miles
Apr. 21	USAC	Champ	Trenton, N.J.	1P	100 miles
Apr. 28	USAC	Stock	Indpls. Raceway Park, Ind.	2½P-RC	300 miles
			Yankee 300		
May 5	USAC	Stock	Langhorne, Pa.	1D	150 laps
June 23	USAC	Champ	Langhorne, Pa.	1D	100 miles
July 7	USAC	Stock	Indpls. Raceway Park, Ind.	⅝P	250 laps
July 28	USAC	Champ	Trenton, N.J.	1P	150 miles
Aug. 31	USAC	Sprint	DuQuoin, Ill.	1D	25 laps
Sept. 2	USAC	Champ	DuQuoin, Ill.	1D	100 miles
Sept. 4	USAC	Stock	Indpls. Fairgrounds, Ind.	1D	100 miles
Sept. 22	USAC	Champ	Trenton, N.J.	1P	200 miles
Oct. 5	USAC	Sprint	Williams Grove, Pa.	½D	30 laps
Nov. 2	USAC	Sprint	Gardena, Calif.	½D	30 laps
Dec. 6	SCCA	Sports	Nassau, Bahamas	4½P-RC	25 laps
Dec. 8	SCCA	Sports	Nassau, Bahamas	4½P-RC	56 laps
			252 miles Nassau Trophy		

1964

Jan. 26	USAC	Sprint	Phoenix, Ariz.	1D	50 laps
Mar. 22	USAC	Champ	Phoenix, Ariz.	1P	100 miles
Mar. 29	USAC	Sprint	Reading, Pa.	½D	30 laps
Apr. 12	USAC	Sprint	Williams Grove, Pa.	½D	30 laps
Apr. 19	USAC	Champ	Trenton, N.J.	1P	100 miles
May 30	USAC	Champ	Indianapolis, Ind.	2½P	500 miles
June 7	USAC	Champ	Milwaukee, Wisc.	1P	100 miles
June 14	USAC	Sprint	Terre Haute, Ind.	½D	30 laps
June 21	USAC	Champ	Langhorne, Pa.	1D	100 miles
July 4	NASCAR	Stock	Daytona, Fla.	2½P	400 miles
			Firecracker 400		
July 18	USAC	Sprint	Mechanicsburg, Pa.	½D	30 laps
July 19	USAC	Champ	Trenton, N.J.	1P	150 miles
Aug. 22	USAC	Champ	Springfield, Ill.	1D	100 miles
Sept. 7	USAC	Champ	DuQuoin, Ill.	1D	100 miles
Sept. 9	USAC	Stock	Indpls. Fairgrounds, Ind.	1D	100 miles
			Started last		
Sept. 13	USAC	Stock	Langhorne, Pa.	1D	250 miles
Sept. 26	USAC	Champ	Indpls. Fairgrounds, Ind.	1D	100 miles
			Hoosier Hundred		
Oct. 25	USAC	Champ	Sacramento, Calif.	1D	100 miles
Nov. 29	USAC	Stock	Hanford, Calif.	1½P	200 miles

1965

July 4	NASCAR	Stock	Daytona, Fla.	2½P	400 miles
			Firecracker 400		
July 18	USAC	Champ	Trenton, N.J.	1P	150 miles
Aug. 21	USAC	Champ	Springfield, Ill.	1D	100 miles

Sept. 7	USAC	Stock	Indpls. Fairgrounds, Ind.	1D	100 miles
Sept. 18	USAC	Champ	Indpls. Fairgrounds, Ind.	1D	100 miles
			Hoosier Hundred		
Sept. 26	USAC	Champ	Trenton, N.J.	1P	200 miles
Oct. 10	USAC	Midget	Terre Haute, Ind.	½D	100 laps
Nov. 13	USAC	Sprint	Gardena, Calif.	½D	30 laps
Nov. 20	USAC	Midget	Phoenix, Ariz.	½D	50 laps
Nov. 21	USAC	Champ	Phoenix, Ariz.	1P	200 miles

1966

| Apr. 9 | USAC | Midget | Gardena, Calif. | ½D | 50 laps |
| Nov. 13 | USAC | Sprint | Altamont, Calif. | ½P | 30 laps |

1967

May 30	USAC	Champ	Indianapolis, Ind.	2½P	500 miles
June 10-11	FIA	Sports	Le Mans, France	8.35P-RC	24 hours
			Co-driving with Dan Gurney		
Aug. 19	USAC	Champ	Springfield, Ill.	1D	100 miles
Sept. 4	USAC	Champ	DuQuoin, Ill.	1D	100 miles
Sept. 17	USAC	Stock	Milwaukee, Wisc.	1P	250 miles
Sept. 24	USAC	Champ	Trenton, N.J.	1P	200 miles
Oct. 1	USAC	Champ	Sacramento, Calif.	1D	100 miles
Oct. 11	USAC	Sprint	Gardena, Calif.	½D	30 laps

1968

May 5	USAC	Stock	Indpls. Raceway Park, Ind.	2½P-RC	250 miles
July 7	USAC	Champ	Castle Rock, Colo.	2.66P-RC	150 miles
July 14	USAC	Stock	Milwaukee, Wisc.	1P	200 miles
Aug. 23	USAC	Stock	Indpls. Fairgrounds, Ind.	1D	100 miles
Sept. 7	USAC	Champ	Indpls. Fairgrounds, Ind.	1D	100 miles
			Hoosier Hundred		
Sept. 13	USAC	Stock	Cincinnati, Ohio	½D	100 laps
Sept. 29	USAC	Champ	Sacramento, Calif.	1D	100 miles
Oct. 13	USAC	Champ	Hanford, Calif.	1½P	250 miles

1969

June 21	USAC	Stock	Indpls. Fairgrounds, Ind.	1D	100 miles
Aug. 3	USAC	Stock	Dover, Del.	1P	300 miles
Aug. 31	USAC	Stock	DuQuoin, Ill.	1D	100 miles
Sept. 6	USAC	Champ	Indpls. Fairgrounds, Ind.	1D	100 miles
			Hoosier Hundred		
Sept. 20	USAC	Stock	Nazareth, Pa.	1⅛D	100 miles
Oct. 5	USAC	Stock	New Bremen, Ohio	½P	200 laps
Oct. 11	USAC	Stock	Sedalia, Mo.	1D	100 miles

P—Paved   RC—Road Course   ½—½-mile
D—Dirt   ¼—¼-mile   1—1-mile

Summary of Indianapolis Record

Year	Car	Speed Ranking	Qual.	Start	Finish	Laps	Speed/Reason Out	LD
1958	Dean Van Lines Spl.	16	143.130	12	16	148	Spun out on SW turn	148
1959	Dean Van Lines Spl.	20	142.648	17	10	200	133.297	200
1960	Bowes Seal Fast Spl.	22	143.466	16	25	90	Clutch failure	90
1961	Bowes Seal Fast Spl.	9	145.907	7	1	200	139.130	200
1962	Bowes Seal Fast Spl.	5	149.074	5	23	69	Wrecked	69
1962R	Sarkes Tarizian Spl. (Relieved E. George 127.147)				17	20	Starter failure	20
1963	Sheraton-Thompson Spl.	2	150.615	8	3	200	142.210	200
1964	Sheraton-Thompson Spl.	6	154.672	5	1	200	147.350	200
1965	Sheraton-Thompson Spl.	1	161.233	1	15	115	Broken gearbox	115
1966	Sheraton-Thompson Spl.	7	161.355	18	26	0	Wrecked	0
1967	Sheraton-Thompson Spl.	4	166.289	4	1	200	151.207	200
1968	Sheraton-Thompson Spl.	8	166.821	8	20	86	Blown engine	86
1969	Sheraton-Thompson Spl.	1	170.568	1	8	181	Flagged	181
12 Races							Total laps	1,709

Summary of Winnings at Indianapolis

Year	Finish	Speedway Prize	Lap Prizes	Accessory Prizes	Grand Total
1958	16th	$ 2,919	$ —	$ 50	$ 2,969
1959	10th	6,125	—	450	6,575
1960	25th	4,120	—	100	4,220
1961	1st	77,625	10,650	29,700	117,975
1962	23rd	5,171	300	250	5,721
1963	3rd	24,100	—	8,514	32,614
1964	1st	97,650	21,900	34,100	153,650
1965	15th	11,641	1,500	7,376	20,517
1966	26th	10,637	—	250	10,887
1967	1st	105,052	9,900	56,575	171,527
1968	20th	10,630	—	500	11,130
1969	8th	19,338	10,223	20,691	50,252
		$375.008	$54,473	$158,556	$588,037

USAC Record

Year	Division	No. of Feature Starts	Point Ranking	Wins	Seconds	Thirds	Top Three Finishes
1956	Midgets	1	109th	—	—	—	—
1957	Championship	5	26th	—	—	—	—
	Midwest Sprints	4	11th	—	—	—	—
	Midgets	35	7th	2	7	2	11
1958	Championship	12	10th	—	1	1	2
	Midwest Sprints	10	2nd	—	3	2	5
	Eastern Sprints	2	14th	—	—	—	—
	Midgets	7	47th	—	—	—	—
1959	Championship	10	5th	—	1	2	3
	Midwest Sprints	9	2nd	2	1	1	4
	Eastern Sprints	2	10th	—	—	—	—
	Midgets	12	33rd	1	—	1	2
1960	Championship	12	1st	4	2	2	8
	Midwest Sprints	12	3rd	—	2	3	5
	Eastern Sprints	8	1st	3	1	1	5
	Midgets	11	10th	3	1	1	5
1961	Championship	12	1st	4	1	1	6
	Sprints	22	3rd	6	4	2	12
	Midgets	13	7th	9	1	1	11
1962	Championship	13	2nd	4	3	—	7
	Sprints	13	4th	2	3	2	7
	Midgets	7	21st	1	—	—	1
	Stocks	16	5th	2	1	3	6
1963	Championship	12	1st	5	3	2	10
	Sprints	13	2nd	5	4	1	10
	Midgets	1	84th	—	—	1	1
	Stocks	13	2nd	4	1	2	7
1964	Championship	13	1st	10	—	—	10
	Sprints	13	4th	5	2	2	9
	Midgets	2	93rd	—	—	—	—
	Stocks	13	4th	3	—	—	3
1965	Championship	17	2nd	5	3	1	9
	Sprints	8	14th	1	—	2	3
	Midgets	3	34th	2	—	—	2
	Stocks	6	8th	1	1	1	3
1966	Championship	12	13th	—	1	2	3
	Sprints	3	20th	1	—	1	2
	Midgets	1	56th	1	—	—	1
	Stocks	2	15th	—	—	1	1
1967	Championship	20	1st	5	3	—	8
	Sprints	3	31st	1	1	—	2
	Midgets	1	113th	—	—	—	—
	Stocks	9	8th	1	—	1	2

1968	Championship	20	6th	4	1	1	6
	Sprints	1	45th	—	—	—	—
	Midgets	1	103rd	—	—	—	—
	Stocks	15	1st	4	3	1	8
1969	Championship	17	7th	1	—	3	4
	Sprints	—	—	—	—	—	—
	Midgets	2	—	—	1	—	1
	Stocks	22	2nd	6	5	2	13

Summary of USAC Starts and Top Three Finishes

Division	Total Starts	Wins	Seconds	Thirds	Top Three
Championship	175	42	19	15	76
Sprints	123	26	21	17	64
Midgets	97	19	10	6	35
Stocks	96	21	11	11	43
	491	108	61	49	218

USAC Money Winnings 1956-69

	Championship	Sprint	Midget	Stock
1956	$ —	$ —	$ 68	$ —
1957	2,171	1,296	5,890	—
1958	15,645	6,377	754	—
1959	17,383	6,044	1,455	—
1960	45,020	11,002	2,536	—
1961	112,818	13,595	5,764	—
1962	39,673	7,540	1,524	9,114
1963	74,515	12,031	192	15,317
1964	156,131	10,498	317	12,502
1965	61,705	4,459	1,453	8,569
1966	24,778	2,116	781	2,743
1967	153,815	2,727	108	9,167
1968	54,079	327	85	27,087
1969	48,376	—	2,305	38,194
	$806,109	$78,012	$23,232	$122,693 = $1,030,046

* Does not include any accessory money or lap prizes.

147

Foyt's
12 Years
at Indy

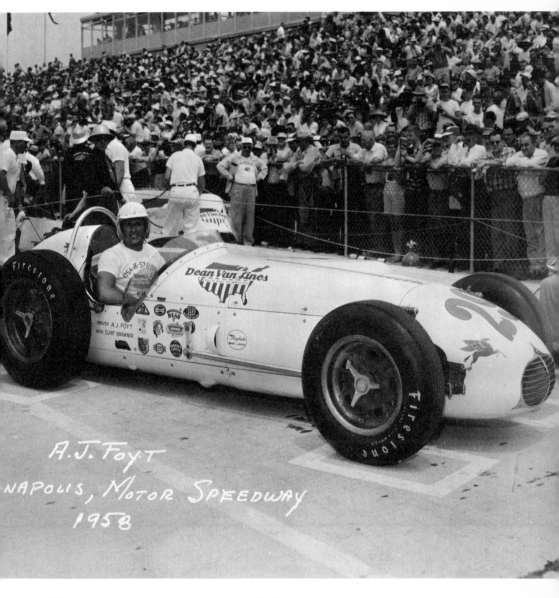

A.J. FOYT
NAPOLIS, MOTOR SPEEDWAY
1958

1958: Started 12th (143.130 mph). Finished 16th.

A.J. Foyt ··· Indianapolis Motor Speedway ··· 1959

1959: Started 17th (142.648 mph). Finished 10th.

1960: Started 16th (143.466 mph). Finished 25th.

WINNER
A.J. FOYT · 1961 · INDIANAPOLIS MOTOR SPEEDWAY

1961: Started 7th (145.907 mph). Finished 1st.

152

A.J. Foyt 1962 Indianapolis Motor Speedway

1962: Started 5th (149.074 mph). Finished 23rd.

A.J. Foyt 1963 Indianapolis Motor Speedway

1963: Started 8th (150.615 mph). Finished 3rd.

1964: Started 5th (154.672 mph). Finished 1st.

1965: Started 1st (161.233 mph). Finished 15th.

156

A.J. FOYT · 1966 · INDIANAPOLIS MOTOR SPEEDWAY

1966: Started 18th (161.355 mph). Finished 26th.

"WINNER"
A.J. Foyt 1967 Indianapolis Motor Speedway

1967: Started 4th (166.289 mph). Finished 1st.

"A.J. FOYT · 1968 INDIANAPOLIS MOTOR SPEEDWAY"

1968: Started 8th (166.821 mph). Finished 20th.

A·J·FOYT 1969 INDIANAPOLIS MOTOR SPEEDWAY·

1969: Started 1st (170.568 mph). Finished 8th.